Solid Waste Management and Recycling

The GeoJournal Library

Volume 76

The titles published in this series are listed at the end of this volume.

Solid Waste Management and Recycling

Actors, Partnerships and Policies in Hyderabad, India and Nairobi, Kenya

edited by

ISA BAUD

University of Amsterdam,
The Netherlands

JOHAN POST

University of Amsterdam,
The Netherlands

and

CHRISTINE FUREDY

York University,
Toronto, Canada

SPRINGER-SCIENCE+BUSINESS MEDIA, B.V.

A C.I.P. Catalogue record for this book is available from the Library of Congress

ISBN 978-90-481-6560-5 ISBN 978-1-4020-2529-7 (eBook)
DOI 10.1007/978-1-4020-2529-7

Printed on acid-free paper

Funded by European Union (ERBIC 18CT 970152)

TABLE OF CONTENTS

FOREWORD AND ACKNOWLEDGEMENTS

The project on which this book is based, results from the interest within the group of staff and junior researchers at the University of Amsterdam on how urban environmental management could be made more sustainable. Earlier studies in India and Peru had given us knowledge on how widespread small enterprises recycling waste materials for profit and people picking waste were in many cities in the South. It also led to a realisation of how important the contribution of such activities was to reducing waste flows, despite the fact that such activities occurred in semi-illegal contexts and received no recognition from governments or middle-class residents.

The debate on how to combine ecological sustainability with socio-economic improvements in the lives of many urban citizens, led to the formulation of the project that lies behind this book. It aimed at improving our understanding of the factors that underlie the dynamics of the provision of a particular urban environmental service (solid waste management), and also at providing policymakers and city managers with ideas about how they can tackle the problems of improving the quality of the urban environment they are dealing with daily.

This project could not have been carried out without the help of a great many people. Our grateful thanks goes first of all to the EU INCO-DC Programme, which provided a generous grant for the research project on Enabling Strategies for Urban Environmental Management in Mega-cities (ERBIC18CT970152). The Consortium formed for the project consisted of researchers from the Amsterdam Global Issues and Development Studies Institute (AGIDS), Department of Human Geography, University of Amsterdam as coordinator under my guidance, the International Institute of Environment and Development (IIED) in London, under the guidance of Dr. D. Satterthwaite, the Centre for Economic and Social Studies in Hyderabad, India under guidance of Dr. S.Galab, and the School of Environmental Studies, Moi University, Eldoret, Kenya, under guidance of Prof. Dr. Th. Davies.

The teams put together from each contributing institute consisted of dedicated researchers, from different disciplinary backgrounds who were willing to listen to each others' language and start to identify a common language. Furthermore, the project could benefit from the synergy arising from joining expertise on various aspects of the solid waste sector. We had experts on the privatisation of solid waste collection, on the organisation of the recycling business, on the role of local communities and their organisations in waste management, on the use of organic waste matter in (peri-)urban agriculture, and on the environmental hazards connected to waste.

At IIED, Dr. David Satterthwaite and Dr. Cecilia Tacoli provided generous hospitality for two workshops and incisive comments on discussions and earlier versions of the manuscript, the latter contributing a background paper on rural-urban interactions. They also generously provided for the time contributed by Chris Furedy to the research project. The Centre for Economic and Social Studies, under the guidance of Dr. Mahendra Dev, provided for the time spent by Dr.S. Galab and Dr. S. Sudahkar Reddy on the field studies done inside and around Hyderabad with their teams, and their reports to the project team. The team in Kenya consisted of people from the School of Environmental Studies, Moi University and researchers already working with staff from AGIDS at the University of Amsterdam. Despite the changing composition of the team, Prof. Theo Davies as coordinator and Anne Karanja and Dr. Moses Ikiara completed the fieldstudies with dedication, formed an integral part of the discussions at the workshops throughout the project, and contributed the chapters found here.

Several contributions were also made by others to the project: Dr. M. Put contributed an interesting commissioned paper on the use of different types of organic manure – including urban solid waste – by farmers in areas around Hyderabad, India, broadening our understanding of the ways in which urban-rural links develop and wane. Anne Karanja wrote a working paper on the informal recycling sector in Nairobi, on the basis of her fieldwork studies for her Ph.D., which illustrated the differences in depth and complexity of the Kenyan and Indian situations. Erwin Koster of the University of Amsterdam carried out a field study on urban farmers in Nairobi, complementary to the work in India, in order to improve our understanding of the ways in which farmers use organic manure and waste materials in their farming strategies. Finally, Dr. Johan Berkhout carried out a transect study in Nairobi, pinpointing geographically areas of activities in SWM within the urban system, laid down in a CD-Rom. R. Dhanalakshmi checked recent elements concerning community-based initiatives in Hyderabad for finalization of the manuscript.

Editing the final manuscript is a process in itself, and I would like to thank my co-editors for the generous amounts of time they made available to bring the manuscript to completion. Particularly, I would like to thank Dr. Johan Post, who provided the main liaison with the publisher and the layout-editor Anne van der Zwalmen during this process. Finally, the assistance of the University of Amsterdam project bureau was invaluable in guiding me through the maze of administrative and financial reporting throughout the project period. Particular thanks goes to Harry van Kesteren, who remained unruffled throughout and coordinated with his financial counterparts from the other contributing institutions, and the desk officers from the European Union, who provided backstopping for this project.

Isa Baud
Project Coordinator
July 2003

ISA BAUD

CHAPTER 1

MARKETS, PARTNERSHIPS AND SUSTAINABLE DEVELOPMENT IN SOLID WASTE MANAGEMENT; RAISING THE QUESTIONS

1.1. INTRODUCTION

The global economic crisis in the 1970s led to significant transformations in international and national institutional arrangements. Major actors on the international stage – countries such as the US and the UK, transnational corporations and the Bretton Woods institutions – strongly advocated the primacy of the market and the retreat of the state. Such neo-liberal ideas on market liberalisation and deregulation were imposed on many countries in Africa, Asia and Latin America under the aegis of structural adjustment programmes. One area governments in the South were strongly advised to withdraw from was that of direct provision of basic services.

However, the results of these reform programmes were less successful than expected. Although state governments reduced spending and economic growth occurred after an initial period in some countries, the late 1980s were characterised by increasing disparities between rich and poor. Within many southern states, urban poverty and informalisation of employment and economic activities grew rapidly, presenting huge problems for local authorities to deal with. In many cities, new forms of collective organisation started to emerge among poor households together with a variety of non-governmental organisations (NGOs) in order to counter poverty and promote community and neighbourhood development.

In the 1990s, the limits of the free market approach were increasingly recognized by even its most fervent advocates. Furthermore, the collapse of state communism and the fall of the Berlin Wall had created an entirely new political climate, one favourable to the democratic reform of state bureaucracies. The difficulties that many countries in the south, but also in the former communist world, experienced in their transition to a market economy also fuelled an interest in the (democratic) institutions that underpin processes of development. Economists have expressed this interest by looking at the role of meso-level institutions and how they influence economic growth[1].

I. Baud et al. (eds.), Solid Waste Management and Recycling, 1-18.

In research on poverty, this is reflected in the new conceptualisation of poverty based on the asset-vulnerability approach that highlights the importance of institutions in mediating access (Moser, 1998; de Haan, 2000). In political and social sciences, ideas on the role of the state have also shifted to seeing the state as 'enabler', a co-ordinating agency working with a variety of other organisations in different forms of partnerships aimed at urban and regional development. The importance of what came to be called civil society organisations and how they could work together with local and national governments as well as private sector organisations became a central topic in the 1990s (Arossi *et al.*, 1994; Mitlin, 2001; Rakodi, 1999; Hardoy, Mitlin, and Satterthwaite, 1992). New interest has emerged as well in the political processes involved in partnership arrangements and how these influence the effectiveness of local and regional governance (Stoker, 2000; Putnam, 1993; Orstrom, 1996; Baud, 2000; Helmsing, 2000).

Each of these debates has as its central concern the changes in the relation between either the state and the market, or the state and civil society, and how their re-alignment affects future paths of development. In this book, our central objective is to look at such processes and patterns of fundamental re-aligning between state, civil society and the market in an integrated manner as it concerns the provision of basic services in urban areas. Urban basic services, such as water, sanitation and solid waste management in the past have been considered by the state to be the responsibility of local or national governments. In the last twenty years, the debate raged to what extent they also should become part of the re-alignment between state, private sector, and community organisations. Rhetoric and practice have differed widely, with expectations running far ahead of changes in practice (Bartone *et al.*, 1991; Batley *et al.*, 1996; Dillinger, 1994).

This has led to discussions on new forms of urban governance, in which the use of partnerships or alliances between different stakeholders is mooted in order to lead to greater effectiveness and sustainable development in urban solid waste management. Partnerships, or alliances, as instruments for more effective forms of local governance are being promoted in a wide range of concerns, but in developing countries have emerged notably in local environmental management[2]. In this book, a basic premise is that the perspectives of local communities, the organisations that deal with them, and support of collective initiatives and small-scale economic actors are equally

---

1. Such institutions include organisations of entrepreneurs at the local and regional level, and supporting legal and training institutions by government or the private sector (Helmsing, 2001). This is reflected in work done by scholars from an institutional economics perspective, such as Storper (1997), Krugman (1997), Hunter and Lewis (1997), and others (cf. Baud *et al.*, 2002).
2. This is in contrast with the situation in developed countries, where public-private partnerships are associated more with large-scale infrastructure and construction projects.

important as the perspectives of urban planners and government policy (cf. Keivani and Werna, 2001; Hardoy *et al.*, 2001).

The authors participating in this study have started from the premise that changes in the provision of solid waste services must be assessed not only by considering criteria of cost efficiency and service effectiveness, but also by considering issues of equality (in access), broad coverage, affordability, and environmental concerns. These criteria are integrated under the broader heading of 'sustainable development'. The basic question posed is the extent to which changes in solid waste management (SWM) systems contribute to (aspects of) sustainable development in urban areas[3]. Our ideas about effective ways of improving urban solid waste management should, to our minds, not be based on broad theoretical generalisations, but should emerge from grounded and comparative studies of existing and newly adopted practices. The use of comparative case studies in two cities, with a common past in British colonial administrative practice but very different current situations, reflects this basic tenet.

The aim of the study is to make an integrated analysis of the different patterns of waste management in order to make an integrated assessment of their contributions to both socio-economic and environmental aspects of sustainable development. This follows international discussions in which the idea of 'integrated waste management' has been put forward by van der Klundert and Lardinois in the context of the UWEP Programme (1995). They try to unravel the concept of sustainable development into its various aspects, and examining the activities carried out by the various actors involved, as well as the changing partnerships around such activities in the light of such aspects. In this way, it is possible to identify the trade-offs between components and how they affect goals put forward by different groups.

The study is designed to come up with a framework which allows researchers and practitioners to make informed choices about the implications of future strategies in urban SWM in terms of the trade-offs among various components of sustainable development. The framework at this stage is based on empirical arguments, as quantifying the different components is almost impossible at this stage of our knowledge. The choices made locally may be politically motivated. However, integrating aspects of socio-economic and environmental assessment allows all the actors involved in such a decision to base their decisions on an explicit analysis and knowledge of the implications of their choices.

3. Waste is defined as materials, which have lost their value to their first owners (Cointreau-Levine, 1984). In this book attention is focussed only on waste that comes into the municipal stream. This can be waste generated by institutions, industries, and households. Waste fractions already separated by firms and households and sold for reuse and recycling means that municipal waste streams do not fully reflect waste generation patterns.

1.2. CHANGING PERSPECTIVES IN URBAN SERVICE PROVISION: THE CASE OF SWM

Research on urban SWM in developing countries has developed from two main concerns; the concern for increasing complexity and costs of waste management, which are proving difficult to manage efficiently and effectively by local authorities, and the concern for environmental impacts of growing waste flows. The latter perspective covers three areas: problems for the environmental health/public health of urban citizens[4], health and safety hazards for those working with solid waste, and problems of sustainable development in terms of resource recovery and recycling of waste materials. These are coupled to the classic concerns of safe disposal of wastes that can be absorbed by local and regional sinks[5].

The first concern has come from the perspective in which local government has primary responsibility for SWM, and carries out its activities from a primary concern with public health issues. Although we currently talk about 'environmental health', many local authorities have limited that view to activities concerning 'public health'. The primary perspective on SWM developed in the course of the nineteenth century in Europe and exported to colonies around the world, was that of public health. Solid waste accumulating in densely populated urban areas posed health hazards, which local authorities sought to control by providing effective collection, transport and disposal services. The organisation of such basic services was carried out through local government, Health departments, in both British and French administration systems. Primary objectives were the effective removal of waste from neighbourhood residential areas, without interference, and disposal sites outside the city boundaries.

The limits to this approach became increasingly clear in industrialised countries during the sixties and seventies as consumption patterns led to sizeable growth of waste flows, whose disposal went beyond the limits of social acceptability and the absorption capacity of local and global sinks (cf. Sachs *et al*, 1997; Mitlin and Satterthwaite, 1997). A perspective aimed at promoting greater sustainable development in the use of resources has influenced solid waste management practices, and is gradually becoming implemented through policy guidelines at national levels in a number of industrialised countries. Guidelines and directives to reduce waste generation, and promote waste recovery are laid down according to the 'waste management hierarchy', in which waste prevention, reuse, recycling and energy recovery are designed

4. The term environmental health is used nowadays instead of public health, as it was felt that public health was linked too much to direct medical provisions (Hardoy *et al.*, 2001). However, SWM has always been an integral part of the Public Health Department in the British administrative system, and still is so in the countries under study. Therefore, we use the term environmental health for the general discussion, but retain the term public health when referring to the specific situation in Nairobi and Hyderabad.
5. These are currently coupled to the concern in the North to reduce use of resources in production, and segregate waste materials at source to increase the possibilities of waste recovery.

to minimise the amount of waste left for final, safe disposal (OECD, 1972; cf. de Jong, 1999).

The perspective giving priority to issues of public health has remained the dominant one until today in developing countries. However, there is more attention of how to deal with the growing costs of managing larger waste flows. Measures taken into account include privatisation and the introduction of cost recovery mechanisms. How such measures should be integrated into a perspective on sustainable development has remained a largely theoretical discussion, in which the public sector has little interest. In international discussions on sustainable development, developing countries have made it abundantly clear that environmental policies should reflect their own priorities and not curtail their legitimate desire for economic growth. They have given priority to issues of pollution (the so-called 'brown agenda') with a predominantly urban focus (UNCHS, 1996), (many of which were already included on their usual agenda) and shifted the environmental focus away from issues of natural resource depletion and resource management[6]. The brown agenda is defined as

> '... the immediate and most critical environmental problems which incur the heaviest costs on current generations, particularly the urban poor in terms of poor health, low productivity and reduced income and quality of life: lack of safe drinking water, sanitation and drainage, inadequate solid and hazardous waste management, uncontrolled emissions from factories, cars and low grade domestic fuels, accidents linked to congestion and crowding, and the occupation of environmentally hazard-prone lands, as well as the interrelationships between these problems' (Bartone *et al.*, 1994: 5)

The focus on pollution problems carries implicitly a conception of sustainable development, which combines environmental aspects with attempts to meet human needs (cf. McGranahan and Satterthwaite, 2000; Drakakis-Smith, 1995). In this perspective, environmental issues are considered in conjunction with quality of life improvements in urban areas, obtained through the changes in the institutional arrangements influencing them. Studies carried out within this framework usually deal with the ways in which various actors contribute to improve environmental conditions as well as increase the effectiveness of urban livelihood strategies (Furedy, 1992, 1997; Pacheco, 1992; Bose and Blore, 1993; Huysman, 1994; Baud and Schenk, 1994).

Although the further emphasis on the 'green' agenda – of preventing waste generation and reducing waste flows – is still weak in current policy perspectives in developing countries, the increases in waste flows make it imperative to focus more attention on this problem as well in the future.

---

6. i.e. the prime environmental worries in the North.

1.3. ACTORS AND ACTIVITIES: PUBLIC, PRIVATE AND CIVIL SOCIETY STAKEHOLDERS

Urban solid waste management as seen from the perspective of local authorities includes the activities of collection of domestic solid waste, either through door-to-door or neighbourhood collection, transportation and disposal of solid waste (usually in dumpsites). A more environmentally oriented view of urban solid waste management includes reuse, recycling and recovery activities, and safe disposal of waste (in sanitary landfills or through incineration): the so-called waste hierarchy, as shown below. This study utilizes the latter framework for analysis, focusing specific attention on the reuse, recycling and recovery processes already existing in many developing countries in the private sector. This allows us to present alternative scenarios to policy makers, community-based organisations (CBOs), NGOs, and local public bodies and private enterprises, showing the kinds of contributions different activities can make in the direction of more environmentally sustainable development in the sector.

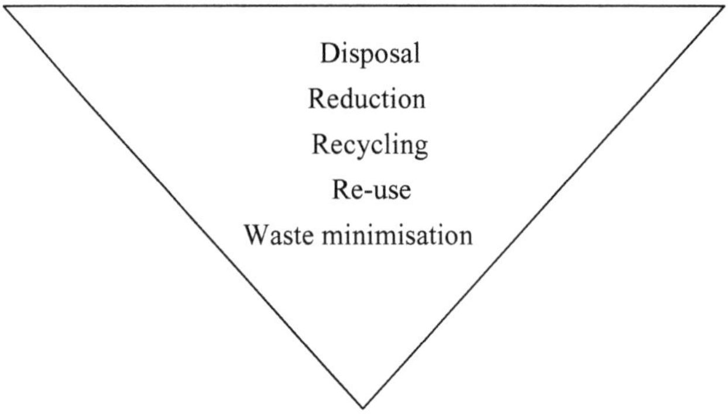

Figure 1.1. Waste management hierarchy

The character of urban waste differs between developing and industrialised countries, and between larger and smaller towns. Cointreau-Levine (1998) has estimated that in developing countries the extent of organic waste is relatively high, constituting between 40-70 percent of solid waste in developing countries. The increasing use of plastics as packaging material, and other inorganic materials has caused the character of solid waste to change composition in recent years (UNCHS, 1996; Schübeler, 1996; Nunan, 2001; Rosenberg and Furedy, 1996). Therefore, because of the potential of organic waste also to be reused and recycled, it was decided to outline specifically what activities were carried out with respect to both inorganic and organic waste flows in the course of this study.

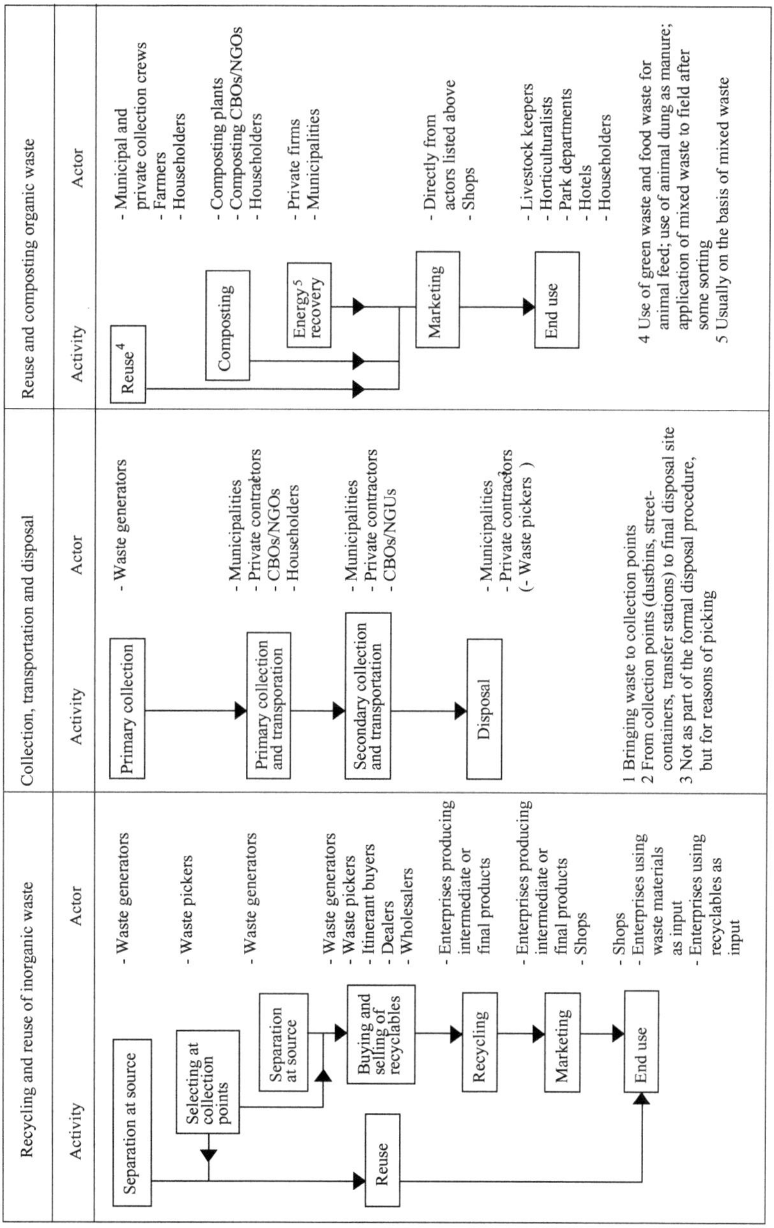

Figure 1.2. Actors and activities in solid waste management

The following figure indicates the ways in which the various activities are assumed to relate to each other, based on previous research in India on inorganic waste (Huysman and Baud, 1994) and on organic waste flows within and outside the city (Rosenberg and Furedy, 1996). The main advantage is that the activities of the municipality are shown in combination with methods of separation and sorting of waste fractions, collection, trade and recycling of waste fractions carried out by other actors Figure 1.2 indicates the activities of municipal and non-municipal actors in urban swm, incorporating the results of the research presented in this volume.

1.4. PARTNERSHIPS IN SWM: POTENTIALS AND LIABILITIES

New ideas about the relations between the local government and other actors in public services initially focused largely on privatisation, but gradually have suggested a new permanent role for the government as enabler of other actors, who should be the direct providers of services (cf. Batley, 1996). From this has followed an intense interest in the idea of 'partnerships', as an important method for providing services not only efficiently and effectively (World Bank criteria), but also equitably. The discussion on what constitute partnerships and under what pre-conditions partnerships function as such, is a broad one, from which some points will be drawn which are necessary for the discussion on urban SWM here. There are different levels of analysis for partnerships: politically as 'institutions' of governance; at the planning level as instruments for public policy; sociologically as forms of social capital; and economically as ways of reducing transaction costs.

In order to understand the way partnerships function, three dimensions need to be taken into account: values, processes, and institutions (Pierre, 1998). The institutional dimension refers to the regularised patterns of interaction between the actors involved. The value dimension includes the broader system of norms, beliefs and objectives that informs interaction, and process refers to the transactions taking place over time between the actors A fourth dimension should be included in such models, as in the end the quality and availability of urban services is the final criterion on which urban inhabitants assess 'partnerships' in service provision (Baud and Post, 2002; Devas, 1999).

There are a number of definitions of partnerships pertaining to urban governance (Devas, 1999; Peters, 1998; Baud *et al.*, 2001; Baud and Post, 2002). Here we draw out the main elements of such definitions, useful for our purpose. Peters has defined partnerships as including at least two principal actors[7], one of whom is public, with more or less enduring relationships and continuing interaction, each contributing resources, and with a shared responsibility for the outcomes of the activities. This last

7. i.e. each partner is capable of bargaining on his or her own behalf.

element can be difficult to achieve in relation with a government-based partner, and also reduce accountability for ordinary citizens.

For our study on SWM, we use the term partnership[8], but in the wider sense of the concept, based on our earlier discussion of the term alliance[9]. Partnerships in this study are used to describe established relationships between actors in the SWM-system (which will be defined in the next section). The distinguishing features of a partnership are (Baud and Post, 2002; Baud, 2000):

- It involves two or more actors[10], although not necessarily a public sector actor;
- It refers to a more or less enduring relationship between the actors (based on a written or verbal agreement) regarding public goods provision;
- The relationship is beneficial for all actors (without assuming equality or equal benefits between actors);
- It finds expression in concrete (physical) activities, in which each actor invests materially or immaterially;
- The bargaining process can include potential areas of tension and conflict as well as co-operation;
- The partnership must regard the provision of public goods (or have a spin-off relating to a public good).

In principle, partnerships provide benefits to each of the actors involved, but this does not imply equality among them, for in most such relationships issues of power are at stake. Although partnerships suggest a degree of stability, they should nevertheless be seen as expressions of people's practices that have an inherent tendency to evolve, adapt and dissolve in response to changing circumstances. The rise to prominence of private waste contractors in a domain that historically was regarded as the domain of the public sector is a case in point. Finally, it is important to map the various partnerships in garbage collection and disposal in order to avoid a preoccupation with the most dominant ones, and to come to an appreciation of the potentials of the others.

---

8. Among urban planners and economists, the term partnership often has the narrower definition of public-private sector partnerships, with large private enterprises being contracted in various ways to provide services.
9. Alliance was used in the article published in Cities (Baud *et al.*, 2001), as an alternative to partnerships, because of the international literature on partnerships where there is certain euphoria about the potential of partnerships. In our definition, as in the further political science literature discussed above, there is no assumption of equality between actors, and possible inequitable divisions of investment in time and money, as well as in benefits.
10. However, we do not want to make the limitation of having at least one actor from the public sector, as a number of partnerships exist between the private sector and communities regarding basic services, where the public sector is not directly involved.

On the basis of previous studies, the following figure indicates a range of possible partnerships in urban solid waste management. It must be considered a heuristic analytical framework, ready to be tested against the empirical situation found in any city. It originates from the models developed not only concerning public sector waste management as seen from the public health perspective, but rather the sustainable waste management perspective (van der Klundert and Lardinois, 1995). It also integrates the specific concerns with the differentiation between the actors involved in diverting organic waste and inorganic waste flows from the municipal streams, as found in the studies by Furedy (cf. Furedy, 1997; Furedy 1998b).

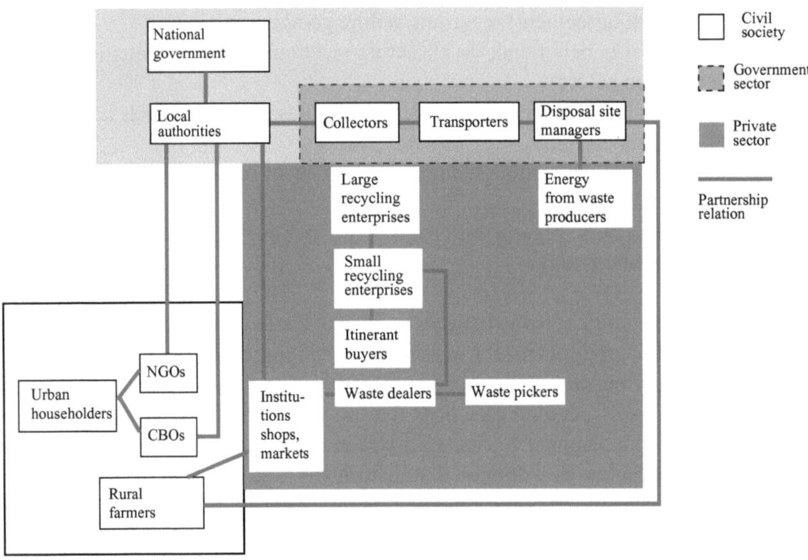

Figure 1.3. Actors and several possible partnerships in urban solid waste management

Public-private arrangements in SWM

Public-private partnerships have received the most attention internationally, as part of the global trends in public management reform. Private sector involvement in service provision raises issues of public interest and acceptability. Governments must still ensure appropriate standards, achieve co-ordinated provision, provide a competitive environment, avoid monopoly control of essential services by non-accountable private providers, and minimise corruption and inequity (Rondinelli and Iacono, 1996; Burgess *et al.*, 1997). Therefore, privatisation in service provision usually implies a form of public-private arrangement, in which the responsibilities of both parties have to be laid down. Although service implementation is contracted out to private enterprises (cf. Dillinger, 1994), governments retain some degree of power in setting conditions and standards. They save on costs, reduce political interference and red tape, and lower levels of coercion (SWM studies in this category include those by Bartone *et al.* (1991), Ali (1993), Fernandez (1993), Lee (1997), and Post (1999)).

In such institutional arrangements, the role of local government officials changes radically from that of implementing agency to that of standard setting and monitoring agency. The extent to which government officials have the capacity to carry out their new responsibilities is a major concern in studies of such public-private arrangements. A second concern is to assess the organisational and financial aspects of such privatisation initiatives (cf. Peltenburg, de Wit and Davidson, 2000).

Governments generally privatise SWM activities to large-scale, formally registered enterprises. Little attention is given to the potential of small-scale, private operators and CBOs to remove solid waste from residential areas. Nor are small-scale waste traders or recycling enterprises of whatever size drawn into privatisation initiatives, despite their existing expertise in this area. Local authorities prefer to link up with formal enterprises. There is an emphasis on strong contractual arrangements, for which informal businesses and communities do not qualify. Although their potential capacity in separation and collection of waste is increasingly acknowledged, few governments have started to include them in their policies (Baud *et al*, 2001).

Private-private arrangements

Studies on SWM covering private-private arrangements focus mainly on waste trade, reuse and recycling and waste recovery activities within the system. Studies deal not only with (semi-) contractual arrangements between traders and enterprises, but have shown a strong concern with labour contracts and working conditions, and the impact of official rules and regulations on private or communal undertakings. Finally, economic evaluations are combined with qualitative environmental assessments.

The first studies on the recovery, recycling or reuse of materials from city waste emerged from the interest in the labour conditions and survival mechanisms of waste collectors and traders (Birkbeck, 1978; Furedy, 1990; Kerkum, 1991; Sicular, 1992; Huysman, 1994). During the 1990s, the fact that waste recovery not only provides income to sizeable groups of the urban poor, but that its value as a commodity also contributed to the ecological aspects of sustainable development of SWM systems became more widely acknowledged (cf. Furedy, 1992; Baud, *et al.*, 1996). Finally, the economic and environmental impacts of international trade and use of waste materials in production has been studied (van Beukering, 1994; van Beukering and Duraiappah, 1996). At the local level, co-operation between local authorities and groups of urban poor involved in 'informal' economic activities is far from materialising. Local authorities usually seek to actively exclude such activities from taking place, as they conflict with their public health perspective on effective collection and disposal

(Cointreau-Levine, 1982; Furedy, 1990). Local residents also perceive street pickers as socially undesirable and seek to reduce their access to local sources of waste.

Waste pickers, itinerant buyers, traders and small-scale recyclers of waste materials carry out their activities in close co-operation and conflict. They depend on each other for credit and informal social security arrangements, but the informality also allows 'free rider' behaviour to go unpunished. All such activities take place in semi-legal conditions, with many enterprises remaining unregistered (van Beukering, 1994; Baud and Schenk, 1994; Jordens, 1996) or only going partially through the process of full registration (Baron and Castricum, 1996).

Finally, spontaneous private-private arrangements occur in collection, transport and disposal. These have emerged mainly in reaction to the lack of effective public sector provision (Mwangi, 2000). They suffer from a lack of recognition of their activities by local authorities, and may also not conform to the public standards for transportation and disposal.

Community-private and public sector-community arrangements

The role of NGOs and CBOs in working with local communities has been discussed widely in the literature, mainly at the level of projects initiated at neighbourhood level. Community-based organisations generally consist of residents of a particular area organizing to improve local waste collection (e.g. clean-up campaigns) and may include composting in their activities, emphasising 'green' aspects of sustainable development (Anand, 2000). They usually do not go much beyond the neighbourhood level in their activities, which can limit their impact (cf. Hordijk, 2000; Lee, 1998).

NGOs, coming in from outside, more often aim at socially vulnerable target groups, such as women and street children picking waste (e.g. Hunt, 1996; Huysman, 1994), and direct their activities towards strengthening their socio-economic capabilities. This includes promoting co-operatives among waste pickers, providing shelter and alternative training, and savings schemes. NGOs may also carry out activities more generally in the area of raising public awareness about sustainable development issues or contribute to developing alternative technology designed to promote recycling and composting activities in a decentralised fashion (Furedy, 1992; Schenk, Bhuvaneswari and Baud, 1998). SEWA in India is a longstanding example of this type of initiative.

1.5. URBAN SERVICES AND CONTRIBUTIONS TO SUSTAINABLE DEVELOPMENT; MAKING AN OPERATIONAL FRAMEWORK

Urban SWM is one of the services that can make strong contributions to sustainable development in cities. However, the underlying concept of sustainable development has led to heated debates and aroused mixed feelings in both academic and policy circles, because of its elusiveness and diametrically opposed interpretations. As a starting point for this study, we will take the position on sustainable development that seeks to combine goals of ecological sustainability with the concern for meeting current human needs (Satterthwaite, 1997; Hardoy, *et al.*, 2001). Striving for ecological sustainability implies that the use of non-renewable resources should be minimised, renewable resources should be used in such a way that regeneration of the resource is ensured, and the capacity of local and global sinks should not be exceeded in either case.

The original model that we followed, emphasised the environmental issues involved, and left relatively undefined what specific institutional arrangements are needed and how to realise those needs[11]. A more recent article by Satterthwaite does discuss the importance of institutional arrangements with relation to environmental issues in cities, suggesting that urban managers need to take into account two areas for which they currently have no mandate. These include: 1) minimising the transfer of environmental costs to inhabitants and ecosystems surrounding the city; and 2) ensuring progress toward 'sustainable consumption' (Satterthwaite, 1997).

It is important to indicate how ideas concerning sustainability are linked with development goals, in terms of sectoral priorities over time. There is tremendous controversy concerning what human needs entail, how their satisfaction relates to ecological sustainability, and what the acceptable levels of trade-offs are. This is apparent, for example, in the contrast between the advocates of the green and the brown agenda in urban environmental improvements. The former emphasize ecosystem health, the impact of cities on rural resources and surrounding regions, and the threat posed by urban consumption to the fulfillment of the needs of future generations. The latter focus on environmental hazards and social justice, and are more concerned with immediate problems at local level, especially those faced by the urban poor (McGranahan and Satterthwaite, 2000). Politically, countries of the South also emphasize the necessity to give economic growth priority at this point in time, before promoting ecological sustainability. In this book we will work from an understanding of sustainable development that combines a developmental perspective with ecological concerns, and makes explicit the trade-offs inherent in the choices that are made.

Attempts to link SWM to issues of sustainable development still need more analytical work. One such attempt is the concept of 'integrated sustainable waste management', developed by van der Klundert and Lardinois (1995). Their concept and related model

---

11. They include: 1) minimising the use or waste of non-renewable resources; 2) sustainable use of finite renewable resources; 3) biodegradable wastes not overtaxing capacities of renewable sinks; 4) non-biodegradable wastes/emissions not overtaxing capacity of local and global sinks to absorb or dilute them; and 5) meeting economic, social, cultural, environmental and health needs, as well as political needs for current and future generations.

is an early attempt to connect environmental issues, with socio-economic and technical delivery issues, on the basis of existing literature and the experiences of the network coordinated by WASTE within the UWEP Programme[12]. It incorporates social, economic, financial, political and environmental aspects. The model is useful in incorporating the many different aspects necessary to analyse to what extent SWM systems are sustainable, but has not yet been applied in empirical studies.

In the model developed in this book, we have linked solid waste management to the discussion on sustainable development, by making operational the three broad goals of sustainable development, *viz.* ecological sustainability, socio-economic equality, and improving environmental health. Solid waste management typically forms part of the brown agenda and its impacts are largely local. This is reflected in the 'localized nature' of the criteria we used to analyze the contributions to sustainable development of various activities and partnerships in SWM, both at the level of actors as well as the urban system.

With respect to the area of ecological sustainability, SWM systems need to work towards the following goals:

- To minimize the amount of waste generated;
- To maximize reuse and recycling; and
- To dispose of remaining waste in a controlled fashion in order to not exceed the capacities of local sinks.

The goal of minimizing the production of waste is primarily a national government (and private sector) responsibility, and can be pursued through production and consumption practices that reduce the input of materials, make more efficient use of these inputs, and increase closed-loop recycling. Whether or not consumers, industries, and institutions contribute to this goal depends on their assessment of the costs and benefits involved, as well as their levels of awareness. The maximizing of waste reuse and recycling can be carried out at primary level – within households, firms and institutions – or at secondary level, i.e. after materials have entered the municipal waste stream. A very important aspect is the extent to which source separation occurs and is officially endorsed and promoted (Lardinois and Furedy, 1999). The contribution to sustainable development lies in the reduction of volumes of waste to be disposed of, and the reduction in use of virgin materials. Controlled disposal is included because the amount of municipal waste that actually reaches the official dumpsites – and, consequently, how much is disposed of illegally and pollutes the urban environment – is an important indicator of the quality of an SWM system. Finally, the method of final disposal – in developing countries, largely through crude dumping or sanitary landfills – determines to what extent ecological sustainability and

12. WASTE is an advisory company for research and practice pertaining to SWM. They run the Urban Waste Expertise Program, funded by the Dutch Ministry of Development Cooperation.

environmental health are impaired through contamination of surface and groundwater or soils by leakage, air pollution by waste burning, and spread of diseases by different vectors

The socio-economic dimensions used to analyze SWM systems encompass both consequences for specific actors and impacts on the entire system (city level). Four criteria are used:

- Financial viability and affordability for the local authorities, consumers, and/or entrepreneurs involved (these may conflict among groups);
- Employment providing a living wage and a certain level of job security to SWM workers;
- Legitimacy from the perspective of the authorities (legal) and the public (social); and
- Effective monitoring and enforcement of standards.[13]

The continuity of an activity ultimately depends on its financial viability, i.e. the assurance that the revenues will continue to balance the costs incurred. Considering the 'public good' nature of SWM, authorities often have to accept a considerable degree of subsidization. However, the financial sustainability of the system depends on the authorities' solvability (their own revenues or grants) and the political willingness to pay the price of adequate servicing. Contributions from residents can help increase the financial viability of waste collection – the concept of allocative efficiency (Batley, 1996: 743) indicates the extent to which charges cover the cost of the service – but if the charges are beyond what they can afford, it will incite them to opt out of the service or to engage in free-rider practices. Within this criterion, we also deal with the issue of productive efficiency, which refers to the operational performance of the service provider measured by such things as labor productivity and costs per ton (ibid.: 743).

The contribution SWM makes to gainful employment is a key aspect of our assessment. It tries to ascertain whether jobs within the sector provide a living wage and a certain degree of job security. In addition, it seeks to compare the working conditions of various social groups. Legitimacy as criterion distinguishes between the legal situation and public attitudes. Legal recognition of a partnership may provide both advantages (access to credit and facilities; absence of harassment) and disadvantages (costs of formalization), and the same applies to non-recognition. Social legitimacy refers to

---

13. In an earlier version (Baud *et al.*, 2001) we spoke of better coordination within the SWM sector, which was discussed in terms of a clear demarcation of tasks and responsibilities (avoiding overlap). To assess this aspect we looked at the existence of policies and bylaws, and capacities for monitoring and law enforcement. Good coordination was supposed to bring about superior system efficiency. In practice, this aspect turned out to be very difficult to test. Therefore, we decided to narrow it down to effective monitoring and enforcement of standards, while discussing the legal framework under the legitimacy criterion, and efficiency under the heading of financial viability.

acceptance in the eyes of the public. The fourth criterion aims to find out whether mechanisms are put in place to monitor performances – in all three dimensions – and sanctions are applied in case agreed norms – output criteria, health standards, labor codes, environmental rules – are violated.

The third set of factors regard contributions to environmental health. The goals are:
- Greater effectiveness in achieving a clean urban environment;
- Minimize occupational health hazards for workers in SWM;
- Minimize environmental health hazards to man and animals related to the use of waste in agriculture.

The contributions to a cleaner environment can be investigated at two levels, that of the neighborhood where the activity takes place, and that of the city as a whole. From the perspective of SWM, cleaner neighborhoods largely depend on the quality of waste collection (service effectiveness), notably the frequency and the reliability of the service. However, pollution produced by local industries dealing with waste materials (air, water, soil) or by collection vehicles (air) also has to be taken into account. At city level, the contribution partnerships make to increasing spatial coverage of collection services is of prime importance. The goal of reducing occupational health hazards is quite obvious. It depends on the level of exposure to waste, especially to dangerous fractions of waste, and can be mitigated by the use of appropriate safety equipment. Finally, when waste – either decomposed or composted organic waste or mixed waste – is applied in peri-urban agriculture, possible negative impacts on animal health, soil conditions, and the quality of food crops also have to be considered.

1.6. TACKLING THE QUESTIONS RAISED

In order to tackle the questions raised in this study, a number of choices had to be made and limitations introduced, which will be briefly discussed here.

To begin with, a number of specific research questions were formulated, which were used as basis for the analysis across the different themes taken up. They were:
- Which are the main actors and partnerships found around the SWM activities taking place?
- How does the regulatory framework (including policies) affect the SWM activities taking place? and
- What contributions (outcomes) do the SWM activities by various actors and their partnerships make to aspects of sustainable development?

The area of SWM includes a wide variety of activities, as indicated in a previous section. Existing literature has often focussed on a specific area of concern, without regard for the whole of the SWM system[14]. However, in this study we wanted to move beyond such sectional and thematic studies. The goal has been to consider three domains within the total range of activities constituting SWM, as parts of a whole system, something, which to our knowledge, has seldom been done. The advantages of doing so is that an analysis can be made of how the different domains affect each other, and what conflicts and trade-offs exist between improvements in the activities in the various domains. The rationale was that in these domains the important parameters considered in the overall research question (social, economic and environmental aspects) could be found, so that an integrated assessment of contributions to sustainable development could be made. The domains chosen are:

- collection and transportation, and privatisation initiatives in this area;
- collection, trade and recycling of inorganic wastes; and
- reuse and diversion of organic materials.

The combination of criteria embodied in the concept of sustainable development as it pertains to SWM, leads to conflicts and trade-offs when improvements in the different domains are attempted. For instance, providing safe and healthy employment increases costs for either local government or private sector employers This may seem to conflict with the idea of reducing costs for SWM activities. However, making explicit the activities of SWM according to operational aspects of sustainable development, allows policy makers, CBOs, and NGOs, to weigh the possible outcomes according to their own preferences, and to make informed choices on that basis. These choices can be political; but the advantage is that this method of analysis allows the choices to be assessed in terms of their contributions (or lack of them) to aspects of sustainable development. At the moment, this is possible in an empirically qualitative fashion, not quantitatively. The contribution to enabling strategies lies in the potential for practitioners and researchers to make explicit the trade-offs between the different aspects of SWM, when specific choices are made.

The questions raised have been studied by carrying out in-depth comparative case studies in two cities: Hyderabad, India and Nairobi, Kenya. Both cities share a common heritage in terms of British colonial administration, but currently differ a great deal in terms of the strength of local government interventions. They differ in size, with the population of Hydrabad being more than twice as large as the population of Nairobi. The population of the capital city Hyderabad is around 4.2 million people[15] (http://travel.indiamart.com/hyderabad/index.html) and the capital city Nairobi counts around 2.14 million people (Republic of Kenya, 2001). The advantage of comparative case studies is that the contrasts found, can highlight the diversity of situations as well as show similarities, which is not possible in a single case.

---

14. For example, there is a great deal of literature on the technical aspects of SWM, and on the chemical hazards involved in unsafe disposal of solid wastes (the journal *Resources, Recovery, and Recycling* is an important source for this kind of literature).
15. As per 2001 census, the population of Hyderabad Metropolitan Area (including the nine municipalities around it) was about 6.5 million.

The methodology used consists of several elements, and was based on a comparative fieldwork approach, carried out by a multi-disciplinary team, in which staff members from four institutions have played a part. They included economists, human geographers, planners, and an environmental scientist. Participating institutions were the University of Amsterdam, which provided coordination as well as participation in research, the International Institute of Environment and Development, which provided international comparative expertise, the Centre for Economic and Social Studies, whose staff coordinated fieldwork and analysis in Hyderabad, and the Kenya team, with researchers from Moi University and the Institute of Social Studies (the Hague).

In order to develop a common understanding of the issues concerned, collaborative elements were built into the research process. At each stage of the research, team workshops were held so that a joint research framework was designed and discussed, the methodology of fieldwork was developed by the Nairobi and Hyderabad teams, compared and finalized, and reporting and analysis of results was done on a comparative basis in joint workshops, where contrasts were brought out sharply.

This approach led to the use of a combination of different sources of data: background literature, fieldwork for primary collection of data (through sample surveys), and a final workshop with stakeholders for feedback on assessments of fieldwork results in Nairobi[16]. The original research design combined an analysis of socio-economic data collected through surveys and qualitative data collected from people with strategic knowledge, with physical and chemical analysis of soil and water samples. However, the physical and chemical analysis was abandoned during the course of the project, because of the extremely high costs of analysing such materials in the countries concerned, and the limited relevance the specific enquiries were found to have in the area of environmental health. Therefore, data on environmental aspects concerns people's perceptions on environmental aspects, rather than physical evidence.

In retrospect, the design of the study held important lessons for future studies in this area. To begin with, the comparative approach brought out contrasts in the institutional and organisational context, which would have been lost in a single case study. Secondly, the joint building up of the research design led to analytical cohesion and built up essential team spirit across regional and disciplinary divides. Finally, the use of both qualitative and quantitative data was found to be essential in making complementary insights possible; neither can do without the other.

16. The specifics of the fieldwork as carried out in each location is the subject of the Methodological Appendix, which explains in detail how primary collection of data and analysis was done.

PART I

COLLECTION, TRANSPORTATION AND DISPOSAL OF URBAN SOLID WASTE

JOHAN POST

CHAPTER 2
EVOLVING PARTNERSHIPS IN THE COLLECTION OF URBAN SOLID WASTE IN THE DEVELOPING WORLD

2.1. INTRODUCTION

In many cities in the developing world – and Hyderabad and in Nairobi are no exception – the major change that can be observed in the collection, transportation and disposal of solid waste is the increased involvement of the private sector either 'spontaneously' in a free market setting or encouraged through local authorities, NGOs or CBOs. Solid waste management is no longer a (local) government monopoly but a domain open to various modes of public-private co-operation. Most studies on the subject start by enlarging on the failures in public servicing, including excessive numbers of workers, low labour productivity, few incentives for better performance, poor cost recovery, and low levels of investment and poor maintenance of vehicles and service equipment. Furthermore, the authorities often ignore the servicing of informal areas, either for principal or for pragmatic reasons or both. Subsequently, different methods of privatisation are suggested for achieving greater service efficiency and effectiveness. The major concern is to evaluate the organisational and financial aspects of privatisation initiatives, and to assess the capacity of government departments and private contractors to perform their new roles.

Although this chapter will start from the general debate on urban (environmental) management and urban sustainable development in the context of the developing world, it will focus on the consequences for policies and interventions with regard to the collection (partly including transportation and disposal) of urban solid waste, while disregarding implications for the other domains within the solid waste management system (see chapters 6 and 9). The analysis starts from a multi-actor perspective and recognises the existence of a multitude of institutional arrangements in solid waste collection. The essence of such 'partnerships' has already been described in chapter 1. The discussion will largely be on privatisation in solid waste collection, notably on the arrangements between local governments and private waste collection companies, as these have become very prominent recently. However, critical remarks will be made about the preoccupation, both in the literature and in policy practice, with the most dominant arrangements, while forgetting about the existence and the potentials of others, usually because of their informal nature. Furthermore, it needs to be said at the

I. Baud et al. (eds.), Solid Waste Management and Recycling, 21-36.

outset that global processes such as the promotion of privatisation policies produce different outcomes in different settings (Post *et al.*, 2001). The case material presented in subsequent chapters emphasises this clearly. In actual fact policies and interventions are strongly influenced by local networks of political and social relations. They shape both the form the privatisation exercise takes and the socio-economic and environmental impacts. The reason why substantial attention will be given to the policy context, both in this chapter and the next ones, is that it helps to understand the diversity of experiences across Africa, Asia and Latin America.

2.2. URBAN DEVELOPMENT IN AN ERA OF REFORM

The re-awakening of the neo-liberal paradigm in response to global economic crisis in the late 1970s and early 1980s also evoked a reorientation of ideas on urban development in the developing world. Under the auspices of the World Bank in particular the emphasis shifted to the economically productive role of cities and to the state as a facilitator or enabler of action by citizens, private firms and NGOs. Cities were once again seen as engines of growth, and the major challenge was to get things organised in accordance with free market principles. Urban management became the watchword and was primarily conceived as arranging for an appropriate division of roles and responsibilities between public and private (both commercial and social) actors, e.g. in which each of them would do what they are (supposed to be) best at (Safier, 1992)[1].

Although governments in the developing world have not always committed themselves formally to the new convention, most of them did accept the neo-liberal reform policies that constitute the basis of the urban management approach. Developing countries have been strongly advised, not to say pressed (through structural adjustment programmes and aid conditionality), to embark on policies to reduce the size of their civil services, to decentralise administrative power, to increase popular participation, and to make their administrations more efficient, transparent and accountable. In many instances these outside pressures were reinforced by local challenges to the prevailing authoritarian and centralist system. The growing inability of many states in Africa, Asia and Latin America in the 1980s to perform a socially relevant function for their citizens seriously undermined their legitimacy. By embarking on policies of decentralisation, democratisation and participation governments have tried to regain part of this legitimacy.

1. The promotion of enablement can be seen as a somewhat belated attempt to bring earlier ideas on self-help housing, community based settlement upgrading, and informal sector promotion to its logical conclusion. The entire structure of institutions, rules, working methods and attitudes must be geared towards working with these ideas. Furthermore, democratic and participatory government structures need to be put in place to ensure effective enablement (Post, 1997; Hardoy *et al.*, 2001).

Both internal and external forces, therefore, have altered the setting for urban development in Africa, Asia and Latin America. For the purposes of the current analysis it is not necessary to dwell upon the meaning and the attributes of urban management or critically examine its ideological roots (for this see Stren, 1993; Werna, 1995; Burgess *et al.*, 1997, Post, 1997). Suffice it to say something on the evolving new role of (local) government and, subsequently, to zoom in on the privatisation debate. However, first we need to say something on how urban management ideas are linked up with the sustainable development debate.

The concern for urban sustainable development in the South gained momentum in the course of the 1980s (see Hardoy *et al.*, 1990 and 1992). The underlying concept of sustainable development has ignited wide debate and aroused mixed feelings in academic and policy circles both because of its elusiveness and diametrically opposing interpretations. The major tension is between goals of economic growth and environmental protection. These are particularly acute in an urban context. Cities are seen as the engines of economic growth and centres of innovation, but they simultaneously draw heavily on natural resources and often threaten to overflow local and regional sinks. It is within this context of conflicting goals that we have tried to take position. As explained earlier, we will adopt an approach that starts from the Brundtland report and was elaborated subsequently, among others, at the IIED by David Satterthwaite. It seeks to combine goals of environmental aspects of sustainable development with the concern for meeting human needs (Satterthwaite, 1997; see chapter 1). Drakakis-Smith has further refined the conceptualisation of sustainable development in the urban setting and distinguishes between the economic component (the productive role of cities and its ability to provide employment and to reduce poverty), the environmental aspect (notably the sustainable use of renewable resources, minimising the use of non-renewable resources and appropriate physical planning), the social dimension (satisfying basic human needs and respecting human rights), the demographic backdrop (urban growth dynamics), and the political sphere (including the role of the state, democratic control and participatory planning). These are the essential components of a comprehensive and integrative programme of urban sustainable development (Drakakis-Smith, 1995: 663-6). What is important for our discussion, however, is the realisation that urban sustainable development, including sustainable modes of collection and disposal of urban solid waste, requires a multi-disciplinary perspective and frameworks of governance and institutional design that enable a constructive merging of these perspectives.

Unfortunately, the marriage between ecology and development has not been a very happy one. The utilitarian and exploitative worldview that occupies centre-stage in the eyes of most contemporary governments and development agencies is fundamentally at odds with an ecological outlook (Rees, 1995; Escobar, 1995). Very often the interdependence between the economic, the social, the political and the ecological system is ignored and academics, policy-makers and activists largely confine themselves to the dynamics of one system (Berkes and Folke, 1998: 4). As a matter of fact, although the activities leading up to the 1992 Earth Summit in Rio de Janeiro have helped to elevate environmental problems to decision-making circles in the South, this has hardly led to a 'greening' of their policies. Developing countries have made it abundantly clear that environmental policies should reflect their own priorities, and, more in particular, that they should not frustrate their legitimate desire to expand their economies (Elliot, 1998). This position is reflected in their efforts to shift the environmental focus from issues of greenhouse gas emissions, natural resource depletion, preservation of biodiversity and resource management – i.e. the prime environmental worries of the northern hemisphere – to issues on the so-called 'brown agenda', e.g. methods to mitigate the health and efficiency impacts of air and water pollution and to improve basic infrastructural services, notably safe drinking water, improved sanitation and appropriate waste management (UNCHS, 1996). To a certain extent these priorities are the 'conventional' preoccupations of (urban) administrators concerned with the basic needs of their citizens.

It is increasingly recognised that a degraded urban environment has a significant negative impact on urban productivity and efficiency as well as on public health, and that these effects disproportionally affect the urban poor (UNCHS/UNEP, 1997; Satterthwaite, 1997; Hardoy *et al.*, 2001). Garbage heaps in residential areas can contribute to the spread of infectious and parasitic diseases among residents, thereby lowering their labour productivity or their chances to become or remain employed. Illegal dumping of refuse into drainage canals can pollute surface water or cause occasional flooding, both with detrimental effects on productive activities. In the early 1990s the need for explicit urban environmental management to combat such harmful impacts was widely admitted. These policies were largely framed along neo-liberal lines, the argument being that the major cause of environmental decay was the combination of misguided economic policies, poor management, inadequate investment in infrastructure, and deficient regulatory and institutional frameworks. The advocated policies tended to rely heavily on incentives and regulations that aimed at establishing the true costs of environmental goods and services. A major proposition was to break up inefficient and ineffective public sector monopolies that provided subsidised services at well below economic costs, and to involve the private sector in service delivery (Burgess *et al.*, 1997: 72).

This is where both discussions, on urban management as well as urban sustainable development, meet: they are structured around exactly the same (neo-liberal) policy reforms. In fact, urban environmental management as practiced in most of the developing world is hardly more than a – largely half-hearted – attempt to merge the concern for improved urban productivity with the goal of environmental protection. The latter is simply added to the agenda, by no means as a primary goal. Although there are occasional success stories of local governments effectively curtailing their environmental problems – Curitiba and Porte Alegre in Brazil being familiar examples

(Hardoy *et al.*, 2001) – this is the exception rather than the rule. Most accounts of urban environmental management attest to the marginality of the environmental concern, particularly the concern of ecological sustainability (Mattingly, 1999: 113). The neglect of the latter may perhaps be caused by the fact that policies with regard to environmental aspects of sustainable development (the depletion or degradation of resources or eco-systems) are usually outside the purview of local authorities. However, the apparent lack of concern in many cities for the environmental hazards their (poor) residents are facing is more a matter of choice. The detrimental effects of a poor and unhealthy environment may constrain urban productivity and efficiency but they are indirect and only partly negotiated through the market. Calculating environmental costs or imposing environmental regulations will confront urban entrepreneurs with higher costs and suppress their productivity rather than enhance it, at least in the short term. The pricing of environmental capital requires strong political commitment, which is unlikely to occur in situations where the constituency is hardly exercising any political pressure to do so. One cannot deny the tension that exists between the drive towards liberalisation and deregulation to allow private operators to take the lead in urban development, and the need for (government) control and regulation of the private sector to effectively circumvent environmental degradation. The inherent qualities of the environment, involving its public goods nature, externalities, and common property problems, turn the market into a deficient institution for its management. These are exactly the kind of contradictions that become manifest in the analysis of the privatisation of urban solid waste collection.

2.3. DECENTRALISATION AND ITS LIMITS

The reason why attention has to be paid to the issue of decentralisation is not so much that privatisation is one of its modalities, but rather that privatisation of public services puts new, qualitatively different demands on local governments especially. In order to match these new demands they need to be empowered, and this is what decentralisation is supposed to realise.

Decentralisation refers to the transfer of responsibility for planning, management and resource agencies to lower echelons of government or to the private sector. Usually, devolution is considered the ultimate or 'real' form of decentralisation, since power (functions and financial means) is actually transferred to sub-national political entities, who in turn have real autonomy in many important respects (Rondinelli *et al.*, 1989; Dillinger, 1994). This is the way in which the concept 'decentralisation' will be used in the subsequent analysis.

In earlier times, decentralisation was largely considered an instrument to improve the efficiency of public administration. In the 1990s, however, the strive for decentralisation was largely motivated by changing views on state-society relations (Helmsing, 2000). As far as the developing world is concerned the crisis years had crippled many central states who were no longer able to perform their social functions nor, for that matter, to award their major supporters the fruits of holding government office. This situation eroded state power and evoked questions, among others, on what institution instead of central government is best suited to deliver public goods and services. In many countries, for example Brazil, Columbia, Ghana, and recently Indonesia, strong domestic pressures were exercised on governments to decentralise and democratise government structures.

The plea for decentralisation was further strengthened by the rise to prominence of the neo-liberal doctrine and the collapse of state-communism. Decentralisation was seen to be instrumental to the desire to roll back the state and create a slim and efficient type of government that would make optimal use of private sector and community potentials (Leftwich, 1994). It became more closely associated with the strive for 'good governance', interpreted as greater accountability (better opportunities for participation and subjecting public officials to popular control), transparency (i.e. a clear demarcation of tasks and responsibilities and more insight into the allocation of resources) and responsiveness (i.e. acknowledging the diversity of needs among the population in terms of policy responses) within a liberal-democratic framework. By bringing government closer to the people, decentralisation is expected to lead to more efficient, realistic and locally adapted development strategies. It will help to mobilise valuable local energies and resources and thereby enhance productivity (Olowu and Smoke, 1992). Finally, devolution was increasingly considered indispensable in an era of globalisation in which the role of the nation state (as a protecting or mediating force) is losing significance and local governments attempt to respond to the restructuring of the global economy (Schuurman, 1997)[2]. In order to maximise their ability to capitalise on their specific locational advantages and compete with other places all over the world, local governments have to be empowered.

Although the overall political-economic climate is evidently pro-decentralisation, there is nevertheless nothing inherently 'good' about the process. Admittedly, it can help to empower lower echelons of government or even disenfranchised groups, but at the same time it can work to extend central authority or strengthen mechanisms of political patronage and corruption at local level (Samoff, 1990). Furthermore, although decentralisation can diminish the gap between the state and civil society, it can also produce greater regional disparities because of the unequal distribution of competitive advantages and institutional capacities (Burgess *et al.*, 1997). Finally, the

---

2. There is some discussion on whether decentralisation is indeed an appropriate regulatory fix for most developing countries. According to Schuurman most of these countries have never experienced a transition to the Fordist regime of accumulation (mass-production) and the concomitant welfarist mode of social regulation. Therefore, '....it is premature to hollow out the state in Third World countries just because current Post-Fordist neo-liberal logic upholds this recipe for the North' (Schuurman, 1997: 156).

relationship between decentralised government and urban productivity remains far from clear. A more efficient and empowered local government might indeed succeed in creating a more attractive business environment by improving infrastructural facilities and reducing red tape. At the same time, however, it can very well lead to a reduction of access of the poor to urban land, and may even, as a result of more aggressive and systematic application of regulations and controls, constrain the room to manoeuvre of informal enterprises (Jones and Ward, 1994: 44). Waste pickers and itinerant buyers of recyclables, for example, may be confronted by more severe harassment despite their contribution to a cleaner environment and a reduction of waste volumes.

Such considerations, however, have not really prevented widespread adoption of decentralisation policies all over the developing world (Helmsing, 2000). In fact, the decentralisation idea was strongly promoted by major donors as well as from within civil society. However, political commitment of governments to a reform that challenges their power basis often leaves a lot to be desired. Therefore, the implementation of decentralisation policies is fraught with difficulties – slowness in reorganisation of central ministries, delay in the design of new procedures, obstruction in the transfer of fiscal powers etc. – and progresses much slower than anticipated (Burgess *et al.*, 1997). Despite decentralisation many local governments are still in a position where they lack the resources, the authority, clear and consistent mandates and sufficiently trained and supported staff needed to effectively enhance the development of their communities. Such shortcomings bear upon the predicaments of public-private and public-community partnerships in urban basic service delivery as local authorities are not automatically capable of delivering their share of the agreement. Negative experiences may even evoke policies of recentralisation, like for example in Accra (Ghana) where the failure of local authorities to adequately address the problem of solid waste collection has incited central government to step in and overrule the former (Obirih-Opareh and Post, 2001).

2.4. PRIVATISING URBAN SOLID WASTE COLLECTION

Governments privatise services for a multitude of reasons. Outside pressures from the Bretton Woods institutions as part of the overall structural adjustment reforms have been reinforced by domestic changes, notably a widespread dissatisfaction with the government's inability to properly manage the economy and provide appropriate services. Now that market principles have been (re-)installed in most national economies, private businessmen are expected to seize the opportunity. In the literature, the private sector is endowed with qualities such as political independence, economic rationality, efficiency, dynamism and innovation, qualities that make it measure up favourably to public sector enterprise. However, it would be extremely naive to take these salutary effects of privatisation for granted. First of all, empirical proof that privatisation actually works is still rather flimsy and largely drawn from experiences in the Western world (Burgess *et al.*, 1997: 81-2; Lee, 1997: 141). Moreover, the argument is based on indirect evidence to a certain extent. The poor track record of many activities run by the public sector is automatically translated into recommendations to work in the opposite direction, making policy into a sort of 'trial and error' process (Ramamurti, 1999). Second, it is far from certain whether the private sector will rise to the challenge at all. Entrepreneurs may hold back for fear of political instability or because they simply do not consider the provision of certain public services profitable. Third, there is often strong political opposition to privatisation from groups that stand to lose from the reform. There is a real danger, for example, that privatisation will lead to a net decrease of employment and that labour conditions in the private sector will compare unfavourably with those in the public sector. Furthermore, in the course of time, complex webs of mutually beneficial relationships have been established between state representatives and private interests. Many stand to lose when this edifice of patronage and privilege is torn down and will therefore resist reform or seek new ways to maintain the status quo, thus eroding the assumed economic advantages of privatisation. Finally, privatisation on the basis of the ability-to-pay principle is likely to exclude those who are beyond effective demand. Similarly, critics fear that private firms will eliminate unprofitable services and will provide inferior services in an attempt to maximise profits. When responsibilities are passed on to the private sector safeguards must be built in to ensure appropriate standards, achieve coordinated provision, ensure a competitive environment and avoid monopoly control of essential services by private providers which are not publicly accountable, and to minimise corruption and inequity (Rondinelli and Kasarda, 1993; Cointreau-Levine, 1994; Rondinelli and Iacono, 1996; Burgess *et al.*, 1997). Therefore, privatisation in service provision usually implies a public-private arrangement in which the government retains some sort of control, while saving on costs and reducing political interference and red tape.

Despite the ramifications mentioned above privatisation has become the political creed of the 1990s and its importance as a policy instrument must be accepted as a matter of fact. Therefore, one must adopt a realistic stance when looking for conditions that have to be fulfilled to allow one to reap the fruits of privatisation, while avoiding the pitfalls. When looking at privatisation, it serves to make a distinction between several dimensions: the form of privatisation, the type of activity to be privatised, the specific nature of public goods, and the main features of the policy context. These dimensions will be briefly discussed for the case of urban solid waste collection.

Starting with the first, each form of privatisation (divestiture, contracting, concession, franchise, open competition) has its own particular set of positive and negative impacts on the various stakeholders In waste collection the two methods most commonly applied are contracting and franchise[3]. In practice there are both situations where the private sector is unwilling to work with the government under contract, usually due to expected risks of non-payment or delays in payment, and situations where entrepreneurs are hesitant about operating under a franchise because they fear high default levels (related to general poverty, a lack of concern for public cleanliness, and the inability to sanction free riders by excluding them from service). Usually, contracting is considered the option that holds greatest potential to developing countries as a way of lowering costs of solid waste collection. It is especially attractive to governments that are eager to retain firm grip on solid waste collection, usually for reasons of public health. This form of privatisation works best if contract periods are not too short, and if tendering procedures and contract specifications stimulate competition and cost effectiveness. The longer the duration of contracts the more likely private operators are to invest in appropriate (cost-saving) equipment as time allows for the depreciation of their capital expenditures. Besides, long-term contracts reduce transaction costs and lessen opportunities for corruption and political manipulation. In case contract specifications are too detailed, e.g. not only on performance standards, but also on working methods, equipment, labour input etc., and control is too tight, this may stifle private initiative and increase supervision costs (Cointreau-Levine, 1994: 21-6). There is also the danger that contracts do not sufficiently recognize the variation in local circumstances, for example the difference between planned and unplanned areas.

Despite the acclaimed advantages of contracting, many advocates of privatisation are clearly in favour of the franchise system because it transfers the risk to the private sector 'where it belongs'. Franchise helps unburden the local public administration and avoids the endemic problems of weak local tax bases and poor revenue collection performance in many developing countries. However, this option will only be successful if safeguards are built in to ensure that privatised operation is sufficiently attractive (e.g. user charges must be commercially viable) and secure (ibid: 27).

With regard to the second dimension, the type of activity, the privatisation of state-owned enterprises is something completely different than the privatisation of urban services. While the former are usually under the jurisdiction of central government agencies and their performance can be measured in purely financial-economic terms, urban services are usually controlled by local authorities and their success is not only governed by economic returns but also by social and political factors (Lee, 1997: 141). In the case of solid waste collection questions of public interest and acceptability are at stake, which implies that privatisation usually requires 'the guiding hand of the state' to become effective. The authorities, for example, will have to arrange for proper monitoring of private waste collectors to ensure compliance to health standards or environmental regulations. This supervisory role is not only very complex and demanding, but also very costly. Studies on privatised waste collection often come to

3. In the case of franchise the local authorities give a private firm both the right and the responsibility to provide refuse collection services to customers within a given zone. The operator is allowed to charge individual households or houses an agreed fee to cover his expenses and to make a small profit.

the conclusion that services are delivered more efficiently than by municipal departments but tend to ignore the additional costs incurred by the authorities for contract management and performance monitoring, not to mention aspects such as land acquisition for disposal or transfer sites (Cointreau-Levine, 1994). Furthermore, it is often necessary for the government to remain actively involved in service delivery, not only for the pragmatic desire to keep in touch with operational reality and to have a fall back position in case of private company failure, but especially while the (commercial) private sector is interested only in servicing 'profitable' high-income and easily accessible areas (Batley, 1996: 744).

Regarding the third aspect, the specific nature of public goods strongly determines what kinds of private sector arrangements are feasible. The familiar classification of public services ranges from purely public goods (also called collective goods) which are consumed jointly and for which it is difficult to exclude people who do not pay (police, fire brigade) to purely private goods which can be consumed by individuals and from which people who cannot or will not pay for them can easily be excluded. In case of collective goods contracting and concession are the most appropriate methods of private sector participation. On the other hand, open competition is the most suitable option for public goods that can be treated as private goods (Cointreau-Levine, 1994: 6-7). Different activities within the solid waste management system fall into different categories. The sale of recyclables, for example, resembles a purely private good, while the cleansing of major roads and public areas falls into the category of collective goods. House-to-house collection of waste is positioned somewhere in between these extremes. It has the nature of a so-called joint use or merit good, which means that the service can be provided on the basis of people's ability to pay. However, charging rates on the basis of full cost recovery may incite many (poor) households either to engage in free rider practices or to opt out of the service (with detrimental effects for public health). Therefore, urban governments in many poor countries will be faced with the continued need to spend substantial parts of their budgets on waste collection services, and perhaps even be required to remain involved in its direct operation (Batley, 1996: 730).

2.5. POLICY CONTEXT

The fourth dimension, the policy context, deserves some more attention. The unique structure of political-economic forces and cultural attitudes to a large degree determines the perception of privatisation (favourable or unfavourable), the preferred type of public-private arrangement (more or less control by the government) and the conditions that require most attention (managerial capacities, limitation of investment risks, community participation etc.) (Rondinelli *et al.*, 1989). Despite strong rhetorical support for privatisation the process moves at a very slow pace in most low-income countries. Very often the preconditions for successful privatisation have not been fulfilled, such as the absence of well-developed capital markets (in case substantial investments are required by local firms), insufficiently adapted legal and judicial frameworks, low per capita incomes, lack of a vibrant private business class, and strong resistance from within the bureaucracy and the trade unions (Rondinelli and Kasarda, 1993; Ariyo and Jerome, 1999 Ramamurti, 1999). In the subsequent review the overall context in India and Kenya will be briefly characterised in order to demonstrate its importance in explaining the vicissitudes of privatisation and partnerships.

India

India's path to development since independence was pre-eminently state-directed with considerable emphasis on rapid industrialisation (through import substitution and with an emphasis on heavy industry). A powerful state apparatus tried to establish a self-reliant economy and to ensure fair distribution of the fruits of development both spatially and socially. However, the complex and excessive system of macro- and micro-economic controls failed to produce a pattern of rapid and self-sustaining growth and by the end of the 1980s the country was in deep economic trouble. In the early 1990s the country switched to a more market-led strategy in accordance with the worldwide prevalence of neo-liberal thinking (Pedersen, 2000). Although average GNP growth rates have been over 6 percent annually during the second half of the 1990s critics claim that the impact of the New Economic Policy has been disappointing, especially in a social sense. They point to the adverse impact of (national) politics and institutions. Economic policy is still largely a function of the whims of politics, notably the anxiety of politicians to stay in power (Khatkhate, 1997). This is an important reason why attempts to undo over-regulation and dysfunctional rules of economic governance are frustrated. The rhetoric of 'equity', according to Drèze and Sen (1997: 186) has often been invoked to justify governmental intervention, but has only strengthened the bureaucracy in its ability to control economic operations, to distribute favours, or to cause obstructions. The alleged beneficiaries of such action, the Indian poor, have got little from it, while those with good access to the bureaucracy fared well.

The peculiarities of India's recent development resonate in the analysis of the problems and prospects of privatising solid waste collection services in the country's towns and cities. Obviously, the idea of transferring this task to the private sector met with considerable opposition. The bureaucracy was largely unwilling because it suspected an erosion of government power. Many of its representatives maintain that only the public sector can guarantee equitable servicing. This helps to understand why privatisation of solid waste management tasks to date has only occurred on a limited scale, despite evidence of substantial cost reduction (Ali *et al.*, 1999: 505). At the same time, however, it looks like more progress is being made in implementing private sector participation projects in the last couple of years, simply as municipalities are not allowed to recruit new labourers whereas their service tasks grow together with their populations and area. The inheritance of state-led development and the power of officials with a traditional 'law and order' mentality also explains why the idea of wider participation of informal actors and community-based organisations in service delivery has not yet moved far beyond the stage of experiments (ibid: 497).

To a certain extent the deeply entrenched belief that the state should manage and control all public affairs could also prove beneficial as the success of privatisation, according to many commentators, depends on the local government providing an appropriate framework. However, this requires that local authorities have the financial means and administrative capacities to act accordingly. In actual fact the devolution of functional and financial power to local bodies in accordance with the Seventy-Fourth Constitutional Amendment 1992 is still being implemented. Critics claim, for example, that gains have been made in terms of 'democratic' improvements in the organisation of local administrative units which, however, are not commensurate with a clearly mandated functional agenda that will enable them to function as units of self-government (Sundaram, 2000; Sivaramakrishnan, 2000). Of course decentralisation is a formidable undertaking that will not produce immediate success. An important obstacle relates to the continued dependence of local bodies on funding by the Central and State governments. Decisions on the division of fiscal resources between the State governments and the municipality are somewhat arbitrary and few attempts have been made to match funds and functions (most important revenues continued flowing to the States). In actual fact a wide variety in approaches exists between different States, which may partly be attributed to their uniqueness but also attests to the lack of guidance at the national level (Sivaramakrishnan, 2000). Furthermore, most local bodies still have insufficient human and institutional capacity to perform their new functions, especially the smaller ones (Sundaram, 2000: 295). The weakness of the local tax imposition and collection system seriously hampers greater autonomy. Finally, many functions that are within the purview of local bodies are still carried out by State governments, leading to confusion and conflicts of competence (Mathur, 1996 and 1998; Singh, 1996). This is especially troublesome in the large metropolitan areas – Hyderabad being one of these – with metropolitan corporations and municipalities zealously defending their respective domains, while the State or Central governments residing in these mega cities simultaneously seek to exercise control (Sivaramakrishnan, 2000).

The imperfections of the decentralisation exercise also trouble solid waste management. Municipalities depend on regional or central state levels to cover investments in landfill sites, composting enterprises or incineration plants, at the expense of their own discretionary power. They may have to call upon the State government to ensure payment of private contractors collecting municipal solid waste. This introduces political elements in the negotiations beyond the domain of local politics and hinders long-term planning and efficient service provision (Ali *et al.*, 1999). Furthermore, the idea of close supervision inherent in contracting out public services requires the concerted efforts and close collaboration of contract managers, performance monitors, information campaigners and public health inspectors Most urban administrations are not really in position to properly arrange all this. Therefore, stories of contractors flouting contract stipulations, corruptive practices, poor complaint handling and inferior labour conditions keep recurring in the analysis of privatised waste collection services in Indian cities.

Kenya

For quite a long time Kenya's development strategy has been profoundly nationalist, attempting to achieve three major goals: to reduce the power of the foreign capitalist class that controlled huge parts of the post-independence economy, to Africanise the economy, and to kick-start the industrialisation process. In order to guide the economy in the desired direction the government became actively involved, and the government executive, notably the president and the cabinet, was given substantial regulatory powers (Gatheru and Shaw, 1998).

As a matter of fact the country's favourable resource endowment ensured a relatively strong economic performance until the 1980s. However, in the early 1990s the economic situation deteriorated sharply due to both internal and external factors (world wide recession, the Gulf War, poor rainfall). It drove the Kenyan government in the arms of the Bretton Woods institutions that imposed a structural adjustment programme. Through the familiar package of deregulation and liberalisation of markets, privatisation of parastatals, and reduction of government spending the economy was expected to be revitalised. But despite signs of recovery the overall performance was disappointing.

Kenya's economic problems can largely be attributed to the peculiarities of the country's political system. Considering the state's firm control on the economy and its role as the most important dispenser of resources, state power is the central preoccupation of politics. The KANU government of president Arap Moi has successfully used the deeply entrenched system of patronage and corruption to build an alliance of minority ethnic elites that constitutes its power base. The current political order, however, is extremely unstable. There are continuous clashes between the government and oppositional groups, culminating in occasional outbreaks of ethnic violence, as well as ongoing rivalry between pro and contra-reform factions within KANU. Furthermore, the government's legitimacy is threatened by people's growing indignation about the skewed distribution of basic services, rising political repression, and failure to turn the economic tide. Forces of change towards a further democratisation of society and a de-emphasising of the role of ethnicity are therefore gradually gaining ground (Southall, 1999).

The Kenyan state is notoriously unaccountable and unresponsive to the needs of the majority of its citizens. Most representatives of the state are understandably reluctant to implement reforms that threaten to undercut existing privileges. The lack of commitment to the idea of scaling down direct involvement of the state in commercial undertakings and service delivery, which are among the stated goals of the Kenya's National Development Plan, is a case in point (Aseto and Okello, 1997). The relationships between the Kenyan state and civil society are tense and characterised by distrust. NGOs and people's organisations have increasingly supplemented the declining state support for basic services since the 1980s. However, the state views this development with unease and has attempted to control these bodies from above and subjected them to the system of political patronage (Kanyinga, 1995).

As far as solid waste collection is concerned most local governments do not have official policies towards the privatisation of these services and actual support to the idea is largely confined to an occasional experiment (UNCHS, 1998). Nevertheless, numerous small and large firms have sprung up trying to fill the gap left by the dismal performance of most public cleansing departments. They are attending to the needs of the upper and middle-income groups who can afford to pay commercial prices (Werna, 1998). However, this spontaneous privatisation takes place without any institutional or legal regulation. Similarly, waste disposal in low-income areas, especially slums and squatter areas, largely depends on voluntary efforts to burn it or bring it to the nearest formal collection point.

The apparent indifference of local governments towards the service needs of their (poor) inhabitants partly relates to their subordinate and feeble position in the administrative system. Kenya is among a steadily declining group of countries in the developing world that have not yet seriously embarked on the road towards devolution. Most decisions of local councils still require ministerial permission. Furthermore, the discretionary power of local authorities is seriously compromised by the constant interference of the provincial administration working through state appointed district commissioners, district officers and chiefs that are primarily loyal to the president and the party. With respect to local finance there is still no system of central grants or revenue sharing in Kenya, which is exceptional from an international perspective. For their budget local authorities rely almost exclusively on land rates, service charges and license fees. However, financial management and resource mobilisation are notoriously weak. Most municipal councils suffer from lack of discipline in financial accountability, nepotism and patronage in recruitment practices, dishonesty of revenue collectors, political pressures on officers to be less aggressive in revenue collection, and high turnover of staff (Gatheru and Shaw, 1998: 142-4; Lewis, 1998: 145). Many of these problems are officially recognised and have made the government embark on a programme of Local Government Reform based on increased devolution. The policy pronouncements, however, still have to be matched with action on the ground.

2.6. UNDER-UTILISED POTENTIALS

Looking at the dominant trends both in the literature on the collection and disposal of urban solid waste and in policy practice several things stand out. First of all, reform in this domain is predominantly motivated by a concern for service efficiency and effectiveness. The former is largely economic – generating higher output from a given input of resources – and leads to cost saving, while the latter is concerned with the quality and coverage of services stemming from the desire to improve the overall public health situation (a concern that has always figured high on (local) government agendas). Therefore, impacts on labour conditions of people working in the sector have been of secondary importance, whereas environmental considerations have been virtually absent. In fact, views and policies on privatisation and partnerships in solid waste collection and disposal are only marginally influenced by the debate on sustainable development. Secondly, in designing new policies for solid waste collection and disposal comparatively little attention has been paid to 'unofficial' practices by informal operators and/or community-based organisations, sometimes supported by NGOs. Very often forms of 'unplanned' privatisation have developed in poorly served urban areas long before privatisation policies became popular (Baud, 2000). As a result of the implicit bias in official policies towards large-scale solutions and formal businesses, city administrations tend to bypass what has already been established on the ground and therefore remain unaware of the contributions these practices can (and often do) make to sustainable development (Baud *et al.*, 2001). A third, closely related observation is that the literature on solid waste collection and disposal in the developing world largely ignores links with the other two major domains in solid waste management, the reuse and recycling of inorganic waste and the reuse of (composted) organic waste matter. The preoccupation of waste authorities with effective and safe collection and disposal takes away their interest in these activities despite the fact that they could help to reduce waste volumes substantially. Besides, significant parts of these recycling and reuse activities take place outside government purview and official attitudes often vary between benevolent neglect and downright harassment. Little is known about possible commonality or conflict of interest between actors working in several domains. Finally, there is an undeniable tendency to think in terms of standard methods of private sector participation in urban solid waste collection and disposal. Whereas in actual fact a multitude of actors and activities can be identified – each having their own rationale – official policy often focuses on one particular method. This does not do justice to the variety of circumstances and needs that have to be catered for, more in particular those of residents living in poorly accessible, usually low-income areas. The current study seeks to correct all these shortcomings to some degree, at least in as much as the two cases allow for that.

A few more words need to be said with respect to partnership arrangements that have only received scanty attention by (local) governments. In case of privatisation they usually demonstrate a clear preference for transferring waste collection services to large-scale enterprises. Government bodies are generally reluctant to create partnerships with small-scale enterprises and community-based organisations (CBOs) despite the fact that they have already proved their value in various domains, notably shelter provision and upgrading (UNCHS, 1996). Most administrators and officials think the inclusion of such small-scale (often, though not exclusively, informal) or communal activities is at odds with their (westernised) convictions on how to run a modern city. The elusiveness of such activities is considered to be a threat to the enforcement of government rules and regulations (including sanitary codes and health standards) and could make effective sanctions in case of malpractice difficult to implement. Besides, most local governments are unable to meet the transaction costs entailed in dealing with a multitude of small firms and community organisations. Only very slowly prevailing negative views are starting to give way to a more positive outlook in accordance with new ideas on urban management and local governance that recognize the potential of these actors as well as the usefulness of collaborating with them (Anschütz, 1996; Haan *et al.*, 1998; Baud, 2000).

As for micro and small enterprises (MSEs) several studies show that they can deliver good quality waste collection services (cf. Arroyo *et al.*, 1999; Obirih-Opareh and Post, 2001). Their advantages include cost savings following from the use of more appropriate (cheaper) technologies such as handcarts and donkey carts, lower wages (albeit it sometimes through severe underpayment of workers), higher flexibility, stronger commitment to the job, closer links with the community, and competition among MSEs (Haan *et al.*, 1998: 12-6). These enterprises are most suited to carry out 'tasks where there are no or few economies of scale, or where the effect of the economy of scale is easily compensated by increased efficiency' (ibid: 20). Sweeping and primary collection of garbage satisfy this requirement. These services may also be organised through some kind of collective effort (cooperatives, community-based organisations). Among the advantages of such collective action are close involvement of the community, which fosters responsiveness to local needs and makes active participation in the service, prompt payment of fees, and direct quality control more likely. Weaknesses are the dependence on the skills and dedication of volunteers, and the lack of a business-like approach to service provision both of which may threaten the continuity of the initiative (Anschütz, 1996). In the current study these 'hidden' potentials are included and their contributions to sustainable development are assessed.

In conclusion, the collection (transportation and disposal) of urban solid waste is no longer an exclusive public sector affair. Several new partnerships have come up either on direct initiative by the (local) government or 'spontaneously' in the private (commercial or non-commercial) sphere. Many cities in the developing world currently demonstrate the existence of pluralistic systems of service delivery. What this system looks like, how it performs, and how it is coordinated largely depends on the societal context. In the subsequent three chapters this will be elaborated for our two case cities, Hyderabad and Nairobi.

S. GALAB, S. SUDHAKAR REDDY AND JOHAN POST

CHAPTER 3
COLLECTION, TRANSPORTATION AND DISPOSAL OF URBAN SOLID WASTE IN HYDERABAD

3.1. INTRODUCTION

Since 1993 the Municipal Corporation of Hyderabad (MCH) has introduced the so-called 'Voluntary Garbage Disposal Scheme' (VGDS). In 1999 the programme operated in over 1,000 residential colonies covering about 175,000 households. The basic idea underlying the scheme is to promote people's participation in solid waste management through their welfare organisations. The MCH provides communities with a tricycle free of charge to enable house-to-house collection of garbage. The welfare organisations employ waste pickers – paid through a monthly fee of Rs 10 per month per household – to collect and transport the garbage to notified vantage points where the MCH collects it for transportation to one of the dumpsites. Households are supplied with two waste bins to facilitate separation of organic and inorganic waste at source. The waste pickers further segregate the inorganic waste and sell valuable materials to waste dealers The organic materials are supposed to be brought to nearby sites allotted by the MCH in order for CBOs/NGOs to convert it into manure through vermiculture (see chapter 10).

This initiative attests to the desire on the part of the local authorities to further improve the performance of the city's solid waste management[1]. The gradual involvement of the private sector in sweeping and waste collection work since 1995 is another token of the MCH's pro-active policy. There were good reasons to try and utilise the capacities of other actors in the urban arena in public service delivery. In the mid 1990s the proportion of uncollected waste was estimated at 25-35 percent in Hyderabad (Omkar and Srikant, 1996; EPTRI, 1997)[2]. At that time the garbage was cleared once in a day in some areas, once every two days in others and in some areas only once a week.

1. Another initiative that can be mentioned is the Clean and Green Andhra Pradesh Campaign that was started in 1998. The State government has issued orders declaring every second Saturday of the month as Clean and Green day. The idea is that residents, municipalities, private contractors and NGOs join hands on a voluntary basis to clean up the city.
2. It is not entirely sure if this figure includes or excludes the illegal settlements within the municipal boundaries.

I. Baud et al. (eds.), Solid Waste Management and Recycling, 37-60.

However, after the introduction of private operators in sweeping and waste collection, especially since the unit system was adopted, the performance has improved considerably both in terms of coverage and frequency. Nevertheless, shortcomings continue to exist, notably in the low-income and illegal sections of the city. For political reasons priority is given to servicing the middle and high-income areas.

In this chapter the performance of the collection, transportation and disposal of solid waste (hereafter SWC) within the MCH is investigated in terms of its contribution to urban sustainable development. The analysis will start by briefly describing the organisational structure. Subsequently, the major waste generators and actors in waste collection will be identified followed by a systematic assessment of the performance of various institutional arrangements in SWC using the indicator system developed in chapter 1. On the basis of this evaluation conclusions may be drawn regarding appropriate policy responses (chapter 13). A methodological note on the research activities carried out in Hyderabad is included in an appendix at the end of this book. More details on the research can be obtained in the reports that constituted the basis for this summary (Sudhakar Reddy and Galab, 2000a, 2000b and 2000c).

3.2. ORGANISATIONAL STRUCTURE

The MCH functions under the control of an elected Commissioner who supervises the various departments charged to implement the municipality's statutory responsibilities. The task of SWC is discharged by the Health Department headed by the Chief Medical Officer of Health. The Health Department has two wings, *viz*. The Sanitary conservancy section (see Figure 3.1) and The Transport section. The sanitary wing takes care of street cleansing and collection of waste, while the transport wing sees to

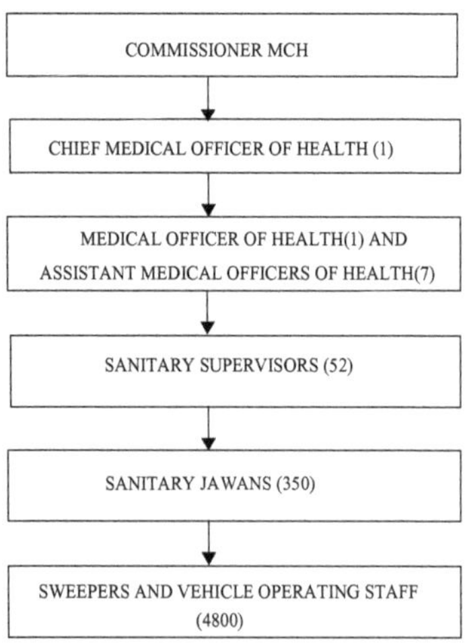

Figure 3.1. Management chart for the sanitary wing of the Health Department

transportation and disposal of waste. The Engineering Department maintains transport vehicles such as tipper trucks, compactors, lorries etc.

The MCH is divided into seven circles (see Figure 3.2) each headed by an Assistant Medical Officer of Health supported by a staff of sanitary supervisors who in turn are assisted by the so-called *sanitary jawans*. The latter manage daily operations by the conservancy staff consisting of sweepers (both men and women) and drivers In the mid 1990s it was reported that the MCH had a staff deficiency of over 50 percent. This problem has meanwhile been partially offset by the involvement of private operators in SWC. The history of the MCH privatisation policy will be elaborated in chapter 5.

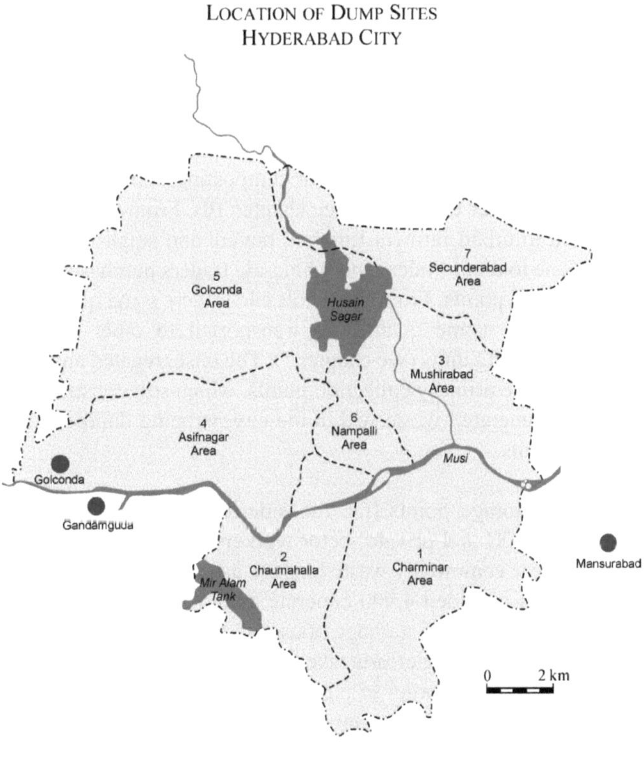

Figure 3.2. Map of Hyderabad's administrative divisions and location of dumpsites

Since 2000 SWC in the city has to abide by the Municipal Solid Wastes Management and Handling Rules issued by the Ministry of Environment and Forests and made mandatory by the Central Government. The municipality is responsible for the implementation of these rules and for any infrastructural development required for appropriate SWC. It may be sanctioned in case of infractions. These rules and their legal implications are forcing municipalities to make serious business of SWM in their judicial areas.

3.3. ORGANISATION OF SWC SERVICES

Collection and disposal

Solid waste is generated in households and institutions. Among the households a further distinction is usually made between the various income groups: high, medium and low. Institutions can be further subdivided in government institutions (schools, offices, hospitals etc.) and private institutions (shops, markets, offices, hotels, private clinics etc.). All these generators store their waste on or near their premises (primary storage). From these primary storage points some of this waste – unfortunately the volume is unknown – is separated by servant maids, tricyclists (in the VGDS scheme), office boys and shop keepers After the above mentioned actors separated the collected waste, some of the organic waste materials is taken by heardsmen to cattle farms, while the remaining fraction is transported to composting units by MCH and private trucks and tricyclists (under the VGDS; see chapter 10). From the composting units, the compost is sold to urban farmers Itinerant buyers and small traders who subsequently sell the waste to retail traders and wholesale traders purchase inorganic waste from primary collection points. From the wholesale traders some quantity of waste is offered for local reuse, some quantity is transported to other states, while the remainder goes to recycling units (see chapter 7). The unsegregated and mixed municipal solid waste from the primary collection points, which still constitutes the bulk of the total volume of generated waste within the city, is being dumped at dustbins or secondary storage points.

From the secondary storage points (i.e. roadside dustbins) the waste is carried to collection points by MCH and private sector workers and it is dumped at secondary collection points. These community waste bins are usually located at street corners In the city, the MCH has provided 4,900 concrete cylinder waste bins (0.6 cum), 420 metal waste bins (1 m^2), and 105 garbage houses (2.5 m^2) covering all the areas[3]. At these intermediary points sorting efforts take place both by the official waste workers and informal waste pickers in order to pick out materials with a potential monetary value, in practice only recyclables, as most reusable materials are already taken out

3. The total holding capacity of the existing collection centres comes to 3,623 m^2 of garbage amounting to 2,170 tons. Thus the garbage holding capacity is higher than the waste generated per day. However, the location and spread of these waste bins across the population and the frequency of collection are more important factors than the garbage holding capacity. The illegal settlements, for example, are poorly provided. For the sake of comparison: while every waste bin in the MCH serves 737 persons, the figure in the nine adjoining municipalities is much more favourable with an average 270 persons per bin.

and the organic substances have become too contaminated. These items are sold to itinerant buyers, retail and wholesale traders From the secondary collection points the MCH trucks transport the waste to one of the transfer stations and from these (in larger trucks) to dumpsites, while the private trucks directly head for the latter. From the transfer stations and dumpsites, some quantity of waste is collected by dumpsite waste pickers and sold to retail and wholesale traders

The MCH currently transport about 200 tons of residual waste per day to Selco International Limited who turn it into pellets (a coal substitute). The pellets are sold on the market to industries using coal-based power generation[4]. In the past, part of the remaining unsegregated waste at the dumpsites was carried to farms and fields by rural farmers to use as compost. However, this practice has discontinued as a result of increasing awareness of negative externalities (see chapter 10). The entire generation, collection and transportation system is summarized in Figure 3. 3.

The MCH currently has 183 vehicles available for the transportation of waste. With an average capacity of lifting 9 tons a day the entire fleet can remove almost 1,700 tons per day. However, in reality a maximum of 900 tons per day is brought to one of the dumpsites. The reason is that almost 20 percent of the fleet is out of order at any point of time. Furthermore, the MCH system is divided into two circuits: smaller (open) trucks transport the waste from the secondary collection points to a transfer station, where it is transhipped into larger trucks that bring it to its ultimate destination.

The most common method of disposing of waste in Hyderabad is open dumping. Within the City, there are three official dumpsites: at the Mansoorabad (Autonagar) dump site, 45 acres located 20 km from the city centre, Jamalikunta (Golconda) dump site with an extent of 20 acres located 8 km away from the centre, and Gandamguda dump site, 22 acres located 10 km from the centre of the town (Figure 3.2). Sanjivaiah dumpsite was closed in 1995 as the park boundaries were being extended. There are other unofficial open dumpsites within the city, such as around Hussiansagar. Some people say that the municipal crew used to dispose of garbage in old, unused wells. Under the current unit system (see below), however, the number of illegal dumps has probably come down because payment of private service providers depends on registration of trips.

Waste characteristics

The high decomposable substance (about 55 percent) necessitates frequent collection and removal of municipal solid waste (table 3.1). Furthermore, the characteristics of

4. The plant's capacity is 500 tonnes per day. Thirty percent of the waste supplied by the MCH is not usable due to contamination by hospital waste but the remainder is converted into pellets. The 30 percent residue is returned to the dumpsite.

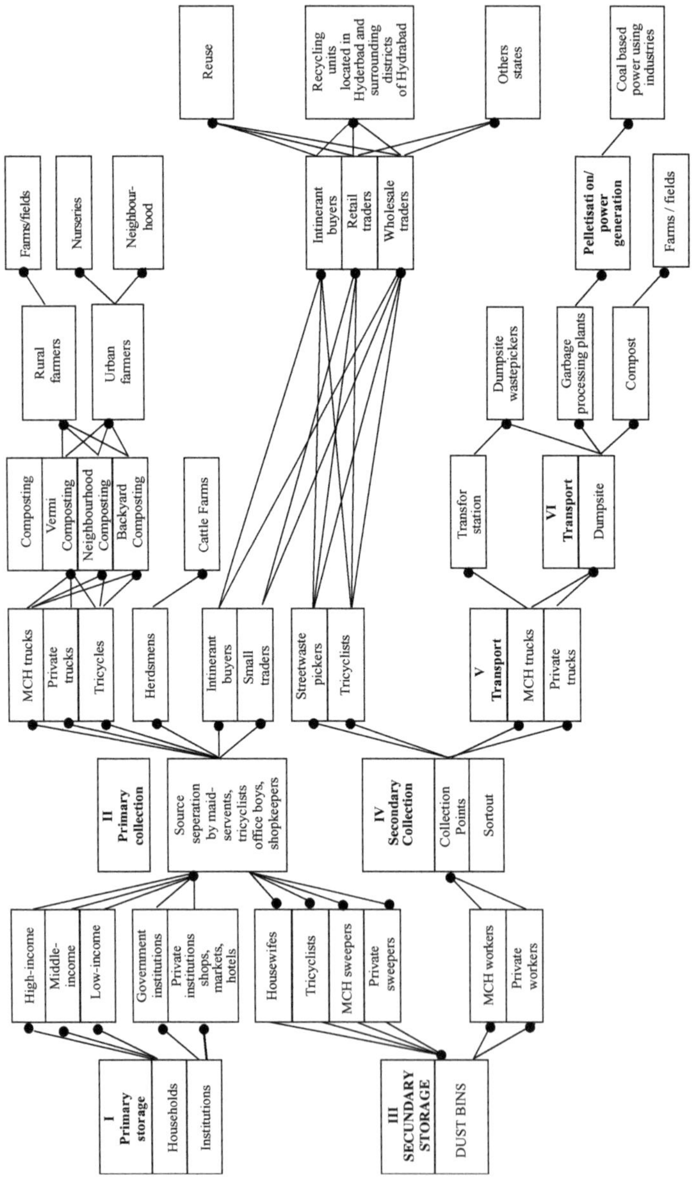

Figure 3.3. Overview of steps in the generation, collection and transportation of solid waste in Hyderabad

the waste have been undergoing change in the course of time: the organic (compostable) fraction is declining, while inorganic material has been increasing. However, the proportion of organic matter is highest among the low-income groups and declining as the levels of income increase, while the inorganic fraction is increasing with rising levels of household income. Household waste mainly includes the following items: i) combustibles (paper, plastics, rags); ii) non-combustibles (glass, metal); iii) compostables (vegetable/food, leaves); and iv) toxic battery cells etc.

Table 3.1 Physical composition of fresh urban solid waste in Hyderabad 1997

No.	*Characteristics*	*Percentage*
1.	Biomass	55.0
2.	Paper	7.0
3.	Rubber/leather	2.0
4.	Plastics	2.6
5.	Rags	8.0
6.	Metals	0.2
7.	Glass	0.2
8.	Sand/fine earth	13.0
9.	Stones etc.	12.0

Waste generators

The quantity of solid waste generated in the MCH is estimated at 1,500 tons/day in 1999 based on the assumption that the average generation rate per capita is around 0.35 kg and the total population is about 4 million. The quantity of waste generated varies from 0.24 kg/day among the lowest income groups to 0.75 kg/day among the higher income groups[5].

The generators of waste can broadly be divided into two groups: bulk generators, both public and private, and small generators, notably households. The former consists of 34 markets, commercial centres and recreation places, 923 hotels and restaurants, 93 function halls, 417 hospitals and nursing homes, 5 slaughterhouses and 30,000 cattle. In addition there are about one million households. Though they generate small quantities of waste individually, taken together they contribute about 75 percent of the total volume (table 3.2).

5. In the nine municipalities adjoining the MCH another one million people are living. The total waste generated in the entire conurbation, therefore, is around 1,850 tons/day. However, these neighbouring municipalities all have their own provisions for SWC. This chapter is confined to the MCH area alone.

Table 3.2 Waste generation by source within the MCH (1997)

Source	*Units (approx.)*	*Estimated volumes (tons per day) Abs*	*Estimated volumes (tons per day) percent*
Households	1.0 million	1050	75
Hotels and restaurants	923	80	5.7
Function halls	93	?	?
Markets	20	60	4.3
Hospitals and nursing homes	417	60	4.3
Slaughter houses	5	20	1.4
Recreation and community centres	35	80	5.7
Dairies etc.	30,000 (cattle)	50	3.6
All sources		1,400	100

Source: MCH, 1997

3.4. SWC FROM BULK GENERATORS

The assessment of the SWC performance in Hyderabad will largely be based on household SWC. To start with, however, a few words will be said on the peculiarities of SWC from various bulk generators using data derived from our survey among these institutions (see methodological appendix).

Hospitals

The number of hospitals, nursing homes and clinics have increased is the twin city Hyderabad-Sekunderabad over the past two decades. They generate waste containing infections and toxic material, which are deposited either in street waste bins or masonry bins built inside the hospital premises. According to national regulations, each hospital should have its own incineration plant. In actual fact few hospitals are equipped with such a facility due to the high costs involved. When they do have an incinerator it is not always possible to use the machine due to regular repairs or power failure.

The present study of eight hospitals has been taken to examine the ways in which hospital waste is disposed of. They include government, corporate and charitable hospitals, as well as a private nursing home[6]. Most of the sampled hospitals have established service links with other hospitals. For instance Vijaya Marie hospital is linked with Appollo Hospital in disposal of wastes from the hospital. All the hospitals, except Medicity hospital, have collection arrangements with the MCH.

There are various methods to dispose of hospital waste. In all government hospitals in our sample anatomical wastes like human tissues, human organs and body parts are disposed of by deep burial. In some other hospitals they are simply dumped in waste bins after being put in covers All other types of hospital waste, ranging from laboratory cultures and biological toxins to outdated drugs and materials contaminated with blood, are mostly dumped in waste bins, although some times these materials are incinerated or autoclaved and then dumped in waste bins. Only three hospitals have incinerators However, in one of them (Osmania hospital) incineration is not functioning due to restrictions imposed by the Pollution Control Board. The remaining five sample hospitals do not have incinerators In all the eight hospitals materials are segregated before being sent to waste bins. Only four hospitals subsequently sell paper, plastic and iron items. All the hospitals are satisfied with the work of MCH workers who clean the waste bins daily or sometimes once in two to three days. Nevertheless, among the eight hospitals included in our sample five are planning to give waste collection to private operators as they feel these will do better than the MCH workers because of their punctuality and regularity. In addition, they think that MCH is not capable of handling biomedical waste in a professional manner.

Educational institutions and offices

For the present study three high schools, three Junior colleges, three Degree colleges, two Engineering colleges and four offices were selected for closer examination on their waste practices[7].

---

6. The hospitals selected for the present detailed study are, Osmania General Hospital, Gandhi Hospital, Government General and Chest Hospital, Medicity Hospital, Medwin Hospital, St.Theresa's Hospital, Vijaya Marie Hospital and Vijaya Krishna Nursing Home. Osmania and Gandhi Hospitals are the oldest and the biggest Hospitals among the sample hospitals with the staff strength of 2,100 and 1,500 respectively. They were established is 1866 and 1851 respectively. Government General and Chest Hospital was established in 1971 with 200 beds and 400 staff. Medicity Hospital was setup in 1992 with staff of 350 and 150 beds. Medwin Hospital, a corporate hospital, was established in 1989 with 750 staff and 350 beds. St. Theresa's and Vijaya Marie Hospitals are run by Charitable Trusts established in 1973 and 1942 with 425 and 60 staff respectively. Vijaya Krishna Nurshing home was started in 1987 with 40 beds and 12 staff.
7. Osmania University College for Women is the oldest college established in 1924 with 480 staff among the sample Institutions selected. The other degree colleges selected are Siddhartha College and Annie Besent College, which were established in 1995 and 1994 respectively. The Engineering colleges selected for the present study are Jawaharlal Nehru Technological University College and DVR College of Engineering and Technology, which were established in 1985 and 1998 respectively. The offices selected for the present study are Office of the Municipal Corporation of Hyderabad (800 staff), ITC Badrachlam Paper Boards Limited (300 staff), Institute of Health Systems (60 staff) and Office of the Accountant General (1,650 staff).

The sample institutions store their waste in plastic containers, gunny bags or polythene cement bags and subsequently dump it somewhere on their compound or in the nearest MCH dustbin. In the latter case their own staff carries this out. As far as the public schools and municipal offices are concerned, these usually take the help of MCH workers to dispose of the waste from their institute. Within the waste generated in the sample institutes, only paper waste is taken out to be sold by some of them. Other materials are rarely sold. In most sample institutions waste bins are cleared daily whereas in the remaining institutions they are cleared once in two to three days, sometimes once in a week and once a month also. Almost none of the investigated institutions plans to give waste collection in their campus to private contractors; the volumes of waste generated in their institutions is usually rather limited and can be disposed of easily by their own staff. Only two institutions are planning to give waste management to private contractors Nevertheless, the majority of them agree with the notion that private contractors do better than the MCH because they feel that private contractors are more accountable and better supervised.

Hotels

Hotels are key generators of waste. For the present study, twelve hotels have been selected to assess the quantity of waste generated and the way of disposal of waste[8]. The majority of the hotels use plastic containers to store the waste while others use polythene covers Moreover, this waste is first dumped inside the hotel premises. All the hotels dispose of their waste – which may amount up to 100 kg per day – on a daily basis. Several hotels sort the waste taking out materials such as liquor bottles, paper, and glass pieces that are sold out. Sometimes edible waste is given to beggars Very often the hotels have some kind of arrangement with cattle farms or piggeries who collect food and vegetable waste free of charge to be used as animal feed. Some of the hotels have contracted out their waste collection to the private sector (Golconda, Paradise-Percis), but most, especially the smaller and medium sized hotels, continue to use the MCH service. The latter do not have plans to give waste collection to a private contractor.

Markets

The city of Hyderabad (including the adjacent municipalities) has 34 markets and 5 slaughterhouses. These are all major generators faced with the problem of waste pilling up and leading to unhygienic conditions on the premises. Among the research

8. The following hotels were included: Golconda, Ashoka, Ramakrishna, Bawarchi, Paradise – Percis, Kamat, Fishland, Rajdoot, Kings 'N' Cordinal Bakery and Confectionery, Dhana Laxmi Mess, Sri Devi Mess and Sri Balaji Mess. All these hotels were established between the years 1953 – 1996. Among all the sample hotels Hotel Golconda and Hotel Paradise – Percis have highest staff strength of 300 and 250 respectively. All the three messes have only 10 staff each.

population two slaughterhouses and eight markets have been selected to assess waste collection and disposal practices[9]. The quantities of waste generated by these units per day range from 20 tons (Monda vegetable market) to some 200 kg (Ramnagar fish market). In most of the markets the waste is stored in metal containers or RCC bins, but some simply dump it in open places. Two markets (Gudimalkapur vegetable market and Gaddiannaram fruit market) have entrusted the responsibility of waste collection to private contractors while all the others call upon the MCH to remove the waste. This is all done by trucks and on a daily basis. The management of several markets and slaughterhouses considers shifting responsibility for waste collection to private contractors, which they expect to provide more regular and reliable service than the MCH.

Among all the markets only Gaddiannaram fruit market has a composting unit. The management of the fruit market has allotted one acre of land to the vermin-composting unit for which technical know-how is provided by the SPEQL (see chapter 10). It is one of the model-composting units in Andhra Pradesh. The farmers and horticulturists from the surrounding villages and districts buy the compost prepared from this unit as well. None of the markets sells any inorganic waste material except for paper.

3.5. SWC IN RESIDENTIAL AREAS

In the remaining part of this chapter the focus will be on household waste and the organisation of its removal. A distinction can be made between the basic SWC system run by the MCH itself or its accredited agents from the private sector, and the VGDS which can be seen as a further upgrading of the basic service.

Basic service

In principle the MCH utilises a uniform SWC system regardless of the status of the area. In actual fact, however, regular servicing is confined to legally recognised residential areas. For reasons of public health, people living in illegal settlements and slums are not entirely ignored but service levels are considerably lower in terms of availability of waste bins and frequency of collection. People are expected to bring their garbage to vantage points (dustbins, garbage houses, containers or open dump-sites). Sweepers see to it that the streets are cleaned and bring the swept waste to the same communal collection points. From there the waste is lifted manually into vehicles that transport it either directly to one of the three city dumps or to a transfer site. The MCH is planning to establish twelve such transfer stations – three are already

9. The markets selected for the study are Gudimalkapur vegetable market, Monda vegetable market, Kothapet Rythu Bazar, Madannapet vegetable market, Uppal vegetable market, Gaddiannaram fruit market, Mojam Zahi flower market, Ramnagar fish market and Amberpet slaughter house and Bolakpur slaughter house.

in operation – in order to save on fuel, and maintenance of trucks, and to reduce time losses as a result of increasing traffic congestion. The entire system is very labour-intensive and wages constitute the bulk of the solid waste management budget.

In 1995 the MCH embarked on the road to privatisation in SWC (for details see chapter 5). The initiative to transfer this task to the private sector came across considerable opposition both from the bureaucracy (fearing an erosion of government power) and the labour unions (protecting the interests of the well-organised MCH-workers). However, as municipalities are not allowed to recruit new labourers[10] while at the same time their service tasks grow together with their populations, they are left with no other option than to go ahead with the programme of private sector participation. Furthermore, promises were made that the existing staff would not be retrenched from their jobs. At first only about 10 percent of the MCH area was brought under privatisation (especially peripheral, newly developed colonies), but gradually more areas were handed to the private sector and in 1998 almost 25 percent of the MCH area was under privatised servicing. Initially contractors were operating under two different systems. The lowest bid system called for tenders in which the private contractors could quote the amount for sweeping and lifting the waste in an entire area on a daily basis. The second system was that of payment on the basis of tonnage of garbage lifted assuming that this would incite the contractor to lift more garbage. Contractors could bid an amount per ton including costs of sweeping. Both systems suffered from serious weaknesses, largely related to a lack of proper monitoring and control by the MCH. In the first system contractors used to get their amounts even if they were not performing well because fixed output norms were absent, and in the second system contractors were resorting to lifting stones, construction debris etc. in their vehicles to increase weight. Furthermore, the contractors were reported to form syndicates, involving MCH-officials, to quote substantially higher prices over the previous rates.

In an attempt to overcome these problems the MCH adopted a new organisational framework and a more detailed monitoring system in 1998. The MCH-area was divided into 266 units each having a total length to be swept of 7-8 km and an amount of waste to be lifted of 7-8 tons per day. Every unit comprises one sweeping unit and one lifting unit each with stipulated labour requirements and specified implements. Payment of contactors under this so-called unit system is based on a uniform calculation including a specified profit margin of 10 percent (which was reduced to 8 percent in 1999). From the 266 units 146 were privatised, while 120 remained under MCH responsibility. Contractors were invited to participate, provided they satisfied certain legal requirements and submitted a Rs 25,000 deposit (US$ 625). Each contractor was

10. It is not entirely sure whether this ban on the recruitment of municipal workers dates as far back as the 1970s already or whether it was enforced in 1984.

allowed a maximum of three units (since 1999 only two). As the MCH received more applications than the number of units, a final selection was made by drawing lots.

House-to-house collection

The MCH is also seeking to establish partnerships with community based organisations and NGOs in order to improve people's participation in SWM. In 1993 it introduced the VGDS that is currently applied in over a thousand residential areas covering approximately 175,000 households. The MCH renders support by making a tricycle – a three-wheel bicycle with a waste reservoir – available to the residents free of costs. The initiative is also extended to slum areas with funding from the British government. Waste pickers who are hired and paid by local welfare organisations collect garbage on a house-to-house basis. The residents contribute to the system by way of a monthly fee ranging from Rs 5-20 depending on the volume of waste to be lifted from each household (joint families paying larger amounts). The scheme is working very effectively because people demand value for money and because the welfare organisations provide strict supervision. Waste bins are no longer required in the colonies participating in the scheme as the waste pickers collect the garbage at the doorstep and bring it to secondary collection points outside the areas.

3.6. ASSESSING SUSTAINABLE DEVELOPMENT IN THE BASIC SWC SYSTEM

The two systems of SWC, although partially overlapping, will each be evaluated in terms of their contributions to various aspects of urban sustainable development. The set of indicators presented in chapter 1 has been used to structure the analysis, albeit in a flexible way to suit the particular circumstances in Hyderabad as well as the data at our disposal. We will start by assessing the qualities of the basic service with special emphasis on the differences between service provision by the MCH and private contractors

Contributions to socio-economic aspects of sustainable development

Economic efficiency and viability
The SWC system in Hyderabad scores low in terms of allocative efficiency (the extend to which charges cover the cost of the service). Although the expenditure for SWC weighs heavily on the municipal budget there has not been any attempt to introduce user charges. The introduction of the private sector has not been accompanied by any form of cost recovery, which may be called exceptional by international standards. The political costs to levy a new tax on the electorate are considered too high even to discuss the issue openly. Similarly, the interviews with contractors displayed a strong dislike on their part to engage in a franchise system because they feel that residents will not pay their dues.

As far as the operational performance of the service providers is concerned (productive efficiency) it appears that the average labour productivity of private sector workers is slightly higher than that of MCH workers (0.24 and 0.19 tons of garbage lifted per day per worker respectively; data taken for the year 1998-99). What is more, the costs incurred to clean the streets and lift one ton of waste is invariably higher in the MCH compared to the private sector. Furthermore, the costs of lifting one ton of waste by local contractors has been declining over the years as a result of improvements in the method of privatisation. The comparative advantage of the private sector, however, largely results from the contract specifications set by the authorities. These force private contractors to pay extremely low wages and to save on transportation costs by using very old vehicles.

Table 3.3. Expenditure incurred on SWC by the MCH and the private sector

	1996-97		*1997-98*		*1998-99*	
Description	MCH	PS	MCH	PS	MCH	PS
Expenditure on sweeping and lifting in millions of Rs per month	16.21	6.00	17.79	14.00	21.25	12.85
Waste lifted in tons per month	25,506	7,200	26,550	29,400	27,210	30,240
Average cost per ton in Rs	636	833	670	476	781	423

Note:
- The year 1996-97 relates to the tender system of privatisation
- The year 1997-98 relates to the unit system of privatisation
- The year 1998-99 relates to the improved unit system of privatisation
- The figures relating to the MCH are estimates using salaries, fuel per vehicle, maintenance of vehicle and other related expenditure to calculate the costs of sweeping and lifting of waste by the MCH
- All the figures in the table are on the basis of data for one year. However, in the case of the private sector under the improved unit system the figures relate to only one month (Dec 1999-Jan 2000).
- The rise in the MCH cost per ton largely stems from increases in salaries and modernisation and strengthening of solid waste management infrastructure.

Source: computed on the basis of MCH figures

Despite the fact that their entrepreneurship has been subjugated and their profit margins adapted downward, levels of satisfaction among local contractors were remarkably high. SWC offers them a solid source of income[11]. From the perspective of productive efficiency, however, the current method of privatisation has two major drawbacks. First of all, the requirements set in the contract are such that the entrepreneur is not at liberty to decide on the mix of production factors The system does not

11. Of course, official profit margins only tell part of the story, as there are numerous ways of gaining more money, notably by underpaying labourers According to the labour unions this is common practice among the contractors In order to avoid deductions as specified in the contracts they bribe the officials charged to control their performance.

allow for (technological) innovations with possible cost saving effects. Secondly, the duration of contracts is only 10 months (with a pre and post monsoon period of 5 months each) resulting in low job security for the entrepreneur and his/her workers These reasons explain why private contractors rarely invest in their activity, as is apparent in the fact that almost all of them hire their vehicles rather than owning them.

A disadvantage of the contracting system adopted in Hyderabad is that it has not contributed to the introduction of new technologies in SWC. Obviously, this is related to the requirement for the contractors to hire specified numbers of workers In addition, they claim that the typical short duration of contracts does not enable them to recover the costs of investment in trucks or mechanized sweeping equipment.

Employment and labour conditions

The total number of labourers engaged in SWC has increased significantly since the introduction of privatisation policies. Under the unit system 3,650 new workers were recruited. A major question is how their labour conditions compare to those of the MCH-workers The average net wage of the latter is *three* times higher than that of the private sector worker. Furthermore, government employees enjoy several non-wage benefits, including pensions, health insurance, leave days, working clothes and boots, and cleaning products. Most of the private sector workers have to do without such fringe benefits. As a matter of course contractors flout contract requirements concerning the welfare of their workers Almost without exception they do not contribute to Employees State Insurance (ESI) and Provident Fund (PF). The average monthly wage of a worker is Rs 1,100, while the contract requires payment of the minimum wage of Rs 1,300. Female workers are supposed not to be engaged in night shifts to protect them from being harassed, yet almost all night workers are women.

Security of employment is another aspect where private sector workers compare unfavourably with their colleagues in government service. MCH-workers are permanently employed, while contract labourers in the private sector can be dismissed any time. The short duration of the service contracts implies that their jobs are guaranteed for only that period. In fact, the position of contract labourers has even become more insecure as a result of MCH intervention. In an attempt to avoid the abolishment of the financially attractive labour contracting system adopted by the contractors, the municipality adapted contract specifications at the cost of the workers (labelling their work 'non-permanent'; dropping the demand for ESI and PF facilities; removing the obligation to wear uniforms). It needs to be said, however, that despite these apparent discriminations the labourers concerned are rather satisfied with their jobs, which compares favourably in terms of remuneration and security to positions they held earlier on in their career.

Most people employed in SWC consider the workload to be very heavy and claim that the job is causing them health problems. However, if the MCH workers are sick they can stay at home and make use of their leave days. Private sector workers cannot easily report sick and almost half of them mentioned that they had worked when they were ill for a period longer than 1 day. For the MCH labourers this figure was 4 percent.

Monitoring

Private sector participation is said to enable local authorities to introduce a more stringent enforcement policy. In India it is probably true that the local government has a better grip on private agents than on their own (highly organised and politically assertive) labour force. Anyway, the private sector is put under strict surveillance, albeit especially with respect to service performance and much less regarding the welfare of its labourers The good quality of services provided by local contractors suggests that monitoring is reasonably effective. Major players in performance monitoring are the sanitary supervisors (52) and sanitary jawans (350). They check the contractors' work daily and register eventual shortages, each of which results in deductions from the contractual amount. Nevertheless, stories of contractors ignoring contract stipulations, corruptive practices, poor complaint handling and inferior labour conditions are very common both in Hyderabad and in other Indian cities (cf. Ali *et al.*, 1999).

A positive feature of the monitoring system is the involvement of the community. In each unit a citizen committee had been formed consisting of 10-15 members Contractors require the signature of at least three members on their performance sheets indicating appreciation of the work, before their bills can be passed for payment. It is not known, however, if this innovation has really improved the quality of supervision, or has developed into a mere formality or even a new avenue for manipulation.

Although it is difficult to calculate the additional costs incurred for contract management and performance monitoring it is a fact that the number of staff positions within the MCH has increased dramatically, whereas the number of labourers decreased. The transaction costs of privatisation are certainly substantial.

System viability

Although exact data on the total cost for SWC within the MCH are missing it is likely that these have increased substantially. Total expenditure on privatised SWC has almost tripled since the latest reform in 1998, while overall costs for the remaining MCH service and monitoring the entire sector have probably not gone down much, if at all. Out of the total MCH budget of about Rs 300 crores in the year 2000-2001, no less than 20 percent is allotted to SWM. According to responsible officials the MCH is having adequate revenue to meet its total expenditures, including SWM. However, the municipality does face difficulties in financing major investments. Under the Mega City Scheme the MCH managed to avail a loan of Rs 28 crores from the Mega City Fund for infrastructure development in the year 1999. Part of this money is intended for the construction of transfer stations and replacement of SWC vehicles. Out of the Rs 28 crores 25 percent comes from the Government of India revolving fund and 25 percent from the State Government as a matching grant. The remaining 50 percent are the own funds of the MCH. The repayment requirements, however, seem to go beyond the current capacity of the MCH and may force the authorities to solicit another Housing and Urban Development Corporation (HUDCO) loan. This indicates that the financial viability of the system of SWC is somewhat disquieting. It is likely to remain an Achilles heel until some kind of cost recovery will be introduced. However, there are currently no plans to charge residents user fees.

Contributions to public health and environmental aspects of sustainable development

Cleanliness of neighbourhoods
Sweeping, collection and transportation of waste is carried out on a regular, daily basis in all units, at least on paper. However, according to the survey among households in areas serviced by the MCH and private contractors, the frequency of sweeping, collection and cleaning of waste bins in their areas is overwhelmingly perceived to be less than once per day. Levels of satisfaction are slightly higher in privately operated areas where frequency and cleanliness have improved in comparison to the situation five years ago when the MCH executed the work. However, service levels in the MCH-serviced units have gone up as well because the municipal labour force, which was about half the normative required strength prior to privatisation, could be deployed in a much more restricted part of the city.

The general attitude among residents towards privatisation of SWC is positive. Most people feel that private workers will perform better due to strict surveillance and higher job insecurity. Consequently, the social legitimacy of SWC by private operators is high. On the level of the entire city improvements can also be observed. Spatial coverage of basic collection services has been extended in the wake of privatisation. This is largely a result of the overall increase in investment in solid waste management in the city. However, privatisation is only implemented in areas with good accessibility and planned layouts (partly as a result of the uniform package). Slum areas continue to suffer from gross under-servicing. Residents in these illegal parts of the city often do not have waste bins in their area (more than half of the selected respondents mentioned this) and consequently many of them resort to arbitrary dumping practices.

Health of workers
The system of SWC in Hyberabad continues to rely heavily on manual labour. Sweeping and lifting is usually done manually and this is extremely demanding. The labour-intensive nature of the work derives at least partially from the strength of labour interests in (local) politics discouraging the replacement of labour by capital. Anyway, labourers are in continuous contact with the garbage and, hence, stand a good chance of being infected or hurt. Although the employers are obliged to provide protective clothing and safety equipment to their workers, both parties often contra-

vene these requirements. There is some reason to believe that private workers, on average, are slightly worse off than their MCH colleagues. In the private sector, loading and unloading of vehicles, for example, is almost exclusively done by manual labour because contractors rarely use trucks with hydraulic lifting devices. The current contracting system does not encourage investment in equipment that will make work easier. Furthermore, contractors are more likely to save on protective gear.

Environmental impact

The SWC policy of the MCH is primarily directed at a reduction of immediate public health hazards. The main goal is to keep neighbourhoods tidy and to have the waste removed from these areas in a controlled fashion. No attention is given to the prevention of waste production or to the minimisation of waste or the promotion of recycling and reuse. Any effects of the policy in these respects, therefore, are accidental. A few indirect environmental impacts may be observed.

First of all, the private contractors are often using very old trucks that usually emit high concentrations of pollutants. According to the contract conditions the trucks of the contractors should be less than 15 years of age. However, in reality the average age of the trucks in use by the contractors is 27 years (!); 52 percent of the trucks is older than 25 years[12]. Furthermore, the trucks of the private contractors are obliged by the MCH to dispose of their loads at the designated dumpsites rather than at one of the intermediary transfer stations. The trucks are very often stuck in traffic on their way to and from the dumpsites, which are situated far away from their unit areas. During transportation littering of waste is quite common due to careless packing without any top cover on the waste. This seems to apply more to trucks used by private operators than to the MCH trucks.

As far as the recycling and reuse of waste is concerned, the privatisation of SWC has probably a marginal impact only. On the one hand the interventions made larger quantities of mixed waste available at secondary collection points and dumpsites, enabling waste pickers to sort out valuable materials. On the other hand those officially charged with SWC will seek to avoid waste pickers having free access to the waste as this impairs their work and may lead to littering.

An obvious positive impact of the privatisation campaign is that the quantities of waste that are being collected and disposed of in a controlled fashion have increased substantially, thereby reducing environmental hazards for people within their residential areas (likelihood of catching an infectious or parasitic disease due to exposure to waste) and for environmental degradation (water pollution or soil degradation due to leakage). At the same time, however, more waste will reach the dumpsites, whose

12. This figure was based on a list provided by MCH specifying the equipment of 44 contractors working in circle 4 and 5.

capacity will therefore be sooner exhausted. As the method of open dumping is adopted (rather than the sanitary land fill), the environmental problems associated with this type of disposal will probably worsen.

3.7. ASSESSING SUSTAINABLE DEVELOPMENT IN THE VGDS

House-to-house SWC is practiced in almost 20 percent of the MCH territory. It has eliminated one step in the basic system,*viz.* the requirement for households to bring their garbage to secondary storage points, and replaced it by doorstep collection. Just like the partial privatisation of the basic service it concerns a major innovation in SWM in the city and therefore its strengths and weaknesses will also be elaborated.

Contributions to socio-economic aspects of sustainable development

Economic efficiency and viability
The scheme is basically self-supporting as its running costs are entirely covered by the households that contribute a monthly fee. Our survey showed that fees range from Rs 5 to Rs 20 per household per month depending on the area (and hence, affordability) and the volume of waste per house. The start-up costs, a tricycle and waste bins for the households, have been financed either by the MCH or the (British) Overseas Development Administration (in slum areas). Replacement of equipment, however, needs to be financed by the communities themselves. Allocative efficiency, therefore, is good, accepting of course that the MCH is still responsible for secondary collection and disposal.

The MCH estimated the economies brought about by the VGDS to be approximately Rs 8 million per year after deduction of the subsidies paid to households. The savings were attributed to the use of fewer trucks and municipal workers Furthermore, it was estimated that the savings resulting from the ODA supported scheme could run up to Rs 13 million per year (Snel, 1997).

Free riding does not appear to be a major problem since most communities succeed in keeping this phenomenon within socially acceptable limits. The welfare organisations were able to recover the fee from most households although some indicated that they also used other revenues to meet a part of the salary of the waste pickers Since it forms a small proportion of the expenditure of the association, they did not really feel the pinch of it. The 60 respondents from our survey in VGDS serviced communities were quite outspoken with respect to deviant behaviour: the vast majority wanted these people to be punished or fined. Affordability, at least in our survey, was not perceived to be a problem[13].

Employment and income

The scheme has generated gainful employment as well as job security to several hundred unemployed youngsters and informal pickers Their wages range from Rs 1,000 to Rs 2,000 per month, but they can earn additional income by selling the recyclables after segregating the waste collected from the households. Those waste pickers engaged in the service in slum areas and low-income colonies, however, are a little worse off because waste generation rates per household are less – requiring them to cover more households – and the volume of recyclable material is smaller than in more affluent residential areas.

Monitoring

Strict supervision by the residential welfare organisations as well as by the residents, who require value for money, ensures proper servicing. The MCH is also very satisfied with the scheme. It has incited the municipal inspectors to regularly interact with the welfare organisations and the residents, both through formal meetings and informal contacts, contributing to effective coordination. The scheme has also increased awareness among the people to ensure a clean environment.

System viability

The long-term viability of the VGDS is expected to be good. At neighbourhood level the running costs are fully recovered. It is not entirely sure if the community will succeed in raising/saving enough financial means to replace worn out equipment. However, the success of the scheme in combination with the enthusiasm among the welfare organisations will probably guarantee its continuity. Anyway, the welfare organisations did not feel this would become a bottleneck. Obviously, the success is also dependent on the ability of the MCH to ensure the necessary means to run the secondary collection and disposal trajectory (see above).

Contributions to public health and environmental aspects of sustainable development

Cleanliness

Satisfaction among residents participating in the scheme is high. The vast majority feels the cleanliness of the neighbourhood has increased. The frequency of collection was high: 87 percent of the respondents indicated that the waste pickers passed by their house every day, while the remaining 13 percent said collection frequency was

13. In some of the slum areas NGOs (such as SUKUKI) have organised door-to-door collection on the same basis. In these areas affordability does seem to be a problem, as many slum dwellers do not regularly pay the required fee.

4-5 times per week. People were very pleased to have their waste collected at the doorstep and no longer having to bring it to communal bins at some distance of their houses. They also appreciated the absence of waste bins – and the related littering – inside their areas.

Environmental impacts

The environmental impact of the VGDS is largely positive. Although there are no effects on the minimisation of waste or on final disposal practices the scheme does contribute to recycling and reuse. Waste is supposed to be separated at household level in organic and inorganic materials. In actual fact, this does not seem to be an established practice yet. The waste pickers try to sell valuable items on the waste market. The organic component of the waste can be recycled in order to be converted into manure through vermicomposting. However, composting at neighbourhood level, despite support by several NGOs, has not yet proved to be a viable option in most cases (see chapter 10). The anticipated benefits of the scheme in this respect have not been achieved.

Another positive environmental impact is that the volume of waste that has to be transported to the dumpsites is somewhat reduced, and hence the capacity of the dumps will last a little longer. Furthermore, collection up to the vantage points on the edge of the neighbourhoods is done manually. The system saves on the use of trucks.

3.8. SYSTEM-WIDE CONCERNS

The investigations of the SWC performance in Hyderabad also pointed to a number of citywide concerns that frustrate or impair proper servicing. A major bottleneck arises from the poor quality of physical infrastructure in large parts of the city, notably in the low-income and slum areas. The basic collection system, for example, requires good accessibility of residential areas by trucks. When it is difficult for vehicles to enter the area, residents, usually housewives or maid servants, are forced to take their garbage to vantage points located relatively far away from their houses. This results in indiscriminate dumping practices. The problem can partly be overcome by introducing a VGDS in the area. However, this requires the presence of a community welfare organisation capable of organising this service, and, if such an institution does not yet exist, community mobilisation. It is a well-known fact that CBOs actively engaged in community sanitation are more often found in middle class areas (Schenk *et al.*, 1998).

The unit system currently applied by the MCH in delivery of the basic service is rather rigid. It implicitly sets minimum requirements to the area's physical layout, especially in terms of road width and places to install waste bins or garbage houses. There seem to be good reasons to adjust the input requirements in terms of labour and equipment to the type of layout. A more flexible approach would probably help to meet the needs of residents living in the poorer parts of the metropolis.

Due to the rapid growth and sheer size of Hyderabad final disposal is becoming a major headache for the administration. The Golgonda site has been used for 20 years and it was closed recently, in January 2000. The site got saturated and it was decided to convert the land into a golf course under the State Tourism Department. Gandamgooda was partially closed in 2001 on the notice of the Andhra Pradesh Pollution Control Board, as it reached its maximum capacity. Now 10 acres are handed over to Selco International for its pelletisation plant (see note 4). The Mansurabad site is still used for dumping although it has long passed its saturation level. Due to its remote location transportation costs are a major worry, especially to the private contractors with their tight contract specifications. There is ongoing protest from the public against the continued use of the site and trucks now require police escort to safely enter the place.

Currently, investigations are carried out into possible new sites. However, even if the authorities are able to identify sites that are technically suitable for being used as dump yards, vehement opposition is expected from the public, and names of potential sites, therefore, were not disclosed. Another problem is that new dumps are likely to be located outside MCH jurisdiction. Currently, two out of three are outside the MCH area. Decision-making on the city's SWC, therefore, depends on collaboration with other administrative units, both neighbouring municipalities and the state government, each bringing in their own political demands[14]. The future location of dumpsites is expected to become a heatedly debated and time-consuming issue. It is very likely that future locations will be at considerable distance from Hyderabad and that transportation costs will become an increasingly troublesome item.

A system-wide concern of entirely different dimension relates to local politics. The early experiments with privatised SWC were severely troubled by practices of corruption and nepotism. Fortunately, local civil society is sufficiently empowered to expose the excrescences of these infractions. Nevertheless, it is common knowledge that officials are bribed to be lenient, for example when private contractors flout contract specifications. Such practices, to be found at all levels within the local bureaucracy, continue to affect service efficiency and effectiveness in a negative way. The rigidity of the unit system, for example, partially results from the desire to minimise opportunities for abuse. At the same time it prevents the MCH from benefiting more from the acclaimed flexibility and dynamism of the private sector. A similar effect is due to the persistent interference of the powerful labour unions, most of which have tried to obstruct privatisation of SWC (see chapter 5). Although they have achieved to protect the interests of organised (municipal) labour, they have done little to support the newcomers in the business. Furthermore, they force the MCH into the utilisation of their own relatively expensive labour force, making SWM much more expensive than

14. The MCH continues to rely on the state for covering an important part of its expenses, giving the latter a major (indirect) say in all decision-making.

it would be under free market conditions. Obviously, this should not be read as an implicit justification of poor remuneration of private sector workers At the same time, however, it needs to be said that the MCH is apparently not at liberty to balance the interests of its workers against those of the general taxpayer.

3.9. CONCLUSION

The MCH has succeeded in bringing about substantial improvement in its SWC system by involving the private sector in service delivery. Private contractors perform comparatively well and at lower costs, compared to the local body. The general public is satisfied with the new policy as it results in cleaner neighbourhoods. The overall environmental impact is positive although this does not derive from privatisation as such but rather from higher levels of spending on SWC. Nevertheless, the system also suffers from a number of weaknesses. The remuneration levels and job security of private sector labourers compare very unfavourably with those of their MCH colleagues. Furthermore, privatised servicing is largely confined to middle and high-income areas, which allow to be operated under the rigid unit system approach. Another shortcoming is that the system does not reap the full benefits of privatisation – dynamism, flexibility, and innovativeness – due to the detailed contract specifications the MCH was politically forced to adopt. Finally, the MCH has not yet grasped the opportunity to design a truly integrated SWM policy. Being preoccupied with the protection of public health, its focus is almost exclusively on controlled collection, transportation and disposal of waste. No concerted efforts are made to reduce waste production, while initiatives to promote recycling and reuse of (in)organic wastes hardly move beyond the support of an occasional experiment.

Another promising step, again in the collection domain, is the introduction of the VGDS in approximately one-fifth of the city. In the areas utilising this system the quality of the service has improved considerably. At the same time it has brought new employment opportunities to a group of people with limited chances on the regular labour market, and it has stimulated recycling and reuse, especially of inorganic matter. Although the success of the VGDS partially depends on the strength and dedication of local welfare organisations, there certainly seems to be room to extend the spatial coverage of this system. Apparently people are willing to contribute financially to an upgrading of the service, albeit that participation in the scheme cannot be taken for granted (in the slum areas affordability does constitute a problem). This knowledge could be utilised to the advantage of the municipal budget.

Among the bulk waste generators, privatisation of SWC is still in its infancy. Nevertheless, the institutions concerned are likely to benefit from private collection, especially in terms of punctuality. The MCH does have to arrange for proper monitoring of these bulk generators to avoid illegal dumping practices. Special attention should be given to hospital waste, which is often not treated carefully enough and ends up being mixed with ordinary household waste. It constitutes a major threat to public and environmental health and should perhaps best be brought under the stricter regime of industrial waste.

MOSES M. IKIARA, ANNE M. KARANJA AND THEO C. DAVIES

CHAPTER 4
COLLECTION, TRANSPORTATION AND DISPOSAL OF URBAN SOLID WASTE IN NAIROBI

4.1. INTRODUCTION

Nairobi is literally under garbage. Of the estimated 1,500 tons of solid waste generated daily in the city, only about 25 percent gets collected. The rest is left in open spaces, markets, bus stops, drains and roadsides forming mountains of rotting, smelly and unsightly waste. Discarded polythene papers of all colours and sizes decorate the city landscape. Solid waste collection, transport, and disposal are thus generally chaotic. It has not been always so, however. As recently as 1977, the Nairobi City Council (NCC) collected almost all the waste generated. Dismal performance started in the late 1970s and persists.

Poor NCC performance stimulated entry of private sector and community actors into the solid waste collection sector. They are, however, operating in a disorderly manner without regulation or guidance. For example, waste pickers board open waste trucks as they transport waste to the dumpsite, pick on top of the moving trucks, toss out whatever they find no use for, and litter the entire route without control. Moreover, waste pickers and dealers illegally control the NCC-owned and operated dumpsite, forcing the NCC and private companies to 'bribe' to access the dump. Lack of control and guidance hinders utilization of the full potential of private sector and civil society involvement.

Marked skewness in the geographical service distribution also reduces social support of existing solid waste collection services. High-income and some middle-income residential areas together with commercial areas are well serviced by private companies and even the NCC. Small private firms are also increasingly servicing some of the relatively better low-income areas. The core low-income areas (slums and other unplanned settlements) where 55-60 percent of Nairobi residents live, however, receive no waste collection service, save for localised interventions by CBOs. Broadly, the Western part of the city is well serviced by the private firms and the NCC while the Eastern part is hardly serviced.

I. Baud et al. (eds.), Solid Waste Management and Recycling, 61-91.

This chapter looks at solid waste collection in Nairobi and its implications for urban sustainable development. It lays particular emphasis on the significance and sustainable development of the activities of recent entrants into waste CTD, the private sector and the civil society. Questions addressed include:

- Which actors take part in Nairobi's solid waste collection and how are their activities organised?
- What is the relative and absolute importance of these activities?
- What are the contributions of the key solid waste collection activities to urban sustainable development?
- What are the problems and potentials of various forms of solid waste collection according to stakeholders?

The discussion draws from literature and data (secondary and primary) collected in 1998 and 1999. Details of the methodology and data used, including the template applied in the analysis of sustainable development, are presented in chapter 1 of this volume and in the methodological appendix.

4.2. HISTORY AND INSTITUTIONAL FRAMEWORK

The NCC, a local authority under the Ministry of Local Government (MOLG), has always provided solid waste management services in Nairobi. Prior to the Environmental Management and Coordination Act (1999), local authorities (LAs) had monopoly control over sanitation and solid waste management services (Mulei and Bokea, 1999). Other agencies required written authority of the relevant LA to handle waste materials or provide solid waste management services.

The NCC experimented with privatisation of solid waste management in 1906 when a private company was contracted to sweep and clean city streets, collect garbage and provide street lighting (UNCHS, Undated). The private company failed to execute the contract satisfactorily and privatisation was abandoned. The NCC provided solid waste collection services satisfactorily until the late 1970s. Deterioration of performance started in the late 1970s, accelerated in the 1980s and 1990s. Starting in the mid-1980s, the appalling NCC performance and demand for solid waste collection services attracted private sector providers Bins (Nairobi) Services Limited and Domestic Refuse Disposal Services Limited (DRDS), registered under the Companies Act to offer solid waste collection services to industries, institutions, commercial establishments and high-income residential areas in 1986-1987. Entry of private sector companies continued in the 1990s. In 1997, following research supported by the Japanese Government, the NCC started a pilot solid waste management privatisation scheme[1] in the form of a management contract to a private firm. The contract ended in 1999 and is yet to be renewed.

1. The details of the pilot scheme are provided in a later section of the chapter.

LAs provide solid waste collection services under the Local Government Act (CAP 265) and Public Health Act (CAP 242). The former empowers LAs to establish and maintain solid waste collection services while the latter requires the LAs to provide the services. The Acts, however, neither set standards for solid waste collection nor refer to waste reduction or recycling. In addition, the Acts do not classify waste into municipal, industrial and hazardous types or allocate responsibility over each type. The NCC has enacted by-laws, under the Local Government Act, prohibiting illegal deposition of waste, specifying storage and collection responsibilities for waste generators, and indicating the Council's right to collect solid waste collection charges. These are not adequately implemented, however. Kenya therefore lacks any comprehensive national solid waste management legislation.

LAs are established as autonomous and independent corporate entities but are emasculated by the Central Government. The example of the NCC best illustrates how LAs operate in Kenya. The NCC is headed by a Mayor, assisted by a deputy Mayor, and comprises of 73 councillors (55 of whom are popularly elected and the rest nominated by the MOLG). The Council manages its affairs through 12 committees (composed of councillors), which vote by consensus. The mayor and councillors have no executive powers Policies made by the committees are therefore implemented by chief officers (headed by the City Clerk), who are accountable to MOLG. These officers are recruited by the Public Service Commission (PSC)[2]. The LAs cannot relieve such staff of their duties or discipline them without PSC approval. This reduces the responsibility and accountability of the staff to the LAs (Gatheru and Shaw, 1998). This structure has wrought enormous friction between the councillors and the chief officers

The Local Government Act empowers the Minister for Local Government to unilaterally establish, abolish, and control LAs (Gatheru and Shaw, 1998). It also allows the minister to remove all councillors, dissolve the council and appoint a commission to run the LA. In exercise of this power, many commissions have been appointed in the history of the NCC, usually in response to political controversies, and often to its detriment. Even though councillors are responsible for policy, ministerial approval is required even for minor decisions. It is not uncommon, moreover, for the NCC to make a decision, only for the Provincial Administration to violate it. While the ministry has exercised excessive control over LAs in some aspects, it has largely failed to play its key oversight role (NCC, 2000). Thus it failed to intervene, and some of its senior staff actually participated, in the plundering of NCC resources.

Since the country adopted a multiparty political system in 1992, political parties compete for the right to form the government, including local government. Like parliament, therefore, LAs tend to have councillors from the ruling and opposition parties. The mayor, being elected by other councillors, usually belongs to the political

2. LAs require ministerial approval to hire even low cadre staff.

party with the largest number of councillors The NCC has had majority of councillors from national opposition parties since 1992. This coupled with the power accorded to MOLG (run by the ruling party) over LAs and the fact that LA decisions are made through consensus rather than majority vote have created fertile ground for sabotage with decision-making being very difficult. Furthermore, the NCC has been bogged down by perpetual competition, among groups of councillors, for the coveted mayoral seat. The councillors whose demands are often anything but altruistic hold the mayor, once elected, hostage.

Strengthening and empowerment of LAs has been a recurring policy intention since the mid-1980s. Recent policy documents (including Budget Speeches and the 'Interim Poverty Reduction Paper for the period 2000-2003') spell out various reforms, such as reduction of the role of Central Government, streamlining of financial management (including revenue sharing programs, and the strengthening of the local revenue mobilisation capacity), and greater community participation in service delivery, planning and project implementation. These policies are at different stages of implementation although the political will is still inadequate. LAs are now receiving a share of the Road Maintenance Levy Fund (Republic of Kenya, 2000). In addition, the Local Authority Transfer Fund (LATF) Act came into effect in July 1999 and the first quarterly disbursements to the LAs were made in January 2000. LATF has received 2 percent of income tax collections (about Kshs 1.2 billion) in the first year but this is expected to rise to 5 percent in subsequent years

New initiatives have also been taken in the legal domain. The Environmental Management and Coordination Act (1999) is the first attempt at a national solid waste management law. The Act entitles Kenyans to a clean and healthy environment, and empowers them to sue for improper solid waste management. With respect to solid waste management the new environmental law:

- Prohibits improper discharge and disposal of wastes;
- Requires licensing (by the National Environmental Management Authority, NEMA) of waste transportation, establishment of waste disposal sites and generation of hazardous waste; and
- Requires waste generators to apply measures for minimisation of waste such as treatment, reclamation, and recycling.

The penalty for violation of the new solid waste management provisions is fairly deterrent, being imprisonment for up to 2 years and/or a maximum fine of Kshs 1 million (US$ 132,000). The new law is, furthermore, laudable as it provides opportunity for licensing of private firms competing with the NCC for solid waste collection. Since implementation of the new law is yet to begin, however, only time will tell how well it works.

With respect to the privatisation of solid waste collection services official attitudes are positive even though no law has been enacted as yet to guide, monitor, and regulate the process (Republic of Kenya, 2000). However, the Local Government Act empowers LAs to contract out services (Section 143) and the Environmental Management and Coordination Act (1999) provides for an institution (NEMA) to license waste transportation, establishment of waste disposal sites and generation of hazardous waste.

4.3. ORGANISATION OF SOLID WASTE COLLECTION SERVICES

Numerous actors are involved in Nairobi's solid waste management system (see Figure 4.1). Broadly, the actors could be grouped into those making policy and those implementing or affecting it as the figure shows. In the subsequent analysis the roles and performance of various actors is elucidated.

Policy-makers

The MOLG is charged with the role of policy formulation, providing technical assistance to LAs, and supervisory oversight and guidance. Moreover, the City Clerk (who is responsible for policy implementation and executive duties) is accountable to the MOLG. NCC councillors formulate solid waste collection policies through the Environment Committee. There is no clear demarcation of MOLG's and NCC's roles and powers, as is evident from the preceding section. MOLG has performed poorly with respect to supervisory oversight, guidance and capacity building and usurps the role of NCC, for example, in rehabilitation of urban roads (NCC, 2000).

The Ministry of Environment and Natural Resources (MENR) also makes policy related to solid waste collection. It is responsible for overall environmental management, which includes pollution and waste management. Over the period 2000-2003, for instance, MENR is expected to implement environmental standards on air, water and land, promote community based waste management programmes, provide incentives for informal sector waste management, and develop environmental partnership with stakeholders such as NGOs and CBOs, among other tasks (Republic of Kenya, 2000). However, the impacts of its policies are still marginal.

The Environmental Management and Coordination Act (1999) has introduced new institutions for environmental management. A National Environment Council (NEC) is envisaged for policy formulation, setting of national environmental goals and objectives, and promotion of environmental cooperation while a NEMA is envisaged for general supervision, coordination, and implementation of all environmental matters and policies. Moreover, a Standards and Enforcement Review Committee of NEMA is planned to advice NEMA on environmental standards, including standards for waste disposal methods and means. The committee will, in addition, issue regulations for

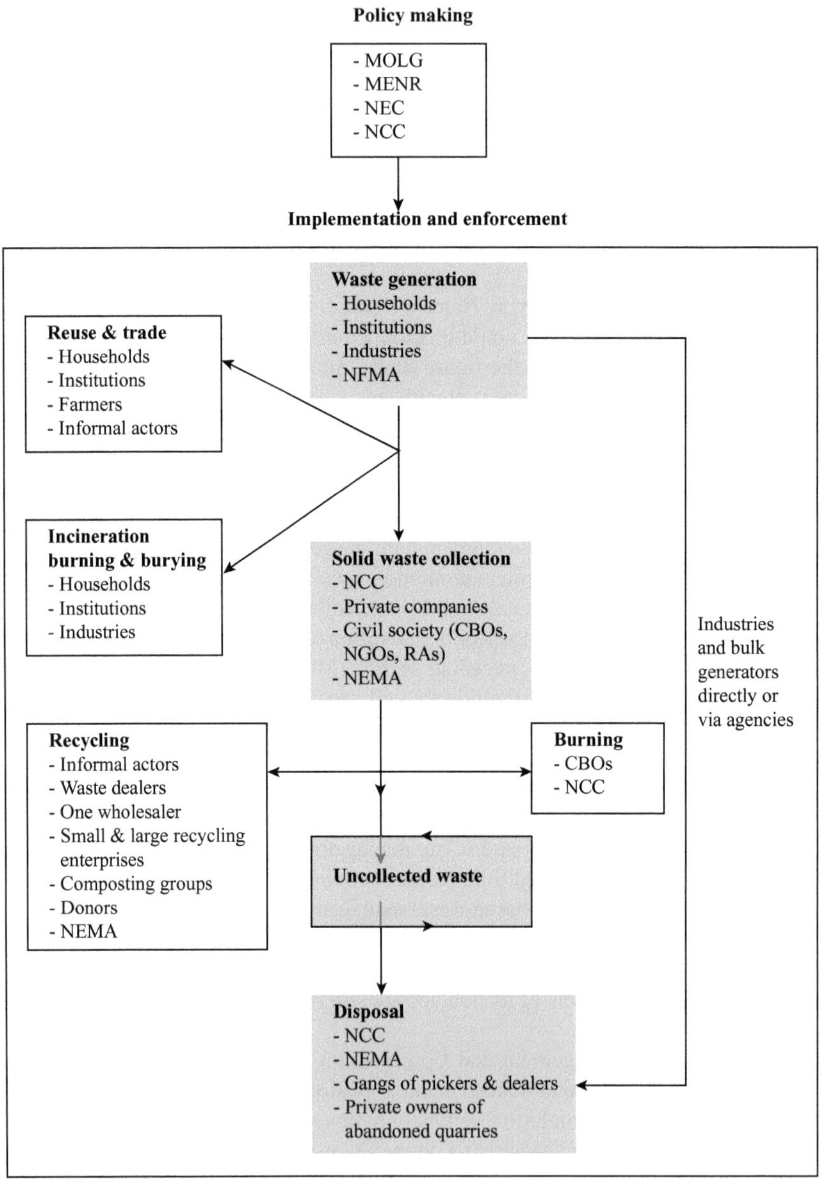

Figure 4.1. Solid waste collection, transportation and disposal system in Nairobi

waste handling, storage, transportation, segregation, and destruction. Once again, it is too soon to tell whether these institutional reforms will sort any meaningful effect.

Waste generators

Households

In 1998, daily municipal (waste) generation in Nairobi stood at 1,530 tons, with households accounting for 82.8 percent of the total, commercial enterprises (6.5 percent), markets (5.8 percent) and roads (4.9 percent) (JICA, 1998). Markets, roads, and institutions like restaurants had the highest generation rates, however (Table 4.1).

Table 4.1. Relative municipal solid waste generation rates in Nairobi, 1998

Type of MSW generator	*Rate (Kg/day)*
Mixed (restaurant)	6.79
Mixed (others)	1.39
High income residential households	3.84
Middle income residential households	3.34
Low income residential households	2.72
Activities in areas surrounding residencies	3.07
1 market	2,425.00
Road (1 Kilometer)	48.30

Source: JICA(1998)

JICA (1998) found per capita generation rates of 0.65, 0.60 and 0.54 kg/day for high, middle and low-income residential areas, respectively. Previous estimates, 1985-1992 for instance, found per capita Solid waste generation rates of 0.35-0.46 Kg/day depending on the socio-economic status of the households (Mwangi, 1990; Syagga, 1992). JICA (1998) data, however, are believed to be better, because the study was very extensive covering even the unplanned settlements, while the others either used the poor NCC data or conducted surveys in the serviced areas only. Waste generation estimates are usually net of the materials taken out by the households for reuse and trade.

Households do not only generate waste, they are also involved in waste prevention or minimisation. About 70 percent of the households we interviewed separate materials for reuse and/or sale. About 15 percent of the sampled households traded with the separated materials. Plastics, clothing materials, shoes, wood from furniture, and other materials are directly or indirectly consumed either by the household itself or by domestic servants. Vegetable wastes are used for fertilizing/mulching kitchen gardens and, increasingly, for livestock feeding, while food wastes are used to feed chickens, pigs and dogs (for more details see chapter 8 and 12).

Because of the inadequacy of solid waste collection services households may convert to indiscriminate dumping and littering, waste burning and/or burying, and some of them even openly acknowledge to engage in such practices.

Institutions

Institutions like schools, hospitals, hotels and restaurants, shops, and markets are also involved in both waste generation and separation of materials at source. They contract private companies for solid waste collection and sell/give waste free to farmers and waste pickers About 47 percent of the 19 institutions interviewed in Esho (1997) handled waste themselves, 45 percent of them by burning it and the rest by selling it to waste pickers About 67 percent of the 18 commercial enterprises interviewed sold their waste to pickers Our 1998 survey indicates that some of the organic waste collected from 43 percent of the markets and institutions in Nairobi is used as animal feed. Kenyatta National Hospital has its own incineration unit and is gearing to sell incineration services to other institutions.

Industrial waste is not considered as municipal waste and the generator is responsible for its disposal. Industries either transport and dispose waste by themselves, contract private agencies or NCC to do it, sell/give it to recycling enterprises, or recycle it themselves. Esho (1997) found that 50 percent of the industries interviewed contracted the private sector to handle their waste while the remainder handled the waste by themselves, 60 percent of them through recycling and the rest through open dumping or selling it to waste pickers

4.4. ACTORS IN SOLID WASTE COLLECTION

The NCC and private commercial companies are the principal providers of solid waste collection service in Nairobi although, as Figure 4.1 shows, there are smaller actors Some industries and bulk generators like Nairobi Airport Services (NAS) store and transport their waste to the dumpsite by themselves.

In 1998, the NCC accounted for 22 percent of the 360 tonnes of solid waste collected in the city per day while the private firm contracted by the NCC to offer solid waste management service in the CBD (Kenya Refuse Handlers Limited) accounted for 46 percent and the other private companies the balance (JICA, 1998). Tables 4.2 and 4.3 suggest the relative importance of collection agencies in the city from survey results[3]. The tables show that private companies and personal initiative are very important. Moreover, many households are not served. About 48 percent of the households we interviewed did not receive any solid waste collection service, generally agreeing with findings of the JICA study that found 26 percent of high-income areas, 16 percent of

3. These results cannot be said to apply to Nairobi as a whole because of unrepresentative sampling and other methodological problems. Nevertheless, they yield valuable insights.

the middle-income areas, 75 percent of the low-income areas, and 74 percent of the surrounding area were not served.

Table 4.2. Relative importance of solid waste collection agencies, 1997.

Solid waste collection agency	*Activity area (clients served)*				
	*Residential**		*Institutions*	*Industrial*	*Commercial*
NCC	1 (1%)	4 (3%)	4 (21%)	-	3 (16.7%)
Private companies	57 (73%)	58 (45%)	6 (32%)	10 (50%)	3 (16.7%)
CBOs	5 (6%)	5 (4%)	-	-	-
Personal initiative	15 (19%)	61 (48%)**	9 (47%)	10 (50%)	12 (66%)
Total	78 (100%)	128 (100%)	19 (100%)	20 (100%)	18 (100%)

Note:
* the second column contains data from our 1998 fieldwork
** the figure refers to those households that indicated that they did not have any collector.

Source: Esho (1997) and own survey (1998)

Table 4.3. The income level of households and source of CTD service used.

Income level	*NCC*	*Private companies*	*No collector*	*CBOs*	*Total*
High income	1	20	8	-	29
Middle income		27	11	-	38
Low income	3	11	13	5	32
Slum	-	-	29	-	29
Total	4	58	61	5	128

Source: own survey (1998)

Communal (or station) and door-to-door are the predominant waste collection methods used in Nairobi, the former being used more by the NCC and the latter more by private companies. In 1998, the communal method accounted for 91 percent and 48 percent of total NCC's and private companies' waste collection respectively (JICA, 1998). The shares for the door-to-door method were 9 percent for NCC and 52 percent for private companies. Private collectors prefer the door-to-door method, as it facilitates collection of charges. The communal method is one where households deposit their waste in a designated area or large container for collection. Though not found in the JICA survey, block and kerbside waste collection methods are also used on a small-scale. Block collection is used in estates with block-shaped structures that have a common gate for several households (Kibwage, 1996). The kerbside method is used in commercial areas and flats and housing complexes where enormous quantities of waste are generated.

NCC

Collection performance

The NCC is still a significant provider of SW collection and transportation service and exclusively provides street cleansing and disposal services. The NCC delivers solid waste collection services through the cleansing section of the Department of Environment (DOE), employing 2,416 people for that task (1998). Responsibilities of the DOE include implementation of solid waste management policies formulated by the Environment Committee, maintenance of public cleanliness, protection of public health and environment, maintenance of aesthetic quality, provision of solid waste collection and treatment services, regulation and monitoring of waste generators, regulation and monitoring of private sector involvement, enforcement of all solid waste management laws and regulations, and coordination of agencies involved in solid waste management. For operational purposes, the city is divided into two divisions, each with three districts. Each district provides SW collection and transportation and street cleansing services. Besides these tasks, Embakasi district manages the Dandora dumpsite also. District sizes vary in terms of staff size, with the Central and Western districts having the largest.

NCC provision of solid waste collection services has been dismal in the last two decades. By 1998, NCC was collecting only 80 tons of solid waste per day (JICA, 1998) compared to 270-350 as recent as 1992 (Syagga, 1992). The NCC no longer schedules services and provides them only when and where most needed. It largely serves core areas like airports, CBD, hospitals and politically sensitive residential areas. Thus, NCC services are concentrated in the same areas that private providers are concentrated in, e.g. those areas and institutions that can afford private service at the expense of areas inhabited by the poor (JICA, 1998). The eastern edge hardly receives any NCC solid waste collection service. Low-income areas receive service only when garbage poses 'a health hazard', when there is public outcry, or during environmental clean-up days. The NCC no longer distributes storage bins although it charges for it in all water bills. Clients served by the NCC use anything available for storage, including plastic buckets and sisal sacks.

Vehicles and equipment

The number and quality of NCC's refuse transport vehicles and other equipment have declined enormously. Refuse trucks, for instance, dropped from 60 in 1969 to only 21 by 1992 while supervision vehicles reduced from 35 to 4 over the same period (Kibwage, 1996). The number of operational waste transportation trucks had fallen further to only 15 by 1998 (JICA, 1998). In February 2001, NCC's Deputy Mayor informed the press that only 22 out of the 202 refuse collection trucks were operational, against a requirement of 235. Most of the trucks used by the NCC, moreover, are in a bad state of repair. Its open trucks have low capacity and cause substantial littering while the imported highly mechanised trucks are difficult to maintain and unsuitable for Nairobi and its waste characteristics. Such trucks are expensive, labour replacing, and require expensive imported spare parts (Syagga 1992, Otieno 1992).

Lack of finance and mechanical problems arising from poor roads, overloading and mishandling by crews, poor service and unavailability of spare parts, and frequent accidents are the principle causes of the poor status of NCC's garbage collection vehicles. Maintenance is often not adequately budgeted for while the Council's procedure for acquiring spare parts is lengthy and time wasting. In addition, the vehicles are illegally used for private transportation.

Disposal

Waste disposal is the exclusive responsibility of the NCC. The city has only one official dumpsite, an open landfill type. Located about 7.5 km east of the CBD (at Dandora; see Figure 4.2), the dump is about 26.5 hectares and filled by approximately 1.3 million cubic meters of garbage after about 14 years of use (JICA, 1998). The dump is owned and operated by the NCC, which charges a dumping fee ranging from Kshs 30 to Kshs 100 (US$ 0.4-1.35) depending on vehicle capacity. The charges are not only low but corruption denies the NCC a substantial portion of the revenue.

Nevertheless, the costs of disposal at the official dump are rather high due to its peripheral location, lack of garbage transfer facilities in the city, and insecurity. Insecurity is associated with gang type control of the dump by groups of waste dealers and pickers Several gangs have formed and divided the dump into territories. Each gang polices its territory to enforce ownership of waste dumped there. Similarly, trucks that regularly transport waste into the dump are "owned" by specific gangs and are usually guided to the appropriate territory for dumping. Truck drivers, including NCC drivers, must cooperate lest the trucks are vandalised. They usually pay for police escort to the dump. However, the police are often helpless against the gangs, some of which are led by heavily armed hardcore criminals masquerading as waste dealers

Insecurity, high cost of SW transports and disposal, and lack of effective monitoring of the activities of private companies create a *milieu* for indiscriminate dumping by private companies. Consequently, illegal dumps (used even by NCC) have sprouted in many places. The demand for dumpsites has induced a few private individuals with abandoned quarries on their land to offer them for waste dumping at modest fees. One of the large private solid waste collection companies has been trying to acquire a licence from the NCC to operate its own sanitary landfill site without success. This will be possible once the new environmental legislation is implemented.

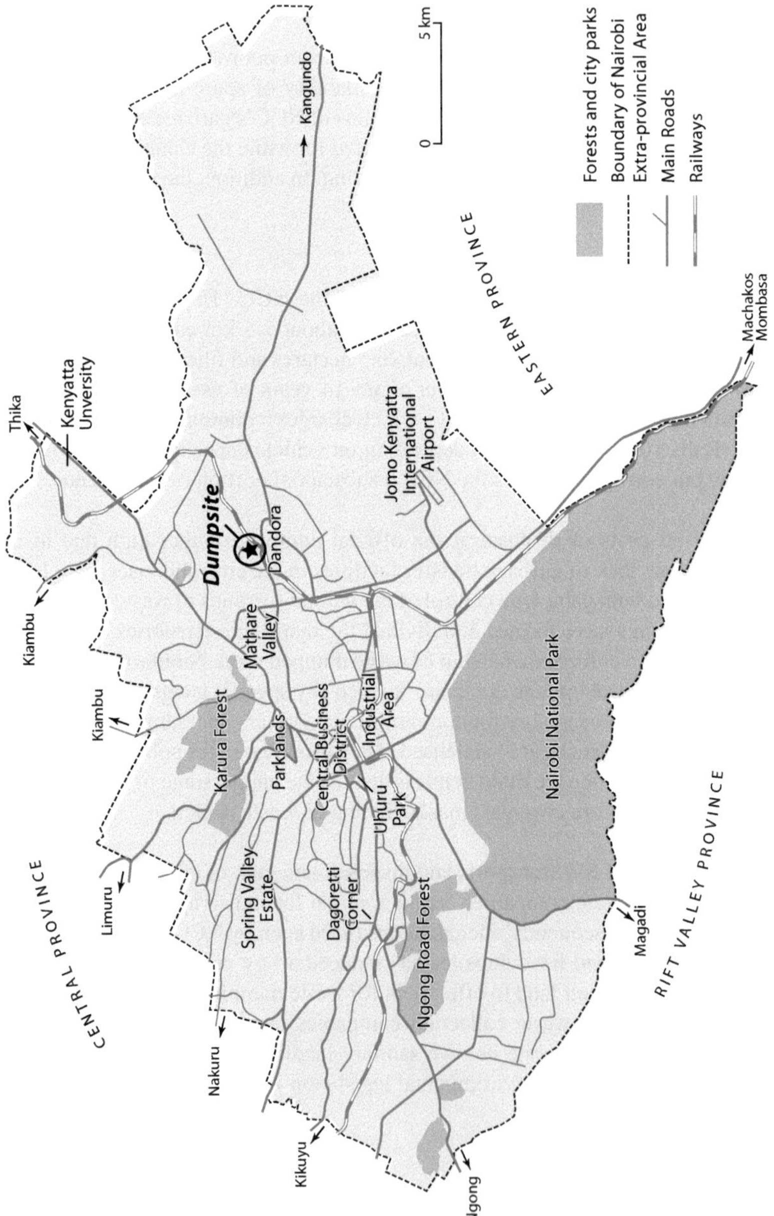

Figure 4.2. Map of Nairobi

Over the years, the capacity of the Dandora dumpsite to receive waste was severely diminished by non-compaction of waste. Only one old and ill-maintained bulldozer is available for waste compression. The Dandora dumpsite neighbours a densely inhabited low-income residential area, indicating poor economic, social, and physical planning. The risk of contamination, spread of diseases, and water/air pollution is high especially because toxic and hazardous materials get into the waste stream. Open waste burning at the dumpsite increases the risk, as do rodents.

Due to secondary pollution, residents of the surrounding area recently (February 2001) demonstrated against continued use of the dumpsite. Although environmental consequences of the dump have been evident for a long time, nothing was done because little or no attention is paid to environmental aspects of sustainable development in the city. On March 7, 2001 it was reported that the dumpsite is being relocated to a site recommended by the JICA (1998) study. The site, Ruai, is about 30 km from the CBD. The decision to relocate was not motivated by these complaints alone, however, as the land on which the dump stands was acquired by a private developer (JICA, 1998).

Explaining dismal NCC performance

NCC's solid waste collection services are poor because of lack of adequate financial resources due to slow down in economic growth[4], indebtedness, and corruption[5]. Proper management is frustrated by the failure of the Government to play the crucial oversight role, under-financing of the NCC in addition to poor revenue collection, insufficiently educated and trained policy makers (councillors), and lack of clear and differentiated duties between the NCC and the Central Government. Politics plays a dubious role in all of this. Many dignitaries (including Cabinet Ministers and officials of the Office of the President) 'play ball' by accepting favours (plots and employment quota allocations, and supply tenders) from NCC officials and, consequently, lose the moral authority to guide and provide oversight to NCC (NCC, 2000). Indeed, NCC's governance and management problems are microcosmic copies of national problems. Public resources designated for the provision of basic goods and services are openly appropriated for personal use. Wrangling over and irregular disposal of Council assets and property are the order of the day, as is nepotism in employment.

The rapid rate of urbanisation and the resultant increase in waste generation have also contributed to deterioration of service. The 1999 national census puts Nairobi's popu-

4. This is shown by annual average GDP growth rates: 6.6 percent over 1964-1973, 5.9 percent over 1974-1978, 4.4 percent over 1979-1983, 4.2 percent over 1984-1988, 2.9 percent over 1989-1993, and 3.0 percent over 1995-1999 (Republic of Kenya, Economic Survey Various Issues).

5. The NCC Extraordinary Inspection Report presents a detailed account of these problems, citing specific cases and the culprits. These include business people, NCC employees, councillors, senior Government officials including Cabinet Ministers, and professionals (lawyers, engineers, architects, etc.).

lation at 2.14 million, 7.5 percent of the country's total population (Republic of Kenya, 2001). The population, moreover, is growing at 4-5 percent annually.

Finally, the solid waste collection bylaws are very general, fail to define appropriate standards, and specify low penalties, while enforcement by the City Inspectorate Department is notoriously weak (JICA, 1998). Empowerment of the DOE to prosecute, regulate and establish environmental standards, and to operate within the new Environmental Management and Coordination Act has been recommended, but is still to materialize (NCC, 2000).

Private commercial companies

Since the mid-1980s, there are now at least 60 private solid waste collection companies in Nairobi (JICA, 1998). Their entry was stimulated by the increasing failure of the NCC service, and the resultant demand for solid waste collection service. The majority of the private companies are either small family ventures or a hybrid between a CBO and a private firm. Typically, a number of young people form a group and start offering solid waste collection service. However, there are about 10 relatively larger firms (see below).

The private companies simply obtain a business license and start offering solid waste collection services, without vetting or regulation. For most of them service commences once a client completes (basically name and address) and signs a form prepared by them. The forms specify the monthly charge, the frequency of the collection service, and the storage facilities to be supplied by the company. Because of increasing competition and cases of unsatisfactory service, moreover, some of the firms include (in the form) a promise to refund for unsatisfactory service. The signed forms are the only "contracts" involved. Some of the agreements, especially those involving small companies, are only verbal. They are also short term. The contracts have no provision for sanctions and there is no legal framework for the companies to deal with payment defaulters or for clients to secure legal redress when service quality is unsatisfactory. The wronged party simply walks out of the relationship. Some of the private companies, however, retaliate for non-payment[6]. There are no bylaws specifying the rights and obligations of the companies and their clients. Private solid waste collection business is operated purely on a willing-buyer-willing-seller basis, with inhabitants in any residential area not obliged to join the service being provided there.

Private companies use superior waste storage technology than NCC. They supply, on a regular basis, either 25-kg plastic bags for individual households or 200-kg drums for high-rise apartments. Some companies, moreover, supply neat hard plastic storage

---

6. One firm reported during our survey that it retaliates by depositing waste collected from other clients at the defaulters' entrance.

containers with fitting lids. Since the efficiency of waste collection is inextricably related to the mode of storage and technical appropriateness of storage containers, it is not surprising that the private companies offer a better quality solid waste collection service than the NCC.

Private companies altogether collect 115-130 tonnes of SW daily (JICA, 1998), almost double the amount collected by the NCC. There is a high concentration in the private solid waste collection sector, with a few of the firms collecting a large proportion of this waste. For example, in 1998 Kenya Refuse Handlers Limited (KRH) alone collected more waste than the NCC: 46 percent of all the waste collected then (JICA, 1998). About 81 percent of the households served by private companies live in the high and middle-income areas (largely the western part) of the city (see also table 4.2 and 4.3).

Privately provided garbage collection services are reported to be regular, reliable and well- organised because the clientele demands quality solid waste collection services. The survey we carried out in 1998, however, indicated that the service provided by small firms is erratic.

Other actors in solid waste collection

Self-provision on an ad hoc basis and outside formal rules is a common response to public service failure, especially when private companies consider market opportunities unfavourable. Communities often organise themselves as providers, especially in the areas of wastewater and solid waste disposal, with NGOs playing the key role of providing technical input during design and implementation (World Bank, 2000). This trend has been observed in Nairobi. Although private companies started joining the solid waste collection sector in mid-1980s they failed to extend into low-income and unplanned settlement areas. Consequently, community-based initiatives in waste collection, transport, storage, trading and recycling started to emerge, especially since the early 1990s. There are now numerous CBOs, in the form of charitable organisations, ethnic associations, welfare societies, village committees, and self-help groups. Residential associations are a special and significant category of CBOs.

CBOs

The majority of CBOs in the city concentrate on composting, which involves solid waste collection as a way of obtaining waste materials. However, the main activity of about 44 percent of the CBOs interviewed was cleaning their neighbourhood while one-third of them were involved in waste picking. Households receiving solid waste collection services from CBOs dump their waste at a central place in the neighbourhood, from where it is collected by CBO-employees or volunteers.

Despite meaningful localised performances, the community in general plays a small waste management role because it is not integrated into the solid waste management system. The World Bank (2000) identifies four barriers to such integration that apply to the Nairobi case: a psychological barrier associated with the expectation that the local government should supply all services, an economic barrier related to the cost of laying community infrastructure, a technical barrier that hampers initiation of self-help activities, and a sociological barrier in the form of lack of trust.

The role that CBOs can play in solid waste collection and other solid waste management domains is limited by lack of official/political support, lack of premises to operate from, lack of protective equipment, lack of finances and capital, low earnings from waste trade, lack of stable markets (therefore price fluctuations) for recycled products and stiff competition from emerging modern waste dealers, lack of transport, lack of local "rootedness" in terms of financial sourcing and political support, and weak leadership (Kajese, 1991; Kibwage, 1996; JICA, 1998; Mulei and Bokea, 1999; and Kwach and Antoine, 2000). Policies on community-based solid waste collection, in addition, have been lacking although the situation is changing. The current objective of the Government is to develop partnerships in environmental management with stakeholders, including promotion of environmental NGOs and CBOs (Republic of Kenya, 2000).

NGOs and international organisations

The involvement of NGOs and external actors in Nairobi's solid waste collection system takes the form of support to CBOs, with training, marketing and provision of tools and equipment being the main forms of assistance. More than half of the CBOs interviewed had been sponsored or facilitated by local and international NGOs and such United Nations agencies like the UNFPA and UNCHS (Habitat). The NGOs cited by the majority of the CBOs and small-scale recycling and reuse enterprises are the Foundation for Sustainable Development in Africa (FSDA), the Uvumbuzi Club and the Undugu Society of Kenya. Of the 15 community based solid waste management groups identified by JICA (1998), 11 were supported by the FSDA. The role of foreign donors is evident not only from their direct involvement in waste management initiatives, but also from the fact that most NGOs working in Kenya are dependent on external funding (King, 1996). UNCHS, in particular, has keenly assisted Mukuru Recycling Project, one of Nairobi's largest CBO complexes.

Residential associations (RAs)

Neighbourhood or residential associations have emerged in many middle and high-income residential areas, stimulated by the success of the pioneering Karen and Langata District Association (Karengata). Karengata, formed in 1940, successfully stopped, through Court, the NCC from collecting service charge from its members before improvement of the quality and frequency of the solid waste collection service and other municipal services. The Nairobi Central Business District Association

(NCBDA) has emerged as another highly organised, resourceful, and influential group. The group is contributing to security in the CBD by donating patrol vehicles and other facilities to the police, and is introducing community policing through training and other support. It has also recently donated garbage storage bins for use in the CBD. It is estimated that there are over 200 RAs registered in the city, engaged in improving security, roads, cleanliness, and other services. The associations have come together under the 'We Can Do It' lobby group, which is increasingly influential, putting the NCC on its toes and contracting, organising, and monitoring private solid waste collection service. Through 'We Can Do It', for instance, about 130 RAs recently signed a memorandum to the Government on the decay of city services.

Farmers
Farmers are increasingly becoming important actors in Nairobi's solid waste collection. The increasing number of urban and peri-urban farmers collect poultry waste, green vegetable waste, and cow dung as well as food waste from hotels, markets and other institutions, and transport it to use either as animal feed or as organic fertilizer (see chapter 12).

Informal actors
Many informal agents are involved in solid waste collection as a secondary activity. They include waste pickers, dealers, itinerant buyers, informal dump service providers and informal recycling enterprises (see chapter 8). These actors are involved in all solid waste management domains, including waste collection, separation, storage, reuse, recovery, recycling, trading, transport, disposal, and littering through discarded non-saleable items. They reduce the waste that has to be disposed of.

4.5. NEW PARTNERSHIPS IN SOLID WASTE COLLECTION

Partnerships between local authorities and other agents (the private sector, NGOs and communities) to facilitate sharing of solid waste collection responsibilities and financial burdens are considered one of the most important ways of achieving more sustainable solid waste management (World Bank, 2000). Such partnerships are only beginning to emerge in Nairobi but support for them is increasing. Broadly, two types of partnerships have come up in the capital, *viz.* public-private and private-private.

Public-private partnerships

The best example of a public-private partnership in Nairobi's solid waste collection was the pilot one-year management contract awarded in 1997 to one of the private companies (KRH) by the NCC. The contract involved daily sweeping of streets, roads, lanes, pavements and markets in the city's CBD, waste collection and transportation from the same area, and disposal of the waste at the Dandora dumpsite, at an agreed monthly rate of Kshs 1,312,500 (at that time about US$ 20,000). The private company did very well initially and the CBD became noticeably clean. Payment problems led to poor performance, however, particularly due to sit-ins by unpaid workers The contract had been financed from NCC's general taxation, as revenue from waste charges was inadequate (JICA, 1998). The NCC does not have contractual arrangements with any private company at the moment. A new management contract for the CBD is being planned, however, as is a pilot franchise system.

The NCC has some relationships with CBOs too. One of these is through the Nairobi Urban Slums Development Project, under which Kawangware Afya Bora CBO was set up. In this project, the NCC, Central Government, NGOs, CBOs, and International Organisations have joined to help people living in Nairobi's slums and other unplanned settlements. The aforementioned CBO, for example, handles more than half of the waste generated at Kawangware market. Similarly, the NCC established a relationship with the City Park Environmental Group (with 16 stall owners at City Park Hawkers Market), under which they allowed the environmental group to use land, kept aside for waste dumping, for composting activities. In exchange, the group was required to collect, transport and control all the waste generated from the market, estimated at 3 tons per day. The sponsor of the market, an Asian Foundation, brokered this informal partnership. The NCC, moreover, participates actively in the environmental cleanups organised by CBOs such as Mathare Youth Sports Association (Peters, 1996).

These examples notwithstanding, the NCC has not forged strong relationships with CBOs, due to several hindrances. First, municipal authorities often do not integrate the facilities of CBOs and NGOs into the mainstream, either because the settlements they serve are considered irregular or because the community-provided infrastructure does not conform to existing codes. Secondly, local governments often lack the technical and institutional capacity to form partnerships with CBOs (World Bank, 2000). Finally, prevailing attitudes among government officials and employees towards these non-state initiatives are still largely negative.

Private-private partnerships

a certain extent one may say that the relationship in which households, commercial entities, and institutions contract private companies to remove their waste on a regular basis qualify as private-private partnerships. Despite the fact that the nature of the arrangement is primarily commercial, it does serve the public interest and, hence, satisfies the criteria outlined in chapter 1. They have already been discussed earlier on.

Collaborative ventures among NGOs and between NGOs and other actors are another expression of private-private partnerships. Examples include collaboration between the FSDA and the Uvumbuzi Club in the 'Garbage is Money' campaign, and collaboration between the Undugu Society and the International School of Kenya in which students of the latter institution collected waste paper and gave it to the Undugu boys in Dandora to sell it and make a living. Mukuru CBO, in addition, has a relationship with waste pickers located at the Dandora dumpsite, in which the CBO strives to secure continued supply from the pickers by buying whatever waste they bring, including materials without re-sale value (Kwach and Antoine, 2000). Once again, the collection efforts inherent to these activities largely stem from the desire to support waste recycling and composting (see chapter 8 and 12).

4.6. ASSESSING SUSTAINABLE DEVELOPMENT IN SOLID WASTE COLLECTION IN NAIROBI

In this section, contribution of the activities of various actors in Nairobi's solid waste collection to urban sustainable development is analysed using, as far as possible, the template developed in chapter 1 of this volume. The subsequent analysis largely pertains to a sustainable development comparison between the two major service providers, the NCC and private companies.

Contributions to socio-economic components of sustainable development

User charges and economic viability

An activity may be regarded as economically viable if it can sustain itself. Unfortunately in this case, it was impossible to carry out a cost-benefit analysis that would allow firm conclusions to be drawn. However, the material does provide us with some clues on the subject. In general, the solid waste collection activities of the larger private sector companies are more viable than those of the NCC. Nevertheless, the economic viability of the private solid waste collection firms, particularly the smaller ones, is also under serious threat from the intense competition engendered by unrestricted entry and lack of control in the sector.

The steady deterioration of solid waste collection activities carried out by the municipality is closely related to the NCC's perpetual financial crises. Financial mismanagement, excessive workforce, corruption, poor revenue collection, and debts adversely affect its solid waste collection operations. The Interim Oversight Board, for instance, estimates that the current debts and revenues due to the NCC stand at Kshs 7.5 billion.

User charges determine to a large extent the viability and profitability of solid waste collection activities. Private solid waste collection companies range from large firms with a clientele of up to 5,000 (Table 4.4) to very small ones with less than 100 clients. Our survey showed that Urban Waste Management Services, for instance, has only 60. There is even a category of 'brief case' businesses comprising of one or two persons, without offices or permanent contact addresses that hire or borrow vehicles for waste transport. Their service is irregular and they charge a minimal fee. They select the areas they service rather haphazardly and may take a long time before returning to the same area. They engage in a myriad of other entrepreneurial activities.

Table 4.4. Relative sizes of private solid waste collection firms in a survey carried out in Nairobi

Firm	*Average charge (Ksh/month)*	*Size of clientele*	*Turnover/month (Ksh/month)*
Bins (Nairobi), large	500	5,000	2,500,000
DRDS Ltd., medium	500	1,400	700,000
Small firm	200	500	100,000

Source: Esho (1997).

Larger firms tend to charge higher rates and to have higher turnovers Medium and large firms have monthly turnovers ranging from Kshs 700,000 to Kshs 2.5 million (US$ 9,350-35,000) while their charge for a monthly service is about Kshs 500 (US$ 6.70). Larger firms concentrate on clients like institutions and industries but also high and some middle-income residential areas, which pay higher rates. The fact that residents in an area are not obliged to contract the private firm offering solid waste collection to other residents in the locality has had serious implications on the cost of the service. Because of it, private firms tend to have few widely scattered clients in different residential areas. The firms charge the few scattered clients as much as Kshs 800 per month to ensure profitability. This high price discourages potential clients and leaves pockets of residences not provided with services. These expensive providers are able to secure the upstream market by offering consistently high quality services. During the fieldwork, households in these up-market residential areas indicated that the private solid waste collection services were affordable. Because of effective demand, therefore, large firms are able to charge high rates and can sustain their services.

Small and some medium-sized firms use competitive pricing so that the charges are determined by the size of bulk load for large generators, the level of competition in the area, and in some cases by the distance of the area from the premises of the service provider (Esho, 1997). Because of greater competition, small firms offer a service that is 60 percent cheaper on average than the service offered by the larger firms. In the lower income areas into which small firms are increasingly venturing, intense competition has driven monthly charges to as low as Kshs 100 (US$ 1.3). 'Brief case' companies in fact charge as low as Kshs 50 for a monthly service, despite widespread willingness among residents to pay substantially higher prices. In a recent survey, for instance, 47 percent of the residents paying less than Kshs 100 per month were willing to pay more than Kshs 200 while 50 percent of those paying Kshs 300-400 were willing to pay as much as Kshs 800 (US$ 10.6) (Esho, 1997). In the absence of official rate setting the tremendous open competition prevents private firms (particularly the smaller ones) from fully exploiting this opportunity, reducing their economic viability.

Another problem that faces private solid waste collection firms, again the smaller firms more than the larger ones, is that of non-payment by some clients and a belief among many Kenyans that the NCC is supposed to provide solid waste collection services. The latter leads to reluctance on the part of residents, especially in less affluent districts, to engage with private providers Privatisation of solid waste collection service, moreover, still lacks political support from NCC stemming from fear that privatisation could lead to loss of employment.

The NCC has a sizeable clientele of institutions and commercial enterprises. In addition, all households with water connection pay a monthly waste collection fee of Kshs 60 each, on average, even though the service is hardly provided. However, in view of its poor performance, the NCC is in no position to adjust solid waste collection service charges upward in tandem with changing service cost. This is one of the areas targeted for reform. The cleansing section has no budgetary autonomy and does not control the waste fees generated through the water bill. It receives only 10 percent of NCC's budget, which is little compared to cities like Bangkok (15.3 percent) and Penang (24.7 percent) (JICA, 1998). Allocations are not based on service needs, but rather on the political fight over scarce local government resources. Consequently, DOE suffers chronic under-funding.

Equipment and transport cost

Economic viability of the solid waste collection service provided by large private companies is also indicated by their larger, well-maintained stock of equipment and facilities. For example, Bins (Nairobi) Ltd. has 11 collection vehicles while City Bins Ltd. has 15. Smaller companies like Urban Waste Management Services and Tacentac Enterprises have 2-6 vehicles. A typical small firm has one vehicle, usually a pick-up (of about 1 tonne) or 7-tonne truck, hired from the open market or family-owned (Esho, 1997). The vehicle is typically dilapidated, without alternative use, or a multipurpose family pick-up. About 60 percent of the small firms surveyed by Esho in 1997 used hired vehicles. The small companies use small vehicles with very low capacity to haul waste.

Vehicles are hired at exorbitant prices and the economic viability of the companies relying on them is therefore low. The poor state of roads and constant traffic jams add to transportation costs. Scattering of each company's clients over different parts of the city also contributes to escalation of transportation costs. For a trip of 4 tons, for instance, private collectors cover an operating distance of 50 km compared with 35 km for NCC (JICA, 1998). Increasing cost of fuel and absence of transfer facilities compound the problem of high transport cost. The unplanned solid waste collection system in Nairobi thus leads to inefficiency in service provision, also negatively affecting the economic performance of private companies.

The NCC has a larger fleet of vehicles than any of the private companies but most of these are not operational because of inadequate maintenance and repair.

Workforce and productivity

One of the principal causes of the financial crisis that perpetually characterises the NCC is its bloated workforce. The growth of population in the city has often been used as a scapegoat for hiring more workers, basically relatives, friends and supporters of councillors and politicians (NCC, 2000). The increase in the workforce has, however, not been accompanied by increasing performance of municipal services. Between 1995 and 2000, for instance, NCC workforce increased from 17,000 to 20,048 but this did not improve the Council's solid waste collection service. The cleansing section had about 2,324 workers in 1998, 22 percent of them involved in solid waste collection and the rest in street cleansing (JICA, 1998). At the time, the NCC was collecting about 70 tons of SW per day, implying that 7.3 workers were required to collect 1 ton. A private company like Bins (Nairobi) Services Ltd. required only 0.62 workers, however, while a smaller company like Tacentac Enterprises required 0.84 workers Moreover, the NCC had an average crew of 4.8 per collection vehicle compared with 4.0 for private collectors (JICA, 1998). A good demonstration of private companies' superiority in terms of labour productivity is the experience of the pilot public-private arrangement between the NCC and KRH. With Kshs 1.3 million (US$ 20,000) per month, the private firm was able to do much better work than NCC used to do with more resources. The municipal authority was able to re-deploy 525 waste workers and vehicles to other areas, and save Kshs 2.6 million per month (excluding fuel and vehicle maintenance costs; Moyo, 1998).

The bloated NCC workforce has had immense implications on the wage bill. More than 90 percent of NCC's annual income is used to meet personnel emoluments, leaving less than 10 percent for operations and maintenance (NCC, 2000). According to the Interim Oversight Board, a sizeable proportion of NCC workers are "ghost" employees (they exist on the payroll only) and only about 12,000 are required. Obviously, this phenomenon is another reason for the comparatively low labour productivity for NCC operations. Unfortunately, recommendations to reduce the workforce to 12,000 in order to improve the revenue position of the NCC meet with fierce opposition. NCC councillors, for example, raised a fury when 273 irregularly employed workers were sacked in January 2001.

Despite problems related to the spontaneous nature of the privatisation process in Nairobi, official policies increasingly admit to the potentials of the private sector (cf. the Government's Poverty Reduction Strategy 2000-2003, the Republic of Kenya, 2000). Moreover, the Interim Oversight Board strongly supports the idea and recommends that NCC's role should change to that of supervision, contract management, and monitoring of private contractor performance. The Board has also unveiled a plan to improve the financial position of the NCC. The plan is hinged upon improvement of the capacity of the Council to collect revenue and debts and net new rate payers, adjustment of Council's charges to bring them closer to market prices, expenditure control through a 40 percent reduction of the workforce, and introduction of a realistic budgeting process.

User satisfaction and public support

Nairobi residents blame the NCC for almost all the problems they face. With respect to solid waste collection the NCC continues to levy charges and rates without delivering any services, leading to substantial dissatisfaction with and lack of public support for the municipal service. Few households now receive the NCC service, which has degenerated to hardly more than crisis management.

There is user satisfaction and public support for private solid waste collection service generally. The households we interviewed felt that the service was regular, reliable and affordable. However, the larger and more organised firms were considered better service providers than the smaller ones. Another indication of user appreciation of privatised solid waste collection comes from experiences with the pilot public-private partnership between the NCC and KRH. City residents felt that KRH had improved cleanliness in the CBD.

Despite these favourable opinions, however, some drawbacks need to be mentioned. First of all, many citizens are reluctant to hire the services of private firms as they consider solid waste collection a NCC responsibility for which they are paying. Secondly, private sector firms have hardly extended services to social groups that are not provided with services (UNCHS, Undated), as they tend to exclude the low-income groups or those associated with payment difficulties for commercial reasons. Consequently, many areas of the city (principally unplanned settlement areas) are still inadequately served or not served at all. These two factors negatively affect the social legitimacy of private solid waste collection.

Employment levels and quality

The NCC employed 2,324 workers, 22 percent of which were engaged in solid waste collection services in 1998. Although salaries are low and purchasing power continues to deteriorate, NCC workers can call on the usual package of fringe benefits related to public sector employment. Besides, a sizeable group of employees does not even bother to show up at work. Those that do are heavily demoralised by delays in payment for up to three months, and lack of equipment and tools. Therefore, their productivity is very low, as demonstrated in the previous section.

Private solid waste collection companies have a fairly large workforce. For example, Bins (Nairobi) Services Ltd. has a total of 62 employees, City Bins Ltd. 65, DRDS 20, and Tacentac Enterprises 21.

All workers in solid waste collection, whether NCC or private, experience a range of similar problems, the most important of which are poor conditions of work, poor terms and payment, heavy workload, health problems, and lack of tools and equipment. Moreover, they suffer low self-esteem. According to over 60 percent of those interviewed in 1998, their waste work is perceived as dirty and lacking respect. About half of the workers, consequently, plan to start their own business or get employment elsewhere. The workers complained about government's insensitivity to their grievances and lack of a proper labour policy. While junior NCC workers belong to a strong trade union that pushes for the payment of their salaries ahead of those of management level staff, the private sector solid waste collection workers are not adequately represented.

Contributions to public health and environmental components of sustainable development

Cleanliness of neighbourhoods

The cleanliness of many Nairobi neighbourhoods, especially the unplanned, low-income areas, leaves a lot to be desired. The NCC's contribution to the promotion of public health through appropriate solid waste collection is insignificant as a result of the virtual collapse of regular servicing. Furthermore, the NCC hardly provides any storage containers, leading to infestation by mosquitoes and other flies, odour, and scattering of waste by animals, according to households interviewed.

The private companies' contribution towards the goal of clean neighbourhoods is much more positive as they usually provide regular (1-2 times weekly) and reliable services to their clients. Some of the private firms, in addition, offer street/compound cleansing and/or sweeping services to their clients as a bonus. Most of the households using private solid waste collection services reported minimal problems in storage. Some households complained, however, that the private companies sometimes fail to collect waste on the scheduled days. The uncollected waste is vandalised by stray dogs and cats leading to littering of streets and surrounding areas. This problem is especially common with the 'brief-case' garbage collection firms.

The most important factor troubling the private sector in performing an even more environmentally friendly role is the lack of government support. There are no by-laws summoning residents to participate in a solid waste collection service offered by a private firm in their neighbourhood. This prevents them from making optimal use of the collection potential in the area and from reducing their rates accordingly (subsequently leading to higher levels of participation, and hence, cleanliness). Furthermore, the NCC fails to enforce private sector adherence to public health and service standards. Since the private entrepreneurs are mainly motivated by the desire to make profit, they are reluctant to comply with cost-raising environmental regulations. The lack of effective monitoring, for example, frequently results in indiscriminate dumping of waste especially by the 'brief case' private companies. Finally, the private sector is incapable of extending solid waste collection services to all areas previously not provided with services. The pollution of these areas produces external effects that negatively affect the public health status of the served areas. On the other hand, the entry of smaller firms that have minimal operational costs has brought private solid waste collection services into reach of some low-income residential areas, particularly those proximate to the city's dumpsite. Thus, Esho (1997) found that 70 percent of the small private firms in his sample mainly served middle-income and some low-income areas. Moreover, about 11 percent of the households we interviewed in 1998 in low-income and slum areas were receiving solid waste collection services from private companies.

Health of workers

The health situation of both NCC and private sector workers gives cause for worry. About 80 percent of the workers interviewed have encountered health hazards in their work, representing a higher percentage than that found among the dump waste pickers Collection technologies are very simple and usually manual labour is applied to haul waste into the vehicles. This has ramifications on the health and well-being of the garbage crew. Ailments commonly suffered by the workers include, in order of importance, injury/cuts, headaches, stomach ailments, skin ailments, respiratory ailments, pains and aches, eye problems, and burns (30 percent). The 'protective gear' used by the workers comprises only of overalls, boots and, in some instances, gloves. Moreover, the lower cadre workers in the private sector are not covered in medical schemes and do not receive regular medical check ups.

Transport and environment

The fact that Nairobi only has one official dumpsite and no transfer facilities makes transportation a troublesome aspect of solid waste collection both in terms of financial and environmental costs. Transportation costs are an important factor determining user charges, and hence affordability of private solid waste collection services. Especially low-income areas located at some distance of the Dandora dumpsite suffer the consequences. Air-pollution caused by collection vehicles is another environmental concern. It applies in particular to the smaller firms serving low-income areas, who can only survive thanks to the use of over-aged vehicles. Finally, it serves to mention that workers of the NCC and the private handling companies have experienced problems with waste pickers jumping on their vehicles using the trip to sort out useful waste materials. Their frantic work usually results in a trail of litter polluting the roads.

Recycling, reuse and composting

Although recycling, reuse and composting are significant sub-sectors within the solid waste management system in Nairobi, collection efforts by the NCC and private firms are not geared in any way towards these concerns. There is no policy or any official support for source separation, recycling, reuse and minimisation/prevention of waste generation. Any relationship that does exist is accidental and the result of initiatives outside the public sector (NGOs, CBOs, international organisations, and some private handling companies). Our survey, for example, indicated that about 5 percent of them are involved in waste separation.

Disposal and the environment

The NCC does not control waste disposal even though it is the only agent authorised to offer disposal services. The method of crude dumping applied at Dandora (a former quarry) poses substantial ecological threats due to leakage of toxic substances. The capacity of the site to assimilate the wastes has been seriously eroded and its life shortened due to the failure to design and implement a solid waste management policy. Currently the NCC uses unofficial dumps due to the absence of transfer facilities, a factor that has seriously affected the cleanliness of the city. The management of the official dump has been so poor that neighbouring residents suffer serious health problems. The upsetting security situation on the dump, which seems to be totally out of official control, helps to compound all of these problems.

As far as the private companies are concerned, even though they remove substantial volumes of waste from residential areas and other parts of the city, only part of it is disposed of in the official dumpsite. There are frequent complaints that most of the private firms (particularly 'brief case' ones) dump waste in open places instead of the official dumpsite in a bid to cut operational cost. Furthermore, private providers have encouraged excessive use of plastic bags for waste storage, which poses new environmental challenges. It is encouraging, however, that some private companies are involved in environmental awareness campaigns.

The planned relocation of the official dumpsite from Dandora to Ruai is likely to improve environmental impacts of final disposal, but may very well worsen the other dimensions of sustainable development. The new site will be a sanitary landfill[7] and will assimilate waste more effectively since its capacity is intact. In the long run, environmental aspects of sustainable development will be determined by the quality of the new dump's management and the rate at which waste is deposited. The dump's location further away from the CBD coupled with lack of transfer facilities, however, will lead more solid waste collection agencies to dump indiscriminately, with serious implications on public health. The move is also likely to negatively affect the viability of private firms. Fewer companies will be able to operate where such long haulage distances have to be covered. This may mean loss of employment for the people working in these companies and perhaps more importantly loss of a means of livelihood for the dump waste pickers and dealers (estimated at about 2,000) located at the

7. A sanitary landfill is a disposal technology wherein solid wastes are placed on impermeable soils that protect groundwater, under laid with impermeable plastic membranes and drainage systems to collect contaminated seepage for treatment, and ventilated with gas management systems.

old dumpsite. The proposed relocation is facing substantial opposition from within and outside the LA.

4.7. CONTRIBUTIONS BY OTHER ACTORS TO SUSTAINABLE DEVELOPMENT

The other actors discussed earlier on only play a very modest and usually indirect role in solid waste collection. The subsequent remarks are confined to those aspects where their support to the goals of urban sustainable development is most pronounced.

Informal actors

Informal actors contribute to the environmental components of sustainable development of the city's solid waste collection service by reducing the waste that has to be disposed of. Moreover, they reduce the waste in critical areas (non-serviced low-income areas), thereby contributing to social equity of service provision. However, waste pickers and dealers throw away non-saleable items indiscriminately, littering the areas in which they operate, and antagonise the NCC and private company workers who have to collect the littered waste. Their contribution to public health is therefore negative. Informal actors are also accused of stealing waste storage containers

CBOs and NGOs

CBOs (and the NGOs that support them) have made individual and localised contributions to sustainable development of the city's solid waste collection service by composting organic wastes, recovering inorganic waste materials, carrying out regular environmental clean-ups, and educating the public on environment and waste management. JICA (1998) cites improved public health, cleanliness of the environment and supplementary income as the localised impacts of composting groups in Nairobi. The Mathare Youth Sports Association organises weekly clean-ups in the Mathare slums. Kitui Pumwani Integrated Project CBO succeeded in sensitising the community on the need to separate waste at generation point (JICA, 1998). In aggregate terms, however, the role of CBOs is still small. There are about 15 community based solid waste groups with a total membership of only 10,300 that treat about 1 ton of municipal waste per day only (JICA, 1998), or less than 0.1 percent of the total municipal waste generated.

Undugu Society of Kenya, a charitable NGO that has been involved in plastic recycling and composting activities since 1981, has contributed to employment and income generation for low income waste collectors and street children (Davis-Cole, 1996). The contribution of CBOs and NGOs is hampered by many factors as discussed in one of the previous sections.

Increasingly, residential associations and their umbrella association, 'We Can Do It' are playing the important role of keeping the NCC and the Central Government on their toes by demanding quality services and greater accountability. They are also monitoring the quality of private solid waste collection service and organising self-provision of not only solid waste collection but other services (such as security and infrastructure) as well. The associations are becoming key change agents towards sustainable solid waste collection, as they have arisen out of need and are led by highly educated and visionary people.

Urban farmers

Urban agricultural activities including kitchen gardening are making a steadily increasing contribution to the economic viability of the city's solid waste collection activities by absorbing a large amount of the organic waste that could have otherwise ended up at the dumpsite (see chapter 12).

4.8. STAKEHOLDER VIEWS ON FUTURE SOLID WASTE COLLECTION POLICY

A stakeholder workshop was held on 8/16/2000. It attracted 31 participants, including representatives of NCC, private solid waste collection companies, recycling enterprises, informal actors such as waste pickers and dealers, Ministry of Health, and University of Nairobi. Stakeholders expressed their concerns and opinions on various issues. Obviously they did not agree on each and every issue. Ultimately, however, the following conclusions were reached with respect to solid waste collection:

- There was need for regular stakeholder meetings and consultations, and the NCC was challenged to convene them.
- Despite the positive contributions of the private sector and civil society organisations solid waste collection was deteriorating. Reuse and recycling of waste materials was deemed inadequate, conditions at the Dandora dump were considered deplorable, and the problem of indiscriminate dumping was seen at a serious threat to public health. Speedy implementation of the proposed dump relocation to Ruai, however, was doubted. Poor infrastructure, in the form of poor roads, traffic jams, and power and water rationing were seriously raising the operational costs of private firms and waste recycling enterprises, thereby jeopardizing their continued contribution to a more sustainable solid waste collection system.
- The NCC's primary role was to facilitate and regulate the other actors and to set a level playing field for them. The NCC was, moreover, obliged to provide an efficient solid waste collection service (either directly or through private providers) as it collects waste charges from residents.
- The private companies had an important role to play. Their involvement, however, required appropriate regulation both to smoothen their operations and to protect the public interest. Although they were making a major contribution already, they were accused of indiscriminate dumping. The private operators

themselves were unhappy with disorganised open competition. They complained that it led to a proliferation of 'brief case' operators who offered very low prices but dumped waste indiscriminately, affecting the perception of the entire private solid waste collection service. There was general agreement on the need to develop and implement standards and guidelines for private solid waste collection service. Private firms, in addition, complained about non-payment. Participants concluded that private sector involvement needed formalisation and organisation to facilitate access by private companies to the judicial system for resolution of such problems. It was also felt that the private sector should be licensed to provide waste disposal services.

- The role of civil society (CBOs and RAs) was acknowledged but also the need to facilitate and strengthen its involvement.
- There was need to raise public awareness of solid waste issues and to study the influence of culture on waste generation and management in order to facilitate the design of effective solid waste collection interventions.

These stakeholder views seem to correspond very well with those expressed during our household survey (table 4.5).

Table 4.5. Solutions to Nairobi's solid waste problem (as suggested by households)

Solution	*No. of households suggesting*	*Percent of all households interviewed*
Improvement of NCC	37	29
Involvement of communities	34	26
Formalisation of privatisation	22	17
Designation of special dumping place	18	14
Increase of government resources to NCC	11	9
Promotion of waste reuse and recycling	5	4
Total	127	99

Source: own survey, 1998.

4.9. SYSTEM-WIDE CONCERNS

The above analysis was largely, albeit not exclusively, framed according to impacts of solid waste collection efforts on the actors involved or the clients they serve. However, it is also important to look beyond the micro-level and to highlight a number of system-wide concerns. This is especially warranted from the perspective of urban sustainable development. Several concerns stand out.

First of all, the low-income areas, principally slums and other unplanned settlements, where about 60 percent of Nairobi residents live, do not or hardly receive any solid waste collection service due to their 'illegal' nature, poor accessibility, weak political voice, and lack of service purchasing power. The NCC, which is concentrating its limited capacity to wealthier areas, is apparently insensitive to the needs of a majority of its citizens and thereby ignores the external costs of poor public health in these areas. The disconcerting lack of equity in servicing is probably the single most important threat to more sustainable solid waste management. The authorities have to come to terms with the phenomenon of illegal settlement in order to make any meaningful city-wide progress

Second, the administration of Nairobi is chaotic, with the NCC and the Central Government often clashing, duplicating roles, and causing confusion. Moreover, policymakers (NCC councillors) are generally poorly educated and lack power to discipline NCC officials and workers The mayor, who is elected by the councillors, owes allegiance to their greedy demands rather than to Nairobi residents. Consequently, mismanagement, corruption, laziness, and general chaos have become the hallmarks of the NCC. This dysfunctional local administrative system has led to the collapse of direct public servicing, but also to the absence of system-wide co-ordination and regulation of other actors in solid waste collection (the NCC's indirect provision role). Private sector involvement in solid waste collection is uncontrolled, adversely affecting effectiveness. Its spontaneous nature implies that it is driven purely by market concerns at the expense of social and public interests.

Third, there is not something like an "integrated" solid waste management system in Nairobi, nor any policy to move in that direction. Relationships between collection, recycling and composting activities, for example, are accidental rather than structural. They do not result from any concerted NCC action, but arise from 'spontaneous' initiatives in civil society. The potential for creating positive spin-offs, therefore, is not reaped. Furthermore, the idea of partnerships between various actors, each contributing their own specific qualities, is not yet taken up seriously. Residential Associations, for example, could effectively supervise and monitor private solid waste collection firms and see to it that they respect service standards. Another distressing aspect of this lack of integration is that various categories of waste are all mixed up. Relatively harmless waste, therefore, becomes contaminated by toxic and dangerous substances, thereby turning collection points and dumpsites into environmental and public health hazards. It still remains to be seen whether the Environmental Management and Coordination Act (1999), which provides for a more integrative approach to solid waste management, will really help to solve these problems.

4.10. CONCLUSIONS

Only about 25 percent of the solid waste generated in Nairobi gets collected. The rest is dumped in open spaces, making the city terribly dirty. The unplanned settlements where about 60 percent of the city residents live do not receive any solid waste collection service. Poor management, lack of accountability and corruption have crippled the NCC, the principal agency for solid waste collection. The Council concentrates its available solid waste collection services on residential areas and institutions that can afford private service, at the expense of low-income areas. Waste disposal facilities and practices are appalling. The involvement of the private sector and actors from civil society has made a difference but exploitation of their potential is hampered by lack of policy support, excessive and open competition, lack of guidance and control, and poor coordination among the agents involved.

The sustainable development record of most solid waste collection activities in Nairobi is rather weak. Private solid waste collection activities of larger companies are the most promising ones, economically as well as in terms of contribution to public health. Even these, however, are suspect as far as social equity of service distribution, working conditions of labourers, and environmental components of sustainable development are concerned. The contributions to urban sustainable development by smaller private companies are adversely affected by open competition and indiscriminate dumping. NCC operations score dismally in all dimensions of sustainable development evaluation. Even the planned relocation of the dump is faulted in terms of socio-economic aspects of sustainable development. The sustainable development scores of solid waste collection activities by actors from civil society and informal actors, although promising in some respects, should not make us forget that their absolute contribution is minimal. Nevertheless, the future of solid waste collection service in Nairobi arguably lies in the rapidly expanding and increasingly articulate residential associations and their uniting bodies. Emerging out of desperate need and led by visionary and highly educated people, these associations are already being felt. They push and sue for quality service from the NCC and value for money service from private solid waste collection providers.

JAAP BROEKEMA

CHAPTER 5
TRIAL AND ERROR IN PRIVATISATION; THE CASE OF HYDERABAD'S SOLID WASTE MANAGEMENT

5.1. INTRODUCTION

Hyderabad, the state capital of Andhra Pradesh in Southern India, has a population of more than 4.5 million, and is ranked the fifth largest city in India. As laid down in The Hyderabad Municipal Corporation Act, the Municipal Corporation of Hyderabad (MCH) is responsible for the collection and disposal of municipal waste in the entire corporation area. Similar to most of the Indian Municipal Corporations, the Health Section and the Transportation Section have the overall responsibility for these activities. The MCH has a total of 12,590 permanent employees of whom 7,150 are attached to the Health Section. Out of these almost 5,700 employees are involved in sanitation work (MCH, 1998a).

Administratively the MCH comprises seven 'circles' that are further subdivided in 56 wards. An Assistant Medical Officer of Health (AMOH) heads each circle. This person is assisted by Sanitary Supervisors who are in charge at ward level and manage a team of Sanitary Jawans. The Jawans are responsible for the day-to-day supervision of the kamatans, kamatees and lorry workers The kamatans (females) sweep the streets and collect the waste, while the kamatees (males) dispose of it at the collection points.

This chapter is based on fieldwork that was undertaken from January until July 1999. The objective of the study was to assess the nature and extent of private sector participation in solid waste collection in Hyderabad as well as its effectiveness in meeting various goals. In the appraisal I have tried to go beyond the familiar concern with service efficiency and effectiveness, and to include the social and political dimension of the privatisation exercise. The following set of questions guided the study:

- What were the motives for the MCH to privatise a substantial part of its solid waste collection services and what are the characteristics of the form of privatisation that was adopted?
- What are the socio-economic impacts of this particular form of privatisation of solid waste collection for the contractors and labourers involved?

I. Baud et al. (eds.), Solid Waste Management and Recycling, 93-112.

- What are the strengths and weaknesses of the current form of privatisation of solid waste collection?

The discussion is confined to the dominant public-private arrangement in solid waste collection, namely the contracting out of sweeping services and solid waste collection in the planned residential areas of the MCH to private firms[1]. The costs of the service are fully met by property tax revenues. Plans are being developed to introduce 'user based garbage collection charges'. However, rumours have it that residents will once again be exempted from paying for political reasons (Interview AMOH, 1999; The Hindu, 2/18/99).

5.2. TERMS OF REFERENCE FOR PRIVATISED SOLID WASTE COLLECTION

The main reason for the MCH to transfer part of its solid waste collection tasks to the private sector arose from the impossibility to increase its current workforce despite the city's continued expansion. Other factors played a role as well. For instance, the Supreme Court stated that sweeping of main roads should be finished before the opening of the commercial establishments in the morning. This obliged the MCH to introduce night sweeping. However, this work could not be taken up by MCH labourers who fiercely opposed the idea of working night shifts and were able to defend their views due to their strong legal position and the resistance by the labour unions.

The first significant attempt to privatise the solid waste collection-services was launched in 1994 when 10 percent of the MCH areas were privatised (MCH, 1998a). In 1996 more areas were handed over to the private sector and by 1998 almost 25 percent of the MCH area was under privatised service (MCH, 1998a). Initially the contractors were operating under either the lowest bid system or the fixed volume system. The lowest bid formula implied that the MCH indicated the areas and was calling for tenders from private contractors Entrepreneurs were asked to quote the rates for sweeping and lifting the garbage for the entire area on a daily base. The second system was based on payment according to the amount of garbage lifted. The contractors were asked to quote the rates for one Mt. of garbage and subsequently, the lowest tender was awarded with the contract for sweeping and garbage lifting in a particular area.

1. A unique feature of the solid waste collection-system in Hyderabad is the involvement of the community under a tricycle scheme. Community appointed waste pickers take care for the house-to-house collection in almost 1,200 colonies, with each colony comprising 100-200 households. The residents pay a monthly fee of Rs 10-20 – the 1999 rate was about Rs 40 for 1 US$ – and the MCH is bearing the cost of the tricycle (Sudhakar Reddy and Galab, 2000; see chapter 3).

In 1998 the MCH acknowledged that the privatisation efforts to date were unsatisfactory. The poor record of the private sector was attributed to three reasons: first of all, the failure to ensure accountability by the private operators; secondly, the lack of a proper monitoring system; and thirdly the absence of fixed output norms (MCH, 1998a). At that time, for example, a local newspaper reported that due to a secret syndicate between private contractors and several MCH officials the quotations increased excessively, by almost 200 percent in comparison with the rates quoted in the previous tender procedure (Deccan Chronicle, 3/31/98). In an attempt to overcome these problems the MCH in September 1998 introduced an entirely new organisational framework and a rational system of monitoring. In order to bring uniformity in the system the city wards were divided in 266 so-called units of more or less equal size. Out of this total, 120 were handed over to the MCH workers while the remaining 146 units were awarded to the private sector. The MCH labourers who used to work in the privatised units were re-distributed among the remaining MCH units. In addition, private contractors took up some new i.e. previous un-serviced areas. Each unit has a total length to be swept of 7 to 8 km and produces around 7 to 8 Mt. every day. Every unit consists of one sweeping and one lifting unit. For each unit the MCH has stipulated the following labour requirements.

Table 5.1. Labour requirements for one unit

Labour requirements			
Sweeping unit		*Lifting unit*	
Type of labourers	No.	Type of labourers	No.
Kamatans (female)	14	Lorry labourers	5
Kamatees	7	Supervisor	1
Sanitary Jawans	1		
Total	22	Total	6

Source: MCH, 1999b

Payment to the private contractors under this so-called unit system is based on a uniform calculation as can be seen in table 5.2. This calculation includes a standard profit margin of 10 percent and fixed implements for workers and vehicles (spades, baskets, brooms, safety equipment etc.). The MCH regards the method of exactly stipulating the contract requirements to be met as a measure to prevent a price-fixing among contractors and to protect itself against under-quoting by contractors, resulting in a failure to perform up to standard.

Table 5.2. Cost calculation per unit

Sweeping unit	*Amount (Rs)*	*Garbage lifting unit*	*Amount (Rs)*
Wages for 21 workers at Rs 1,800.-	37,800.-	Rent for the vehicle with driver, oil etc.	18,150.-
Wages for 1 supervisor at Rs 2,500.-	2,500.-	Wages for 5 lorry workers at Rs 2,300.-	11,500.-
Implements for male workers at Rs 130.-	910.-	Wages for 1 supervisor	2,500.-
Implements for female workers at Rs 70.-	980.-	Implements for vehicle	100.-
Safety equipment at Rs 10.-	210.-	Safety equipment for workers at Rs 35.- per head	175.-
		Disinfectants	1,275.-
Sub-total	42,400.-	Sub-total	34,000.-
Profit 10 percent	4,240.-	Profit 10 percent	3,400.-
Gross cost per month	46,610.-	Gross cost per month	37,400.-

Source: MCH, 1999b

After the unit rate was fixed contractors were invited to reflect through advertisements in three daily newspapers (a leading Telugu, Urdu and English newspaper). The time span between the call for tenders and the actual start of the work was rather short, only two weeks. Although there were no demands with regard to the type of company – meaning that both proprietary firms, partnership firms, registered companies, Labour Contract Societies as well as individuals were invited to respond – the MCH did indicate its preference for companies satisfying the following requirements: having their own vehicles, Labour License, Profident Fund (PF) and Employees State Insurance (ESI) code number, and good record of previous performance. Labour Contract Societies were specifically mentioned as a contractor category and incited to reflect. The MCH also had a preference for firms or persons willing to buy Dumper placer vehicles on a loan basis. Contractors were restricted to a maximum of three units each. Despite all these demands the MCH received more applications than available units. Final selection, therefore, was done by drawing lots.

The applicants were requested to make a down payment of Rs 25,000 (US$ 625) for each unit. This amount, the Earnest Money Deposit (EMD), is meant as an assurance for the MCH and calculated as 2.5 percent of the annual contractual amount. In case the contractor, even after repeated warnings, fails to perform well, the MCH is permitted to terminate the contract, forfeit the EMD and blacklist the contractor.

With respect to the contract requirements it merits to mention one particular incident that has had far reaching consequences for the contracting approach adopted by the MCH. In its decision of 3/30/99 the Supreme Court of India awarded a request for the absorption of contract labourers working for a private contractor into government service because they had been found 'state employees in disguise'. Although it was unclear whether this ruling would also be applicable to the unit-system in Hyderabad, the MCH anticipated with some adjustments in the terms of reference of the contracts. As one MCH official put it, these 'little manipulations' were intended to prevent abolition of the contract system. These corrections, however, have made the contracting out of public services less secure, especially for the contract labourers (shorter contracts, abolition of PF and ESI facilities etc.).

5.3. IMPLICATIONS OF PRIVATISATION FOR THE MCH

Since the introduction of the unit system in September 1998 the amount of money channelled to the private sector has increased sharply. According to the revised budget estimates 1997-98 (before the introduction of the unit system) the total amount spent on privatisation of solid waste collection services was 57,39 lakh Rs per month.

Table 5.3. Expenditure on privatised solid waste collection in 1997-98

Works	*Expenditure per month (in lakhs)*
a) Night sweeping	18,37
b) Day time sweeping of wards	32,15
c) Sweeping of small areas in day time	6,87
Total	57,39

Source: MCH, 1998b

The monthly amount spent on privatisation after the introduction of the unit system can be calculated as follows:

Table 5.4. Monthly amount spent on privatised solid waste collection after introduction of the unit system

Work	*No. of sweeping units*	*Monthly amount for one sweeping unit*	*Monthly amount spent on sweeping units*	*No. of lifting units*	*Monthly amount for one lifting unit*	*Monthly amount spent on lifting units*	*Total*
1. Day	142	46,640.-	6,622,880.-	121	37,400.-	4,525,400.-	11,148.280.-
2. Night	50	73,695.-	3,684,750.-	21	45,898.-	963,858.-	4,648,608.-
Total	192		10,307,630.-	142		5,489,258.-	15,796,888.-

Calculation based on: MCH, 1999b

The total expenditure on privatised solid waste collection almost tripled after the latest reform. Unfortunately, it proved impossible to discover how the rise in expenditure was paid for (see also chapter 3).

Another major consequence of privatising solid waste collection for the MCH, especially since the introduction of the unit system, was the restructuring and upgrading of the supervision and monitoring system. Advocates of privatisation always claim that it is easier to check the performance of private operators than that of government agencies. In Hyderabad 52 Sanitary Supervisors are in charge of the day-to-day supervision of the private contractors However, their work suffers from several drawbacks. First of all, the workload of the sanitary inspectors varies tremendously, with some covering only a few units and others many more. Secondly, the supervisors have to register any shortfalls in the service provided by the contractor and to register these at the Circle Office. On the basis of these reports the Commissioner of the MCH can calculate the deductions from the contractual amount due to the entrepreneurs depending on the type and frequency of the violation. However, the supervisors primarily pay attention to task performance – proper street cleansing and removal of waste from the collection points – while requirements with respect to labour conditions are neglected, partly because they think this surpasses their responsibility. Thirdly, files of the infractions made by individual contractors are not kept. Therefore, the MCH is not capable of making a distinction between well and poorly performing contractors at the time of awarding new contracts. Finally, the entire system is sensitive to corruption. It is common knowledge that deals are made between Sanitary Supervisors and contractors in order to prevent reporting of shortfalls. It is interesting to note that the current system also attributes a monitoring role to the community. In almost every unit a citizen committee was set up, comprising of 10-15 members After the contractors have finished their work they have to collect a signature of at least three members, indicating approval of the work. The bills of the contractor will not be passed without these signatures. Although these requirements certainly contribute to better performance, they are not free from manipulation either.

5.4. Position of the contractors in the new unit system

This section discusses the results of a small survey among the private contractors In total 28 interviews were carried out of which 10 had a profound character. Initially the contractors were contacted on the basis of a list that was provided by the MCH, after which, often with the assistance of the Sanitary Supervisors, appointments were made. In most cases the interviews were held at the house of the contractor.

Socio-cultural background

The vast majority of the interviewed contractors are Hindu (90 percent) and many of them are Reddys. Together with Kammas and Velamas they belong to indigenous agricultural communities, which can be classified under the forward communities[2]. The remaining 10 percent of the contractors are Muslim. Among the interviewed contractors three are female. All of these work under the umbrella of a women co-operative society. Almost one third of the contractors was born in Hyderabad whereas the others were born in one of the neighbouring districts. However all the contractors were born in Andhra Pradesh. Most contractors are well educated. Only 15 percent of them has no or a maximum of a few years of primary education, whereas 52 percent has at least a bachelor degree.

In the literature describing the antecedents of contractors in solid waste collection services, it seems that they are often engaged in truck-related activities such as transportation and construction work (Cointreau-Levine, 1994; Coad, 1997; Broekema, 1998). In Hyderabad, however, the contractors constitute a rather heterogeneous group performing a broad array of different activities with a strong variation in dependence on solid waste collection activities. Among the interviewed contractors, for instance, there were civil contractors, labour contractors, real-estate agents and small-scale factories. Some of the contractors can be labelled 'unconventional' since they are Labour Contract Societies[3]. On the instigation of the state government the MCH enabled these organisations to apply for a number of contracts. Most of them are co-operative societies for women, but also various welfare organisations of the labour unions are participating (see box below). Although these organisations are labelled co-operative societies, in reality they operate under exactly the same conditions as the purely private contractors

Box 5.1. A Co-operative society in solid waste collection

Labour Welfare Organisation of the MCH Employees Union

The MCH Employees Union is a labour union affiliated to the CITU that, in turn, is politically linked to the CPI(M) the communist (Marxist) party. Its labour welfare organisation is involved in solid waste collection since 1998. As a Labour Contract Society it is exempted from paying EMD. The organisation claims to provide employment to the children of retired and current MCH labourers, since they do no longer have the opportunity to enrol in MCH-service. There is a certain degree of controversy with the other unions about their participation in Hyderabad's solid waste collection services, ranging from principal objections to becoming an employer to accusations of profit sharing among leaders The union claims that no profits are made due to deductions and additional costs, i.e. management expenditure.

2. These agricultural communities can be considered as the traditional ruling class in Andhra Pradesh. Until recently they have also dominated the political spectrum in the state.
3. A Labour Contract Society primarily aims to provide and secure employment to its members

Starting up business

Most contractors started working in solid waste collection with the introduction of the unit system in September 1998. They applied for a contract after reading the tender notice in the newspaper. Sometimes friends informed them. Most respondents motivated their choice to enter the profession by saying that 'it is just another method to earn a living'. In addition, some contractors stated that they already had a truck on their own, which made the decision easier, while others pointed at the investments and administrative requirements being low. None of the contractors had acquired special skills or training before they started to carry out solid waste collection services. However, none of them considered this to be a handicap. Some argued that the MCH had provided them with sufficient information while others regarded the work as comparatively simple or stated that they could call upon experienced labourers who had worked for other contractors in the past. In addition, some contractors referred to the principle of 'learning by doing'.

Considering these optimistic views it is interesting to see with what kind of starting problems the contractors were confronted with. One third of the contractors claimed not to have experienced any specific problem in the start-off period. The most frequently mentioned problem was gathering the funds needed for investment and EMD. Other, less frequently cited problems were lack of specific working experience, recruitment of labourers and labour problems (in terms of supervision).

Recruitment and supervision of labourers

A majority (90 percent) of the contractors did not face problems with regard to the recruitment of labourers Most contractors use the snowball method and ask their labourers to introduce new labourers to them. Furthermore, the contractors or their supervisors inquire in the area where they will start their work. They may also contact local labour leaders at the '*addas*' (square) to see if he can introduce them to good workers The requirements the labourers need to fulfil are stated in very general terms: they should preferably be healthy, young, and experienced, and be residing in or near the area where they have to work. In case a contractor takes over from another one they recruit among the labour force of the former operator. The majority of contractors emphasised that the labourers themselves prefer to continue to work in their current unit(s). With regard to new recruitments contractors usually stick to a maximum age between 40 and 45 years Nevertheless, it seems age is not an important selection criterion, contrary to what is often said in the literature (referring to the higher productivity of younger workers). Many contractors seem to believe that the labourers do not have a very heavy job as they only work 6 hours per day.

Use of vehicles

It is interesting to observe that a large majority (90 percent) of the contractors hire their truck. None of them have purchased a truck when they started to work in solid waste collection. Despite the preference of the MCH for contractors who own their own truck, most contractors consider the purchase of a truck economically unviable, even if it is more than 20 years of age. The contractors who own a truck have used it for other purposes, either for transportation or for renting it out. According to the contract conditions the trucks of the contractors should be less than 15 years of age. However, in reality most trucks are much older. The average age of the trucks in use by the contractors is 27 years; 52 percent of the trucks is older than 25 years[4]. Again this can be attributed to the law of economics: newer trucks are simply too expensive.
The contractors have made different arrangements with truck owners Some contractors only hire a truck, while others hire a truck as well as a driver and petrol. The amount that the contractors spend on renting a truck ranges from Rs 9,000 to over Rs 30,000 per month, depending on the type of arrangement, (with or without driver and petrol), the age and condition of the truck. In case the truck breaks down the contractor is responsible for replacement.

Whereas all the trucks of the MCH are provided with hydraulic lifting devices, the trucks of the contractors usually have to be unloaded manually. From this point of view one could argue that private sector participation, contrary to the general belief, has led to a reduction in productive efficiency[5], not to mention the decline in working conditions.

Networking and co-operation

The degree of co-operation among the contractors is limited. Although they often come across each other at the Ward or Circle Office where they submit their bill, they do not really engage in mutual exchange of equipment or joint buying of materials. Only a few contractors indicated that they occasionally provided lime powder (for disinfecting the dustbins) or rented a truck to colleagues, often after a request from the Sanitary Supervisor. Nevertheless, the contractors see themselves primarily as a unified business community rather than a disjointed group of competitors All the participating contractors are members of the Private Contractors Association (PCA).

4. This figure was based on a list provided by MCH specifying the equipment of 44 contractors working in circle 4 and 5.
5. In addition, vehicles of the private contractors are excluded from the use of the transfer stations. They are supposed to bring their waste directly to the dumpsites. This is another negative effect on efficiency. Finally, whereas the MCH uses a variety of vehicles each specifically suited to the requirements of various parts of the city, the private contractors overwhelmingly use one type of vehicle, the open truck. All these 'qualities' attest to the comparatively poor environmental record of privatised solid waste collection in Hyderabad.

The organisation is frequently organising meetings to discuss problems presented by the contractors It also acts as the official body representing the contractors in meetings with MCH staff or the (special) Commissioner. All the contractors demonstrate satisfaction about the functioning of the organisation. They see the association as an important tool to appeal against harmful MCH policies, since individual claims make no sense.

Problems encountered by contractors

Although the terms of reference of the unit system are described in detail, it still remains to be seen if the initial perceptions among the contractors accord with their actual experiences. Earlier on the problems faced during the start were examined. Obviously, it is equally important to map the structural problems the contractors are faced with in the execution of their job. Successively, the major problems during the work, the unexpected tasks and the unexpected costs will be discussed.

Almost half (46 percent) of the interviewed contractors stated that they did not experience specific problems during their work. Among the remaining group problems related to poor infrastructure (especially overflowing of drains) and lack of co-operation from the residents were raised most frequently. A problem that deserves to be mentioned explicitly concerns working in nightshifts and the risks related to that. One contractor – the only one who is carrying out nightshifts – pointed at the problems female labourers face while working at night. Often (drunk) men trouble these labourers Furthermore, experiences have shown that there is a high risk that these labourers become subject to traffic accidents. Besides, employment of female labourers in the night-time is at odds with the 1970 Contract Labour Regulation and Abolition Act.

Most contractors (85 percent) were faced with one or more unexpected costs. The most frequently cited additional expenditures came from the removal of debris, the breakdown (or accident) of the truck, sickness of labourers, cleaning of drains, peak periods of waste production, and extra efforts to redress malpractice by the former contractor. Some of these additional costs or tasks stemmed from MCH directives, i.e. removal of debris and cleaning of drains. Both types of tasks are not part of contractual requirements, but contractors feel obliged to satisfy these requests even though they are not or only partially compensated by the MCH. Occasionally there is mention of pressure being exercised on contractors to carry out informal tasks, for instance cleansing works in the compounds of local leaders

An interesting aspect concerns the participation in all kinds of special state programmes, such as clean-up campaigns. In Hyderabad once every two months a special, thematic day is organised on which both labourers of the MCH and the contractors are expected to participate. Although some contractors argue that these days led to extra cost, none of them considers this problematic.

Most contractors emphasise that their labourers hardly ever create problems. Obviously, if they did they would be removed from the job quickly. The most common problems in this respect are late arrival and absenteeism. However, the contractors say that the labourers themselves usually send somebody else to replace them (either friends or family members). Many contractors have made arrangements to this effect with the labourers in order to avoid deductions by the MCH. Another problem according to some contractors is 'pressure' by labourers to get paid. Often payment by the MCH is overdue and therefore labourers also receive their pay too late.

Contractors generally do not fear organisation of their labour force. Only 12 percent thinks labour unions might get a grip on their workers However, most contractors refer to their own association and believe that they can jointly resist the danger of organised labour. Furthermore, some contractors engage in practices to reduce the likelihood of labour organisation by frequently changing labourers, offering only short-term contracts, and dismissing of labourers who are members of labour unions or demonstrate activism.

5.5. POSITION OF PRIVATE SECTOR LABOURERS

In order to get a comparative picture of the labourers active in solid waste collection, 170 interviews were carried out in all the circles in Hyderabad. Both types of labourers, i.e. MCH and the private labourers, were interviewed with permission of the Sanitary Supervisors

Table 5.5. Number and type of interviewed labourers

Type of labourer	*Kamatan (female)*	*Kamatee (male)*	*Lorry worker*
MCH	67 (78.8%)	18 (21.2%)	4 (4.5%)
Private	74 (83.1%)	11 (12.4%)	-

Source: fieldwork 1999

Background and economic position

Both categories of labourers share the same socio-cultural characteristics and can be categorised under the same socio-cultural class. In general they are Hindu and belong to the lowest casts and have none or only a few years of primary education. Most of them were born in Hyderabad, however, with a sizeable group of second generation-migrants. More than 65 percent of the private labourers who were born in Hyderabad have parents who came as migrants to the city, while this figure is 45 percent for the MCH-workers A large majority of the migrants (including the second generation) is from Andhra Pradesh, predominantly from one of the neighbouring districts of Hyderabad.

A significant difference exists between the two categories of labourers in terms of overall household income[6]. This difference can be attributed to the wide gap between the salary of the MCH-workers and the private contract labourers (more on this will follow). In 38 percent of the households of the former group, the household is exclusively dependent on the salary earned by the MCH labourer. In the households of the private labourers this dependence is much lower (13 percent) and more household members have an income. Usually the involvement of more household members in income generating activities can be regarded as an indicator of poverty.

Only in about 20 percent of the households of the private labourers the salary of the labourers constitutes more than 50 percent of the family income. For the MCH labourers this figure is much higher: in 94 percent of the cases their incomes constitute more than 50 percent of the total family income.

In Hyderabad the average age-difference between the two groups is 11 years This seems to support the general belief that private contractors prefer to recruit young labourers because they are most productive (Cointreau-Levine, 1994). However, this explanation is not entirely satisfactory. One has to remember the ban on the recruitment of new MCH labourers since 1990. If one looks at the age at the time of first employment the difference is three years, to the advantage of the MCH labourers (an average recruitment age of 28.1 years versus 30.9 in the private sector). Therefore, in Hyderabad contractors do not seem to attach great value to the productive gain of recruiting younger labourers However, this might be related to the particular system of private sector participation in place. Considering the rigid contract specifications there is no incentive for contractors to strive for higher labour productivity.

Job history

As explained earlier, the scale of privatisation has increased considerably since 1998. This development is reflected in the job history of the private labourers A majority (almost 70 percent) of the labourers started their job as private kamatee/kamatan when the unit system was introduced in September 1998. Only about 30 percent of the labourers had worked as a kamatan/kamatee before that date.

Although the short working period of most private labourers does not allow drawing firm conclusions on mobility patterns, there is one aspect that deserves to be

6. Here the household income is defined as the sum of the incomes of all the household members regardless of the actual distribution of the income among the members

mentioned. The main reason for those who were working before the introduction of the unit system to quit their former boss was the expiry of the contract with the MCH. Although most of these contractors obtained a new contract elsewhere, most labourers preferred to work in the same or an area nearby for another contractor. Apparently, labourers are not very attached to their employer. When asked if they would join their boss if he gets a new contract elsewhere, a majority says that they prefer to stay in the area were they are currently working.

Both categories of labourers were mainly active as daily labourer (coolie) or as maid-servant before they entered the solid waste collection sector. The reasons for the MCH labourers to quit with their former jobs are obvious. They were able to change their casual, low paying jobs for a permanent and more rewarding one in the public sector. For the private labourers the reasons are less pertinent. However, half of them referred to the lower salary of their previous job, while more than a third pointed at the high irregularity of their former engagements.

Most MCH labourers are recruited through official channels, i.e. via an employment agency, or on the basis of the well-established rule that a family member of a deceased worker can take over his/her job. The recruitment channels of the private labourers are different. Informal networks play a major role here. Very often friends, neighbours or family members of the labourers have helped, either passively (informing about vacancies) or actively (recommendation/introduction). It is interesting to note that most of these relations were themselves not working for the particular contractor. The rather indiscriminate nature of current recruitment patterns seems to be related to the fact that the large-scale privatisation efforts in Hyderabad are of recent date. It is quite possible that social networks will develop further in the future.

Income and labour conditions

It is often said that privatisation leads to a net decrease in employment as a result of the private sector's acclaimed higher efficiency. However, in the Hyderabad case the opposite is true. The total number of people active in solid waste collection has increased significantly since the introduction of the unit system. Of course, this is related to the substantial rise in government expenditure on solid waste collection. In terms of employment the major question in Hyderabad is "how do labour conditions compare between private contract labourers and their MCH colleagues?" As demonstrated in the box below a wide gap exists between both types of labourers.

Box 5.2. Labour conditions on paper

Labour conditions of MCH labourers

Salary: the wage structure consists of the basic pay rate augmented with dearness allowances, fitness benefit, HRA, and CCA. The basic pay is dependent on the number of years in service and ranges from Rs 1,360 to Rs 2,360.

Allowances/goods: there are a number of additional allowances that are not included in the basic salary. Every labourer receives Rs 30 dirty work allowances per month, and Rs 15 washing allowances. In addition every worker receives 2 pieces of soap per month, 1 bottle (975g) hair oil per 3 months, two towels, 2 sari's (shirts) and one pair of shoes per year.

Leave days: there are 15 casual leaves and 7 optional leaves. If they are not used these leaves will be lost. In addition there are 30 leave days that can either be used or exchanged for money. A total of 240 days of leave will be paid after retirement. Furthermore, there are 15 medical leaves per year. If they are not used, half of these are paid after retirement.

Maternity leaves: every pregnant labourer has 90 days of maternity leave. There must be a two-year interval between the pregnancies. There will be an increment if the labourer decides to engage in family planning (provided there are no more than two children).

Pension: the pension is calculated as half of the basic pay including allowances. If the labourer dies the pension will be given to the partner until his/her death. In case the partner also deceases the pension will be given to the son until the age of 24 years, if it is the daughter, up to her marriage.

Provision of bus pass: *all the MCH labourers are provided with a bus pass. The MCH is compensating 50 percent of the costs and the labourers are paying Rs100 per month.*

Health Insurance: if a labourer is subject to an accident he will be paid for treatment.

Safety equipment: hand gloves, nose masks

Labour conditions of private labourers

Salary: the contractor shall pay the labourer minimum wages as fixed by the Commissioner of Labour, i.e. Rs 1,300 per month.

Provident Fund (PF) and Employees State Insurance Scheme (ESI): the contractor should contribute to the E.S.I. and Provident Fund every month for each worker engaged by him at the rate of percentage fixed by the E.S.I. and P.F. authorities.

Leave days: the private labourer should have 3 leave days per month

Safety equipment: hand gloves, nose masks

In reality the differences are even more pronounced. The average monthly salary of the interviewed private and MCH kamatans – the vast majority of labourers is female – for example, was Rs 1,120 and Rs 3,360 respectively. On average the private labourers received Rs 200 less than the stipulated minimum wage. Furthermore, within the private sector there was a slight difference in payment between kamatees and kamatans. Despite the fact that the MCH calculates the same amounts for male and female labourers, the former earned between Rs 60 and Rs 100 more per month.

As far as the secondary labour conditions are concerned none of the interviewed contractors contributed to the ESI and PF. Although the MCH gave preference to companies with a PF and ESI code, almost none of the contractors have one. Another condition that was frequently violated by the contractors is the condition to recruit additional labourers to give other employees the opportunity to take their leave days. Partly as a result of this, labourers usually work more days per month than stipulated.

A striking example of the inability or reluctance of the MCH to acquit itself of its regulatory role is related to the position of the female labourers According to the 1971 Andhra Pradesh Contract Labour Rules female contract labour should in principal not be employed by the contractor before 6.00 a.m. or after 7.00 p.m. In fact, most sweepers working nightshifts are female (just like the situation during daytime).

Organised labour

According to the Indian labour laws the labourers have the right to organise. The MCH labourers have taken advantage of this right: a large majority (74 percent) of them is registered as a member of a labour union. Most of them (53 percent) belong to the Andhra Pradesh Municipal Sahakar Mazdoor Union (INTUC), which is politically affiliated to Congress. Another 40 percent is member to the MCH workers union (AITUC). All the unions oppose the municipality's privatisation policy and appeal for the removal of the ban on new recruitments of MCH labourers However, the unions are strongly divided and look with suspicion at each other's strategies (cf. the involvement of two union welfare organisations in contracting). While some of the unions expressed their willingness to absorb the labourers of the contractors, others regarded this as an implicit recognition of the privatisation campaigns of the MCH. In practise all the unions tend to focus on the MCH labourers and their labour conditions.

The labourers of the contractors are not organised. Only one of the interviewed labourers stated that he was a member of a labour union. In addition to the fear of losing their jobs, many labourers were simply unaware of labour unions and their activities.

Perceptions about the job

Most of the private labourers regard their current jobs as better than their previous ones, both in terms of security and pay. To illustrate this, more than 70 percent of the labourers said that they were not actively looking for a new job. Furthermore, about 78 percent of them expressed a desire to continue with this kind of work in the future. About one fifth of this group thinks that involvement in the sector as contract labourer enhances their chances for a permanent position at the MCH.

Health problems

A majority of the labourers – 68 percent for the MCH workers, 70 percent for the contract labourers – believes their work is causing them health problems. The most frequently mentioned health problems are: fever, muscle/back pain and headache. Many respondents came up with more than one type of health complaint.

If the MCH labourers get sick they stay at home and make use of their leave days. On average they have stayed 7.2 days at home due to sickness over a one-year period. Their colleagues in the private sector actually do not have a leave facility. As a result the average number of days of absence was only 3.4 (extrapolated on the basis of a half year period). Moreover, about half of the private labourers claimed that they had continued working while sick (for the MCH labourers this figure was only 4 percent).

5.6. Assessment of the privatisation of solid waste collection

In this section the views of contractors and labourers on the strengths and weaknesses of the current privatisation policy are presented. The emphasis will be on the perceptions of the former as they have been the MCH's contract partners

Opinion on the tender procedure

Despite the short duration of the tender procedure – between the publishing of the tender-notice and the actual start of the work a couple of weeks later – only 3 contractors (12 percent) considered this problematic. An overwhelming majority was satisfied about the transparency of the procedure, the method of announcing the work, and the information that was provided by the MCH on the tasks and requirements. Only one contractor considered the EMD to be too high and a barrier to entry. The other said that the EMD had never prevented them from applying for a contract. Of course, many potential contractors may have backed out due to the amount, but this cannot be substantiated.

Opinion about punctuality and regularity of payment

At the end of the day it is common to see many contractors coming and leaving the compound of the central MCH building. The main reason behind this ritual is to speed up the processing of their bills. Although the bills have to be submitted at the Circle Office, they pass through a number of steps and eventually arrive at the main MCH office. Nevertheless, only a minority of the contractors complained about the punctuality and regularity of payment by the MCH. Usually the contractors submit their bills on the 2nd of each month and are paid between the 10th and 15th. Occasionally, delays in payment create problems with regard to the payment of the labourers.

Opinion about contract conditions

Almost a third of the contractors stated not to have problems with the contract conditions. The most troublesome condition is the fixed amount of waste that has to be collected and disposed of each day. Many contractors mention that in some periods (especially summer) they are unable to collect the stipulated volume, and, consequently, that they get deductions on their contract amount. Some contractors would like to have the amount of waste to be fixed on a monthly rather than a daily basis. Another unfavourable requirement according the contractors is the fixed number of trips per day, and the fact that the MCH is not paying them for any additional trip. Finally, about half of the contractors objected to the system of deductions itself, which they think is unjust. They may say, for example, *"if we collect more than the fixed amount we don't get paid extra, but if we collect less we are charged with deductions"*. The MCH is accused of not taking into account the conditions under which the contractors have to work *("they just deduct blindly")*. Also many contractors find it unreasonable to deduct two times the salary if a labourer is absent due to sickness. Nevertheless, half of the respondents feel the MCH is usually lenient.

The unit system adopted in Hyderabad is based on the assumption that the units are more or less homogeneous. Most contractors are satisfied with the size and demarcation of their area. Only 10 percent of the contractors thought that their area was larger than the specified standard. The contractors were also asked what they thought about the number of calculated labourers Again 90 percent of them was satisfied.

Contract period

In general private entrepreneurs prefer to have long lasting contracts, which assure them a proper return on investment. Needless to say that the likelihood of private sector investment all depends on the terms of reference under which it has to operate. At first sight the situation in Hyderabad does not seem to be very favourable. The system is strongly regulated with a low degree of entrepreneurial freedom. Contract periods are extremely short (currently two successive periods of six months each). Most contractors have a strong preference for contract periods longer than 1 year, and the average preferred contract period is 3 years and 3 months. The arguments to support this choice are that it offers more security (work for a longer time) and will enable them to operate more efficiently. With regard to the latter they say that they have to get used to the area. At the start of the contract period they are still unfamiliar with the unit and suffer from higher deductions due to malpractice. In the course of time they can improve upon themselves. None of the contractors came forward with the investment-argument. Most of the contractors are rather reticent when they are asked about investment opportunities under the current system. Those who showed a willingness to invest, however, demanded an upward adjustment of the contract amount. Apparently the current profit does not allow for investments.

Role of MCH

Although the majority of the contractors is content about the way the MCH manages the out-contracting of solid waste collection services, they also see opportunities for improvements. Only 12 percent of the contractors are completely satisfied with the MCH in its regulatory role. The question how the MCH could improve the system gets a varied response. Most recommendations boil down to a more liberal contracting approach, allowing entrepreneurs more freedom of choice and reducing government interference with the organisation of production. Improving communication is also on the shortlist of suggestions. Although the majority was satisfied about information they received at the start of the privatisation campaign many would nonetheless like to receive more feedback from the MCH in the form of advice, suggestions and training. Another frequently cited recommendation concerns the attitude of the MCH. According to some contractors the MCH should be more responsive to the complaints and suggestions of the contractors Finally, some contractors would like to see the MCH rent out trucks and equipment.

Form of privatisation

The contractors who were active in solid waste collection before September 1998 used to work under a system of lowest tender. In principle this model allows the entrepreneur more room to manoeuvre. Nevertheless, all the interviewed contractors prefer the unit system. According to them the current system allows more contractors to participate, reduces possibilities for corruption, guarantees a fixed profit margin, and uses standardised norms applicable to all contractors (size of area, fixed numbers of labourers and fixed amount of waste to be collected). They unanimously agree that the unit system is an improvement in comparison with the old system. This finding attests to the high risk avoiding mentality among contractors They are remarkably mild in their critique on the unit system, although some raise objections to the system of deductions, the heterogeneity of the units, and the fixed number of labourers to be employed. The contractors who have worked under the previous system showed some dissatisfaction with respect to the lower profit margins and the system of deductions (which existed before but is applied more stringently today).

Franchising as a mode of private sector involvement is not popular among contractors Most of them are convinced that they will never be able to recover their costs from the residents. To substantiate their view they point at the rickshaw-schemes. According to the contractors many of those schemes have failed because residents were reluctant to pay, despite the relatively low rates. Only one contractor (a women co-operative society) was willing to take up a franchise scheme, provided they could select their own areas.

Pointing at the recruitment stop for MCH labourers, the tight MCH budget, and prevailing positive views on privatisation within MCH circles, a large majority of the contractors expects that the new role of the private sector in solid waste collection is an irreversible development. However, they do expect some adjustments in the nearby future regarding the terms of reference, notably an enlargement of the areas (i.e. merging of units) and a downward adaptation of profit margins.

Willingness to continue

Considering the high degree of satisfaction among contractors, it is not surprising that most of them like to continue. Only one contractor said he would step back in a next contract period due to the low profit margin. The majority would like to take up more areas in a new contract period. However, the number of contractors who responded negatively to this idea was sizeable (almost 30 percent).

Opinions of labourers

A large majority (83 percent) of the MCH labourers disagrees with the MCH privatisation policy. Only a tiny fraction (6 percent) adopts a positive view arguing that *'they work well'* or *'at least they create some employment'* More than 10 percent of the labourers has no opinion on this matter, or is merely stating that *"it is a government decision"*. Although the role of the private sector in Hyderabad's solid waste collection has grown fast, most of the MCH labourers rest assured of their strong legal position and do not seem to be scared of losing their job. When, for instance, they were asked if they perceive increasing privatisation of solid waste collection as a threat to their own position, a vast majority (87 percent) answers 'no'.

Both categories of labourers seem to be aware of the difference in labour conditions under which they are working. Obviously, none of the MCH labourers would like to work for a private contractor, the main argument being that they are extremely pleased with the permanency of their job (the other benefits of their current employment such as salary, pension and other non-wage benefits are not mentioned). Private labourers, on the contrary, are very eager to work for the Corporation. The main arguments relate to the permanent character of the job (61 percent), the higher salary (21 percent) and because it is a 'government-job' (17.4 percent).

5.7. CONCLUSION

The nature and course of the privatisation process in Hyderabad are largely determined by the increasing government concern about solid waste management and the power of various labour interest groups, notably the unions. The MCH was under considerable political pressure from the general public as well as higher levels of government, c.f. the new Municipal Solid Waste Rules issued by the Ministry of Environment and Forests in 2000, to pay more attention to the sector. Privatisation became the only possible avenue to improve solid waste collection services both because of the ban on recruitment of additional personnel, and because of fierce opposition to any restructuring of public sector labour conditions (such as the introduction of flexible working patterns). After some experimentation the MCH committed itself to the unit system of privatisation, which is characterised by rigid contract specifications and minimal entrepreneurial freedom. The choice for this particular model was motivated by the social concern to create new employment opportunities – the system is very labour intense – and the desire to combat corruption. Furthermore, the system keeps the MCH firmly in the driver's seat, which is probably one of its hidden objectives.

Overseeing the socio-economic impacts of the privatisation of solid waste collection in Hyderabad it can be qualified as a mixed blessing. Unmistakably, the quality of solid waste collection at the city level has improved since the introduction of the unit system, notably in terms of spatial coverage. Furthermore, more stringent methods of supervision have been adopted, which have stimulated proper service performance. However, the privatisation campaign alone cannot be credited for all of these improvements, considering the fact that the MCH has also raised its total expenditure on solid waste collection tremendously.

On the negative side the big gap that exists in terms of labour conditions between the MCH labourers and those working for private contractors is certainly the most outstanding flaw. The latter receive substantially lower wages, their work is pretty insecure, and they have to do without numerous other facilities, including those they can legally claim. Although the people concerned do not perceive their position negatively, it is nevertheless distressing to note that the position of the contract labourers in Hyderabad has even worsened as a result of the very laws and regulations that were designed to protect them. The MCH fails to perform its role as a protector of labourers (by not enforcing the labour provisions specified in solid waste collection contracts). The unions, on their part, do not seem to bother much about the fate of unorganised labour and continue to focus on the rights of the labour elite.

Service efficiency, usually one of the major driving forces behind privatisation policies, does not seem to have played a significant role in the Hyderabad case. On the contrary, the MCH has been willing to compromise on the productive efficiency of privatised solid waste collection in order to achieve its other goals. The unit system does not provide contractors with sufficient incentive to work more efficiently. In comparison with the MCH, for example, private contractors are using much older equipment. Moreover, they are unwilling or unable to invest and innovate. Nevertheless, substantial gains in productive efficiency have been achieved due to privatisation. Similar to what can be observed elsewhere in India (cf. Ali *et al.*, 1999) the costs per tonne have gone down considerably. This can exclusively be attributed to the lower labour costs within the private sector, which have been set by the MCH. Therefore, the government has gained by trampling on the workers By doing so the state is forsaking its traditional social obligations.

PART II

REUSE, RECOVERY AND RECYCLING OF URBAN INORGANIC SOLID WASTE

ISA BAUD

CHAPTER 6
REUSE, RECOVERY AND RECYCLING OF URBAN INORGANIC SOLID WASTE; MODALITIES, COMMODITY CHAINS AND SUSTAINABLE DEVELOPMENT

6.1. INTRODUCTION

In the second part of this book, the focus is on inorganic solid waste reuse, recovery and recycling[1]. Trade and recycling activities, and the use of secondary materials by enterprises as sources of raw materials for production, is widespread in many developing countries, although it varies by region. The importance of making solid waste management[2] more environmentally sustainable by increasing waste minimisation, reuse, recovery and recycling by the private sector is well recognised in industrialised countries. Discussions in developing countries, in contrast, have centred on co-operation between governments and civil society organisations in promoting recycling activities and have largely ignored the existing private sector in this area. If new initiatives lead to reduced access to waste by existing recyclers, the economic viability of private sector activities may be endangered (Blore, 2000). In contrast, CBO activities are often dependent on the volunteer work of community groups, and are not economically viable.

In this chapter, we will briefly discuss the different perspectives on waste recovery and recycling which have guided policy makers, and how the socio-economic, public health and environmental perspectives can be integrated for the transition to sustainable development. Secondly, existing modalities of private sector recovery and recycling will be briefly reviewed and contrasted with public sector-community initiatives in this area. Thirdly, we will briefly map the partnerships between the various actors involved and their potential contributions to aspects of sustainable development, as laid down in our model (see also Baud *et al.*, 2001).

---

1. The main emphasis will be on waste recovery and recycling that takes place through a 'market', rather than the reuse taking place within households.
2. The waste included here is municipal solid waste and waste that would generally go into the municipal waste stream, and does not include waste materials from production that companies sell directly to other companies for further transformation.

I. Baud et al. (eds.), Solid Waste Management and Recycling, 115-131.

6.2. INTEGRATING SOCIO-ECONOMIC AND ENVIRONMENTAL CONCERNS

Although recovery, trade and recycling of solid waste materials are widespread in most countries, opinions about its value have differed widely, and are derived from several perspectives. Classically, local governments, whose responsibility it is to collect and dispose of solid waste, have focused their strategies on improving public health conditions. Solid waste should be quickly removed and disposed of in such a fashion that it no longer presents a danger to public health. Any activity deviating from that process should be avoided. Therefore, government attitudes towards reuse, recovery and recycling of waste materials have usually been very ambivalent. This attitude is still prevalent in many developing countries, and is reflected in rules and regulations that prevent material recovery and recycling at early stages of the collection, transport and disposal cycle (e.g. in Peru).

The public health approach came under increasing pressure in industrialised countries in the 1960s and 1970s, as consumption patterns changed. The growth in solid waste flows outgrew the capacity of local and regional waste sinks, and became environmentally and socially unacceptable. As a result, strategies to reduce waste flows and dispose of them in alternative ways became a major concern. This held true both for municipalities dealing with domestic waste, as well as for industrial enterprises, dealing with rising costs for removing waste from production processes (E&U, 1999; Frijns *et al.*, 1997). End-of-pipe solutions were followed by the more integrated chain approaches to promote environmental aspects of sustainable development. In the early 1970s, the waste management hierarchy became the internationally accepted standard for environmental assessment of solid waste management systems as a whole (Blore, 1999) (see Figure 1.1).

Waste minimisation potentially offers the best possibilities for environmental aspects of sustainable development, by reducing resource use and preventing waste generation. However, this leads to a focus on changes in production systems, which lie outside the scope of this volume[3]. More commonly, reduction of waste flows through source separation, reuse and recycling (material and energy recovery) systems have been introduced by governments in developed countries such as the Netherlands, which has been active in this area (de Jong, 1999). These systems were based on households separating materials voluntarily with little or no compensation, and subsidies by governments to reuse and recycle materials (from domestic waste)[4]. Implicitly, this system was based on households' agreement with the goals of waste reduction, and their active co-operation[5].

---

3. (cf. Tukker, 2000) on the use of LCA in considering alternative ways of producing a certain good.
4. Industries increasingly recycled either bottling through financial incentives to consumers, or by separating and re-using resources used in production processes (these had to be commercially viable, or made mandatory by government regulation).

This situation is in stark contrast to that in developing countries. Waste minimisation is, to our knowledge, not generally applied as a matter of public policy[6]. Waste reuse, material recovery and recycling on a private commercial basis are extensive, despite the fact that most local and national governments have not supported such activities, or actively forbidden them[7]. The economic viability of the trade and recycling sector varies widely across regions and urban areas, and depends on several factors It is viable in local urban economies, where wage levels are low and alternative employment is relatively scarce, where high quality waste is sizeable and readily accessible, and where there is a (created) scarcity of raw materials, which form a major cost to local enterprises (van Beukering, 1994; Bose and Blore, 1993; Dhanalakshmi and Iyer, 1999; Lardinois and Furedy, 1999: 18-9). Regional differences exist: in Asia, such systems are well established, whereas in Sub-Saharan Africa, they are much less extensive (Karanja, 1999; chapter 8 in this volume). There is also some evidence that they are related to the size of the cities concerned, and the density of urban agglomerations. In Southern India, waste materials from small towns were transported to central locations in large towns for further recycling. In Kenya, recycling companies have waste materials collected from all over Kenya for further recycling (Karanja, PhD thesis, in preparation).

Nevertheless, private sector activities are susceptible to change in external conditions. Waste recovery and recycling decreases in situations when wage levels go up and alternative employment increases, waste is mixed and not readily accessible, when collection costs are high, or when alternative sources of raw materials become cheaper and widely available,e.g. when import duties are reduced (Karanja, pers com.; van Beukering, 2001).

Public sector discussions of trade and recycling in urban environmental management have largely ignored the existing private sector commodity chains transforming waste materials. These activities have been organised into a variety of commodity chains. They include those with street pickers at the lower-income end, dealers, and owners of enterprises using recycled materials as raw materials for production at the highest-income end (the figure in chapter I illustrates the chain)[8]. Alternatively, itinerant buyers buy waste material and send it through dealers up the commodity chain (Furedy, 1993).

5. There are differences between countries in the relative contributions by industry and households separating material.
6. The director of the NIUA in New Delhi confirmed that only one large company in India is known to voluntarily be introducing waste minimisation measures.
7. It has consisted of informal collection, trading, and recycling of materials that have an economic value to others than the original owner.
8. In the next section, we will discuss the modalities of such commodity chains.

Instead, local governments have supported the creation of community-based micro- and small-scale enterprises for collection, transport and disposal, and supported separation and trading activities as an additional source of income for these enterprises (Arroyo Moreno *et al.*, 1997/9; Lapid *et al.*, 1996; Haan *et al*, 1999; Schübeler, 1996). However, from a commercial point of view, such initiatives have had mixed success (Baud *et al.*, 2001; Arroyo Moreno *et al.*, 1997/99; Lardinois and Furedy, 1999), as large numbers of micro-enterprises collapsed without government subsidies.

From an environmental point of view on the SWM system as a whole, the extent of waste flows separated from domestic waste flows through voluntary and community efforts is very small in comparison with the size of the waste flows diverted by private commercial commodity chains[9]. Therefore, governments should avoid undermining the existing environmental efficiency of SWM by taking up initiatives which will reduce private sector access to waste, an issue which is still not being addressed by local authorities in many developing countries.

The major question raised in the current study was how to integrate the socio-economic and environmental perspectives in urban SWM, such that strategies for new models in SWM contribute to urban sustainable development. Although 'integrated sustainable waste management' is a concept discussed in international conferences, in local practice in developing countries (van der Klundert and Lardinois, 1995), there has been little integration of the socio-economic, environmental and public health aspects concerning reuse, recovery and recycling. This can lead to negative effects for the SWM system as a whole, as we will argue in the following sections.

6.3. MODALITIES OF RECOVERY, TRADE AND RECYCLING

From an environmental perspective, primary importance should be given to minimising waste flows, by changing the way goods are produced. In a number of industrialised countries, businesses are becoming involved in several such processes. These include extended product responsibility, eco-labelling, and waste exchanges. Extended product responsibility (EPR) places the responsibility for reducing the environmental impact of the product on the producer throughout the life cycle of the product (Hoorweg and Thomas, 1999; Davis *et al.*, 1997). Some European countries have introduced comprehensive frameworks for EPR measures in this area (Germany, Sweden, the Netherlands). However, such systems are mainly voluntary. Although waste exchanges are found in Latin America, these types of initiatives are rather rare in developing countries and are left out of the discussion here (Lardinois and Furedy, 1999).

---

9. Low-income households may reuse waste from high-income households, but this flow has hardly been documented.

Recovery, trade and recycling activities comprise the next step in the waste management hierarchy, and involve a larger and variable chain of actors (see chapter I). Maximising waste diversion toward recovery and recycling is the primary environmental goal at this point in the waste management hierarchy. In developing countries, however, the environmental perspective at this point is less well developed among both local authorities and citizens (cf. Chaturvedi, 1998). Rather, these activities take place primarily in a private market context. Therefore, assessing their economic value is currently the only method we can use to predict to what extent recovery of waste materials is likely to increase when the incentives are changed.

An important distinction in assessing the value of such waste flows is that made between unmixed sources of waste, which retain a higher economic value, and mixed waste, which presents a much less attractive source of raw materials for trade and recycling enterprises. Therefore, some authors suggest that more emphasis should be put on increasing source separation in order to maximise clean, homogenous waste materials, whereas mixed wastes should be phased out of an integrated system (Lardinois and Furedy, 1999). This might enhance the size of both organic as well as inorganic waste flows and their economic value, if other factors remain constant. It also has implications for which actors remain involved, which will be discussed in the following paragraphs.

Assessments of the size of municipal waste flows often implicitly limit the system borders to the urban locality[10]. However, trading and recycling commodity chains are not necessarily closed local systems. Regional, national and international trade flows of waste materials exist, in which large volumes and sales turnovers are generated (Dhanalakshmi and Iyer, 1999; van Beukering and Duraiappah, 1996; van Beukering, 2001)[11]. Van Beukering argues that international flows of waste materials can add to the quality of waste material flows in developing countries, if 'high-quality' waste is exported from industrialised countries (van Beukering, 2001). This obviously does not apply to hazardous waste, for which developing countries may be less well equipped to deal with the dangers they pose.

Locally, many entrepreneurs are less than willing to discuss their commercial activities, and the smaller enterprises often operate in a semi-illegal context (Baron and

10. Many municipalities in developing countries only estimate the size of their waste generation. If the volume is actually measured, it is usually done so at the dumpsite through a weighing bridge, which means that the major part of the recyclable waste has been removed from the stream before.

11. In developed countries, a discussion has raged whether waste materials should be transported to developing countries (especially hazardous wastes). However, non-hazardous waste materials originating in developed countries, such as paper, are often considered sources of superior raw materials by small entrepreneurs in developing countries (Dhanalakshmi and Iyer, 1999; van Beukering and Duraiappah, 1998).

Castricum, 1996; Dhanalakshmi and Iyer, 1999). The result has been that size of material flows, turnover and profit levels are rarely established in studies on recycling and recovery[12], which makes discussions on their commercial and environmental contributions very difficult. However, rare field surveys show that the final links in the commodity chain, consisting of wholesalers and enterprises using recycled materials as (a portion of) their raw materials, have sizeable turnovers in both volumes and sales levels (cf. Dhanalakshmi and Iyer, 1999). In Chennai, the largest wholesaler in broken glass buys materials from all over Southern India, and has a turnover of Rs 2 million per month. In Bangalore, plastic recycling enterprises on average process 40 tonnes of material per month, two glass factories use 861 tonnes per month, and metal producers some 9 tonnes monthly (van Beukering, 1994).

Although their existence argues for the commercial viability of trade and recycling enterprises using inorganic waste materials, their existence can be threatened by externalities. The use of waste materials as raw material can become less competitive, if the price of imported virgin materials goes down or is no longer subject to high import duties (as has happened in Nairobi with plastics; Karanja *et al.*, this volume). Alternative waste flows of higher quality waste materials from abroad also push lower quality domestic waste streams in developing countries out of the market. However, if the demand for the materials remains higher than the supply, waste flows can continue to grow (van Beukering, 1994). Economic viability is also influenced by other external factors Lack or failure of power supply is a frequent problem, and many smaller recycling enterprises and traders maintain a degree of 'informality' to avoid high tax costs and time-consuming administrative regulations (van Beukering, 1994; Baron and Castricum, 1996).

The large enterprises producing final goods at the end of the commodity chain determine the final composition of their product, and influence prices in the rest of the commodity chain. When they reduce the price of waste materials, the negative effects are fed back through the commodity chain to other more vulnerable groups.

Employment and income

Trading and recycling commodity chains also have an economic importance in the employment and income created. At the collection stage, such activities have provided a basic livelihood and a source of consumption goods for large numbers of poor urban citizens picking waste materials off streets and dumpsites (Furedy, 1990, 1992; Hunt, 1996; Huysman, 1994; Baud and Schenk, 1994). These have included immigrants to urban areas, (male) street children, and women without other employment opportuni-

12. An exception is van Beukering, who found profit levels to increase up the commodity chain, with recycling enterprises earning average monthly profits of around 25 percent in Bangalore, India. Traders lower in the chain had profits ranging between 4-6 percent (van Beukering, 1994).

ties. An equally important group are the itinerant buyers, who buy materials straight from households and institutions, using small amounts of capital (Furedy, 1993; Dhanalakshmi and Iyer, 1999; Chikarmane *et al.*, n.d.). The waste materials collected in this fashion are sold to (non-specialised) dealers, who further sort and clean the materials, and sell them to large(r) wholesalers specialising in one type of material. Small dealers employ family members, and provide informal insurance and credit facilities to waste pickers working for them. Larger wholesalers mainly employ casual workers to sort and clean the materials they sell (van Eerd, 1996; van Beukering, 1994). Small recycling enterprises transform waste materials into intermediate goods. They employ a combination of permanent and daily wageworkers The end of the commodity chain consists of large enterprises making end products, using waste materials as inputs.

A study from Pune, India indicates that there are 4600 waste pickers, and 370 scrap traders known there. The number of enterprises using recycled materials is relatively small and they operate regionally rather than city-wide (Chikarmane *et al.*, n.d.). The Pune study indicates that the majority of waste pickers are scheduled caste, and women. Among the women, a quarter to one-third are divorced or abandoned. The itinerant buyers tend to be men (60 percent), and to have some primary education. Mean monthly per capita incomes in pickers' households are Rs 650 (50 percent above the local poverty line), although the distribution is uneven across the whole group. Among waste traders in Pune, men predominate (women are only 20 percent of all traders). The retail traders have a social background similar to the waste pickers, although they are better off economically. In contrast, the wholesale traders come from a Muslim or higher caste background.

Attention in research has been largely focused on actors at the beginning of the commodity chain: the waste pickers A large number of studies has been done on their contributions to diverting waste from municipal waste streams, and the difficult circumstances under which they live and work (Huysman, 1994; Poerbo, 1991; Hunt, 1996; Furedy, 1992; Chikarmane *et al.*, n.d.). Actors at the intermediate income levels of the commodity chain (itinerant buyers, neighbourhood dealers, small and large wholesalers) have received much less attention in research[13]. Nevertheless, they are crucial nodes between the pickers and itinerant buyers and the recycling enterprises buying waste as raw materials, in both economic and spatial terms, organising the transfer of small amounts of waste materials to the enterprises. Van Beukering (1994) has indicated that small dealers do not only work with pickers, but also obtain materials from institutional sources and directly from households. Their numbers are large in Bangalore, in contrast with the large wholesalers of whom there are only a few.

---

13. Exceptions are the studies by Sinha and Amin, describing the complete commodity chain in Dhaka (1995), and the compilation of studies in Bangalore (Baud and Schenk, 1994) and Chennai (Dhanalakshmi and Iyer, 1999). Furedy also carried a .... on IB in Bangalore (...).

Their income levels are higher than those of pickers, but remain susceptible to adversity. Many small dealers become pickers again if their businesses fail. Wholesalers are fewer in number, but make higher profit levels by maintaining an oligopolistic position. The study in Chennai showed wholesalers with distinct differences in size of turnover and market control (Dhanalaksmi and Iyer, 1999).

Such networks were also found in the plastic waste recycling trade and small enterprises processing waste materials as intermediate products for the enterprises using secondary materials as sources of raw materials mixed with virgin materials. Jordens (1996) indicates that vertical family networks are set up in the plastic trade and recycling sector, in order to keep a firm control over quantities and prices of the waste materials collected and utilised for processing. Such small enterprises are the only 'true' recycling enterprises, as the large enterprises always combine virgin materials with secondary materials to produce their final products.

Medium and small enterprises or co-operatives collecting waste in Latin America have been formed by people from slum neighbourhoods. In a number of countries, they perform basic collection services (cf. Arroyo Moreno *et al.*, 1999). A number of medium scale enterprises/co-operatives have also carried out trade of secondary materials. They have different financial models, receiving payments either directly from their customers, or through municipalities. These models limit their profit margins, as municipalities set limits. They also face a restricted market for their recovered materials, in which large buyers and end-users set the price (Arroyo Moreno *et al.*, 1999). Co-ops working directly with private customers show the best results financially.

Very little is known about the enterprises using secondary materials for their production. This constitutes a major gap in our knowledge, as these enterprises determine the extent to which secondary materials are utilised again, and the prices paid for them to other economic agents in the commodity chain.

The commodity chains described here can be spread geographically within a city (waste pickers, itinerant buyers and retail traders), but also include units that cover much wider areas (wholesalers collecting scrap from a whole region of the country). However, the base of the recovery sector can also be termed an 'industrial district', in that local recyclers work together pro-actively within a limited geographical area to promote their businesses. In Dhaka, Maqsood Sinha and Nurul Amin have studied neighbourhoods where small recycling industries are heavily concentrated, and have argued that this constitutes an 'industrial district' (1995). In fact, their study also indicates linkages with businesses outside the city putting in orders for recycled materials.

Social aspects

The majority of people working in trade and recycling of waste materials, do so in informalised work settings, which leave them vulnerable to poor working conditions, low and irregular wage levels, and harassment from outsiders (van Eerd, 1996; Bromley, 1978; Huysman, 1994; Beall, 1997). This holds true for all waste pickers, but also in large measure for itinerant buyers, and the daily wageworkers hired by dealers, wholesalers and recycling enterprises. The women and children face particular health hazards from injuries, inhaling toxic fumes, or contamination from infectious waste (see chapter 11).

Concerns for their health and working conditions have led to opposite views on the social acceptability of trade and recycling activities. Some commentators feel that informal collection of waste by pickers should be banned, whereas others suggest that working conditions should be improved to support the efforts of people undertaking such activities (Hunt, 1996; Furedy, 1992). Furedy (1995, 1997a, 1997b) has suggested that people working in the most hazardous and degrading conditions – street and dump pickers – should find alternative employment, and that others in the commodity chain – such as the itinerant buyer, trader and wholesaler – should have their work supported by removing existing barriers to their activities. Karanja supports the idea that social and working conditions of pickers should be improved, as few alternative employment opportunities are available.

The pickers themselves face a great deal of social ostracism, which sometime leads to social isolation and a lack of channels through which to improve their conditions. NGOs working with waste pickers have concentrated on women waste pickers and street children picking waste (Lapid *et al.*, 1996; Vincentian missionaries, 1998). In several places, women pickers have been assisted in organising co-operatives, in attempts to increase their incomes and power vis-à-vis their customers (Bhuvaneshwari, 1998; Lapid *et al.*, 1996; Chikarmane *et al.*, n.d.). Youngsters, mainly boys, have been approached as street children, often by religious orders Attempts have concentrated on getting them off the street, providing them with alternative skills and accumulating savings, or improving their health conditions (Hunt, 1996). In Nairobi, about 60,000 street children are estimated to survive on reuse of waste materials at some time (Karanja, pers com.). Although NGO initiatives are important, they are not widespread, and the majority of waste pickers have to survive on their own.

Medium scale enterprises in Latin America provide employment for their workers at minimum wage level and up to twice the legal minimum wages (Arroyo Moreno *et al.*, 1997/1999). The recovery activities pay less than the activities involved in collection and disposal. In Peru, particularly women are the active associates in the enterprises involved in recovery and recycling activities, whereas in the other countries this tends to be a male activity. The percentage of men and women varies within the region: from a low of 17 percent women in Costa Rica to a high of 90 percent in Guatemala (Arroyo Moreno *et al.*, 1997/1999: 77). Such micro-enterprises offer the pickers and sorters previously working individually, the possibility to organise themselves into large units to demand better contracts and conditions. However, this generally requires the assistance of NGOs to act as advocate of such groups in negotiations with municipalities.

Traders and recycling enterprises have generally not organised themselves into associations. The Pune study found an interest to exist among retail traders to do so, but very little interest among wholesalers, who feel powerful enough to maintain their businesses individually.

In conclusion, this sector can be analysed as any other industrial sector, focusing on types of enterprises, employment, turnover and incomes. The sector is organised into a variety of commodity chains, with the enterprises using waste materials as inputs determining the economic viability throughout the chain. The sector is particularly vulnerable to changes in prices of alternative materials that can be used as inputs, or changes in import controls. It needs to be included in analyses of SWM as a matter of course, in order to understand shifts in material flows.

6.4. Changing context for recovery and recycling of solid waste materials

Three important processes are changing the context of the ways in which waste is recovered and recycled. They reflect the re-alignment of state, civil society, and the private sector in the last two decades. They are privatisation of public sector services (as described in chapter 2), decentralisation within government and changes in autonomy of local authorities and funding flows, and the initiatives undertaken by NGOs/CBOs to substitute for market and government, or to fill the gaps left by them.

Effects of privatisation

In many developing countries local authorities have recognised their lack of capacity to provide environmental services directly to the entire population of large cities[14]. In order to increase the provision of services, many local authorities have begun to contract out the delivery of services to private sector contractors The conditions specified for the quality of their service provision often include effective collection of waste at the neighbourhood level. When collection is indeed effectively done on a house-to-house basis, the amount of waste that is left in public spaces is reduced. It implies the closing off of access to solid waste in public spaces for informal recovery

14. They also explicitly exclude certain informal and illegal settlement areas of the city from such services.

and recycling, and an increasing competition among waste collectors for remaining waste materials (Blore, 2000).

From a public health perspective, cleaner neighbourhoods are a crucial benefit. From an environmental perspective, there can be negative impacts as the size of the waste flow going for disposal is increased significantly, and currently disposal sites in most cities in developing country are already filled to capacity. It is also negative in terms of recovery and recycling of materials, if alternative ways to retain or improve current levels of recovery are not taken[15]. From a socio-economic perspective, more effective collection without separation reduces employment in the recovery and recycling sector, eliminating already scarce employment possibilities for vulnerable groups on the labour market. It also has proved in practice to lead to conflicts between public sector employees providing waste services, and private sector employees, who are often paid far less, and have less employment security (Baud *et al.*, 2001; Blore, 2000). This has also proved to be the case in Surabaya, where source separation led to strikes by collection crews, who lost access to economically valuable materials (Furedy, pers com.). If indeed there is increasing reliance on source separation by households, it may necessitate major changes in cultural attitudes towards waste handling, although this problem is found very unevenly across social classes and in different countries.

This discussion suggests that privatisation efforts need a regulatory framework that provides incentives for source separation by households, and allows or makes mandatory recovery of materials commercially. In India, until recently, local authorities had only rules applying to SWM under the Health Departments, largely originating from the British system, with no higher authority co-ordinating the overall legislative framework. This is currently under revision, with state and Central Pollution Control Boards being allotted new tasks in setting national standards for SWM. However, this does not yet include incentives for separation and recovery of waste materials[16]. In Latin America, legislation generally does not clearly mandate one jurisdictional authority to formulate and enforce legal requirements (Arroyo Moreno *et al.*, 1997/1999). A number of public sector organisations at different levels are involved, and contest each other's responsibilities.

Decentralisation processes

Recent decentralisation policies being carried out in many countries have promoted the possibility to involve other actors in SWM. In Latin America, such policy changes

15. More effective methods of collection now include containerisation, which closes off access to picking materials by street pickers
16. A recent court case has led the Supreme Court to mandate separation and recovery; however, with no provision for incentives.

have allowed CBOs in Bolivia to participate in the provision of public services by monitoring, control and evaluation of public or private sector SWM activities (Arroyo Moreno *et al.*, 1997/1999). In Colombia, a law now allows recycling co-operatives to bid for public contracts for carrying out solid waste management activities.

In India, the effects of decentralisation have been different. There was already a local responsibility for direct provision, and almost no policy development concerning SWM at state and national levels. Court cases by concerned citizens have led to more attention now being given to developing standards for SWM by centralised government levels (to be developed further through the Pollution Control Boards).
Decentralisation as a result of changes in the Constitution in India, has led to large financial flows going to the state governments, and from there, being earmarked for local authorities (such as in Tamil Nadu). This means that more funds are now in principle available for investment in SWM. Although this is usually not taken into account, there have been some initiatives to have such investments include waste recovery through conversion to energy (see also chapter 9). Such initiatives are all large-scale, and ignore the recovery and recycling systems currently in place.

Effects of NGO/CBO initiatives

Local authorities have shown themselves more willing to work together with NGOs and community-based organisations in developing new models in the area of SWM. This has often taken the form of promotion of micro- and small enterprises for primary collection of waste, but in which recovery and recycling also take place by the members of the enterprises. Moreno has called this a form of social privatisation (cf. Arroyo Moreno *et al.*, 1997/99).

An inventory of initiatives that NGOs/CBOs undertake in SWM in India shows that there are several types of civil society organisations. They include those that are interested in adapted technology, those working with women waste pickers to form co-operatives, those working with street children picking waste to 'rehabilitate' them, and those providing public education on SWM issues (Bhuvaneshwari, 1998). Their perspectives differ: some work from a social justice viewpoint, others from a 'cleaner living environment' perspective, and others from an ecological perspective (Schenk *et al.*, 1998). A different inventory of community-based solid waste and water projects, carried out by WASTE (Anschütz, 1996), also indicates that recovery of (in)organic waste materials is often part of a community-based SWM system. However, the report also shows that the micro-enterprises are the ones interested in promoting resource recovery, because it can provide added income. The micro-enterprises involved find it very difficult to motivate households to carry out source separation effectively, which would make resource recovery more economically viable.

What these inventories show, is that CBOs operate mainly at the neighbourhood level, whereas NGOs can undertake recovery activities over a wider area. They also indicate that NGOs/CBOs are able to work with local authorities in developing new models of SWM at the local level. In fact, this is seen as crucial in obtaining a long-term viable system that can be replicated fairly rapidly, although it is difficult to interest municipalities initially (Lardinois and Furedy, 1999). However, they also indicate two basic weaknesses in these new models of co-operation. First, CBO/NGOs depend heavily on voluntary labour from their members, which has to be continued over a long period of time, and which such organisations find very difficult to expand (Lee, 1998). Secondly, where NGOs/CBOs have set up co-operatives for waste pickers to improve their earnings, they have often found difficulties in being sufficiently business-like to achieve better prices and contracts for the pickers involved (in India, e.g.)[17]. In the Philippines, access to credit and increasing the range of materials to be collected has been carried out by an NGO (Lardinois and Furedy, 1999). Thirdly, those working from a 'cleaner neighbourhood' perspective, have to depend on the co-operation of households specifically for source separation of waste materials, which requires a basic understanding and commitment to material recovery. Such initiatives may also have unintended impacts on the access of existing waste pickers and traders to saleable waste materials, because they usually ignore existing private sector initiatives in doing so (Baron and Castricum, 1996).

A final issue is how existing levels of source separation can be made more effective. Households have different reasons for carrying out source separation, related to economics and concern with the environment (Furedy, 1992; Lardinois and Furedy, 1999). In developing countries, economic reasons tend to dominate among most households, although the patterns of source separation differ according to income levels[18]. The type of materials to be separated also presents a problem: inorganic materials are recovered more extensively than organic materials (see chapters 9-12). This seems to be related to the lack of an effective collection system for organic materials: households use organic materials mainly for feeding animals or fertilising gardens.

---

17. Waste pickers themselves have also organised in 'unions' (in Latin America), which is a different model.
18. Households are interested in some degree of source separation, according to their levels of income. In high-income household, servants generally take care of separation of waste for selling themselves. In middle-income households, housewives tend to be more involved in separation, and in low-income households, waste will be both reused and the final residue discarded.

6.5. PARTNERSHIPS IN INORGANIC MATERIAL RECOVERY AND RECYCLING: CONTRIBUTIONS TO SUSTAINABLE DEVELOPMENT?

In this section, I turn to the question of what contributions the reuse, recovery and recycling of inorganic materials can make to the different aspects of sustainable development as applied to household solid waste management in developing countries.

In chapter 1, the different aspects of what constitutes sustainable development were unravelled in their component parts, as applied to urban domestic solid waste management. The main areas of contribution were covered under nine headings: the environmental aspects, which included minimisation of waste, recovery and recycling (as part of other processes), safe disposal of wastes, and socio-economic aspects, which included good co-ordination, financial viability, safe and healthy employment, clean urban environment and legitimacy. Here, I examine the question what different kinds of partnerships (or what we have termed elsewhere alliances (Baud *et al.*, 2001)) can contribute to specific aspects of sustainable development.

The major type of 'partnership' found in recovery and recycling are the forward and backward linkages within the variety of commodity chains found. The contributions of any type of commodity chain to the extent of recovery and recycling depend highly on the economics of the firms using waste materials as raw material inputs. This means that the fluctuations in the price of secondary materials for such firms determine in large measure the size and composition of the 'recovered materials' flow. Fluctuations are influenced heavily by alternative sources for (and quality of) raw materials, and prices for imported materials (Karanja *et al.*, chapter 8). As these factors lie outside the purview of municipal authorities and outside those of national Departments of the Environment, unexpected changes in the amount of recovered materials can take place outside the knowledge and influence of these organisations (cf. van Beukering, 2001). This means that further research is necessary on such external factors, in order to understand those parts of the system. Also research on the use of secondary materials in different industrial sectors should be extended in order to obtain a better understanding of firm strategies with regard to the use of secondary materials as resource.

In the area of socio-economic aspects of sustainable development, the recycling commodity chain makes fewer contributions. There are few contributions to better co-ordination with the official urban SWM system, as the local authorities usually refuse to deal directly with the recycling commodity chain in an effort to increase the environmental aspects of sustainable development of the system. The creation of employment by those in the commodity chain is obviously widespread, but of varying quality. Particularly the question of safety and health aspects remains an area in which much could be improved, with the situation being worst for the street and dump waste pickers, and best for the permanent workers in the recycling firms. The legitimacy of the commodity chain is also a vexed question, with the most vulnerable groups in the chain (waste pickers) having the least legitimacy usually, and the firms using secondary materials being part of formal sector production (Baud *et al.*, 2001; Arroyo Moreno *et al.*, 1997/99).

The second major type of partnership found is the triad between local government, NGOs, and waste pickers (women and youth mainly). This triad is usually based on a combination of social justice and economic considerations, with the NGO as nodal point between local authorities and waste pickers and dealers This type of partnership contributes to environmental aspects of sustainable development, by supporting existing patterns of collection and sale of recovered materials, by the creation of co-operatives and/or contracting with local authorities for neighbourhood collection schemes. It also contributes by the improvement of waste collection and separation at neighbourhood level and thus to cleaner neighbourhoods. Because of their small-scale, it is not yet clear whether these initiatives actually contribute to larger and more varied flows of recovered materials at city levels (see also chapters on organic waste diversion) (Furedy, 1997b).

In the area of socio-economic issues, partnerships between local government, NGOs and waste pickers contribute highly to better co-ordination between the informal and formal sectors of the urban SWM systems (cf. Dhanalakshmi and Iyer, 1999; Lapid, 1996). Studies in Chennai, Manila and Lima indicate that the recognition of their activities makes waste pickers much less vulnerable to harassment by police and residents, and improves their working conditions because more energy is spent on safety and health aspects (Baud *et al.*, 2001; Lapid, 1996; Furedy, 1997a). Such partnerships also contribute to the financial viability of the SWM system as a whole, because more materials can be recovered in a cleaner state (increasing their economic value), as well as increasing the returns for the waste pickers themselves. Finally, such partnerships contribute to the reduction of flows for final disposal (Baud *et al.*, 2001).

The assessment of partnerships according to specific aspects of sustainable development is still a new area in studies on the situation in developing countries. The Urban Waste Expertise Programme (UWEP) is the only other programme, to our knowledge, that also uses the integrated sustainable approach to solid waste management for studies in developing countries. The paper by van der Klundert and Lardinois (1995) was an early effort at developing the concept of integrated sustainable solid waste management. They define sustainable development in waste management as incorporating social/cultural criteria, environmental criteria, institutional/political criteria, financial criteria, economic criteria and technical criteria. 'Integrated' waste management includes the use of 'a range of different collection and treatment options', based on the use of the 'waste hierarchy' described at the beginning of this chapter. Therefore, in the UWEP studies, sufficient information is provided on both collection and disposal, as well as on recycling and recovery issues.

6.6. LESSONS DRAWN FROM THE CASE STUDIES

The next two chapters describe in detail the recovery and recycling commodity chains of inorganic domestic waste in Hyderabad, India and Nairobi, Kenya. Here, I will draw out several lessons from these case studies on the basis of discussions held in several workshops during the course of the research programme carried out.

Waste generators, both households and institutions, sell the most valuable and unmixed inorganic materials to private sector traders in both locations, in contrast to the situation in many industrialised countries where this process is less developed except for the second-hand goods trade. Mixed waste is collected by socio-economically vulnerable groups of people at the lower end of the commodity trading chain. The trading chain in Hyderabad (like elsewhere in India) is more complex than in Nairobi, Kenya in terms of the number of actors involved. Nevertheless, both locations show similarities in trading chains, which grow in size of material flows (and turnover) and specialisation of materials as these move up the commodity chain. In both locations, the commodity chain is not limited to the city concerned, but is more regional (Hyderabad) and even national for some materials (Kenya). The final link is the production unit, using waste materials as raw materials.

The production units using recovered materials as input, do so on the basis of technical and economic considerations. Technically, waste materials are almost always used in combination with virgin materials for producing end products; therefore, such enterprises cannot be truly called 'recycling enterprises'. This term has to be reserved for those enterprises producing intermediate products solely from recovered materials for the production units making end products (such as the plastic lump making units found in Hyderabad). This distinction can become important when deciding what type of recycling unit to support in its efforts to divert waste flows. The intermediate units often operate on the border of informality in paying taxes, which support programmes would have to recognize.

Secondly, trade and recycling activities are vulnerable to a number of external factors, which can influence them quite negatively. The end production units switch from recycled to virgin materials as and when the costs of virgin materials become similar to the levels of recycled materials. In Kenya, where the prices of imported virgin plastic went down sharply, the bottom dropped out of the market for locally recovered plastic materials. This suggests that the competition from imported virgin materials and imported waste of higher quality can remove the incentives to recover waste materials locally. This is not to say that the international trade in waste materials cannot make a contribution to international levels of material recovery, as van Beukering has recently suggested (2001). However, localised effects should be recognised and dealt with accordingly.

Incomes earned by street and dump pickers in both locations are below survival wages. The wages of the itinerant buyers are also low, and fluctuate a great deal. The incomes earned by retail traders, wholesalers, and production units are substantial and allow them to survive and build up assets (land and housing in Kenya). In both locations, women and children are heavily represented among the groups of pickers The working conditions of pickers particularly are unsafe and unhealthy, with harassment from residents and policy as a common problem. This suggests that support efforts to pickers should be directed towards alternative sources of employment, and social support to improve their living and working conditions.

Informal social security systems exist in both locations between the actors at different levels of the commodity chain. The most vulnerable groups of pickers and itinerant traders have informal access to credit from the traders to absorb sudden shocks through illness, death or rites of passage. Although these security systems are also designed to create dependence on the lender, they often provide the most important form of social security and credit to pickers

The lack of co-ordination between the formal government-led system of collection, transport and disposal and the recovery and recycling commodity chain reduces the chances of maximising recovery of inorganic materials from households. Therefore, efforts should be made to promote source separation and house-to-house selling in ways that are locally culturally adapted and include private sector actors

These lessons feed back into the discussion on more effective ways of co-operating between government, civil society and private sector, which will be discussed in the final chapter of this book.

S.GALAB, S. SUDHAKAR REDDY, ISA BAUD

CHAPTER 7
REUSE, RECOVERY AND RECYCLING OF URBAN INORGANIC SOLID WASTE IN HYDERABAD

7.1. INTRODUCTION

This chapter discusses the way that the inorganic waste generated in Hyderabad is reused, recovered and recycled. The agents participating in these processes include households, servants, sweepers, street waste pickers, dumpsite pickers, itinerant waste buyers, retail traders, wholesale traders and recycling units.

The aim of the chapter is:
- to indicate how different waste fractions (paper, plastic, metal, and glass) move through trade and recycling channels as they are recovered, reused and recycled;
- to analyze the organization and economic viability of production in recycling enterprises, and the links with retailers, wholesalers, and waste pickers;
- to examine the nature of employment of the different people working in the sector; and
- to analyze to what extent recycling inorganic waste contributes to socio-economic and environmental aspects of sustainable development in urban areas.[1]

Households and institutions generate inorganic waste that in principle goes into the municipal waste stream[2]. Primary access in India to separating inorganic waste for sale is by servants or family members, or lower-level personnel in institutions (Furedy, 1992; Baud and Schenk, 1994). After separation of inorganic waste at source, the remainder is put into the municipal stream where it is mixed with other waste. At this point, waste becomes a public good (or 'bad'), and access becomes more general. Further separation takes place of waste that is no longer uncontaminated[3]. Access and separation takes place at dustbins, collection points, and transfer stations by tricyclists,

1. For the research methodology, see appendix at the end of this book.
2. The level and type of inorganic waste generated by households differs according to income, with more inorganic materials being generated the higher the income (Baud, Dhanalakshmi, Baron, and Castricum, in preparation; Anand, 1999). Institutions include public sector as well as private institutions, shops, markets, and hotels; they also differ in the types of inorganic waste they produce (Snel, 1997).
3. As such, it loses part of its sale value.

I. Baud et al. (eds.), Solid Waste Management and Recycling, 133-159.

MCH crews, street waste pickers, and municipal crews. At the dumpsite, inorganic waste is separated and sold by dumpsite waste pickers.

Household members, servants, office boys and tricyclists sell waste mainly to itinerant buyers and retail traders Street waste pickers, dumpsite waste pickers and some tricyclists, itinerant buyers and small traders sell to retail traders and wholesale traders Retail traders sell some waste to wholesale traders in other states, and the recycling units located in Hyderabad and surrounding districts. Some waste from retail traders and wholesale traders is reused. Wholesale traders also sell the waste to other states and recycling units. Although this is the general picture found in Hyderabad (see Figure 7.1) the flows vary in complexity according to the type of materials involved[4].

The waste flows generated in Hyderabad are presented in diagrams 2 to 5, according to type of waste material. Comparing the diagrams brings out several differences. First, the flow of metal scrap and glass bottles is more complex than the flow of paper, plastic and glass waste, since reuse in this case as well as recycling are involved. Secondly, the flows of waste are not limited to the Hyderabad Corporation area and the surrounding municipalities, neither in terms of generation nor in terms of recycling units to which waste is finally sold[5]. This indicates clearly that the city is not a closed system in terms of recycling flows. It also makes it very difficult to assess the contribution that waste taken out of the municipal stream makes to reducing city waste flows, because neither the full quantity of waste generated in Hyderabad is known nor the amounts purchased from outside the city by the recycling enterprises.

4. This also applies elsewhere in India, as described in Dhanalakshmi and Iyer, 1999; Baud and Schenk, 1994; van Beukering, 2001; Snel, 1997.
5. Metal waste is also purchased directly from the building industry and other manufacturers

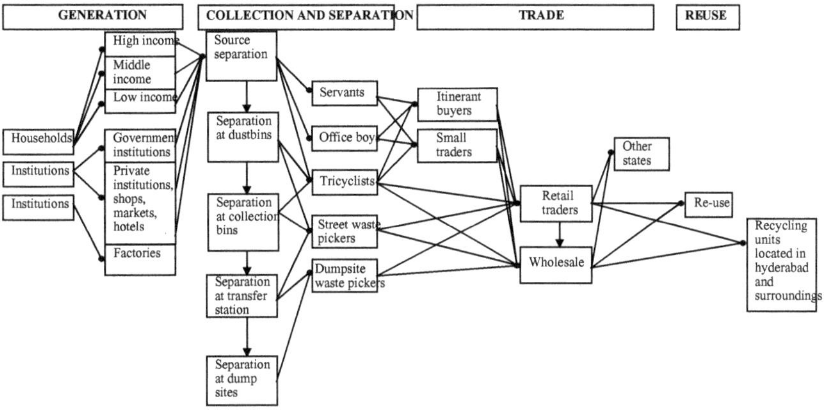

Figure 7.1. Inorganic waste flow

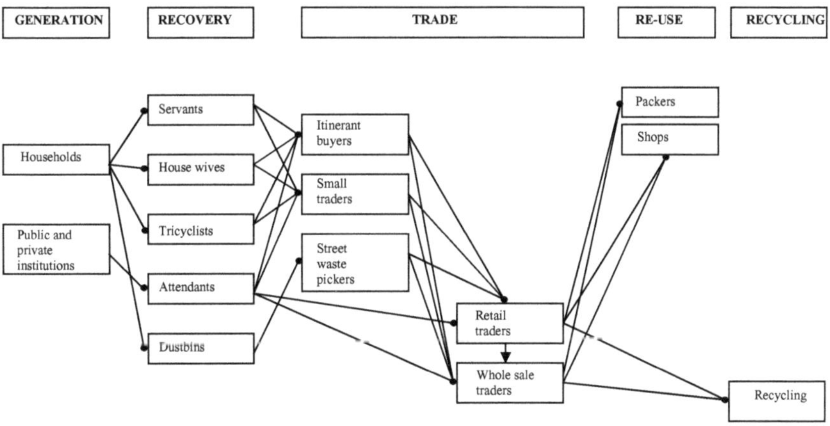

Figure 7.2. Paper waste flow

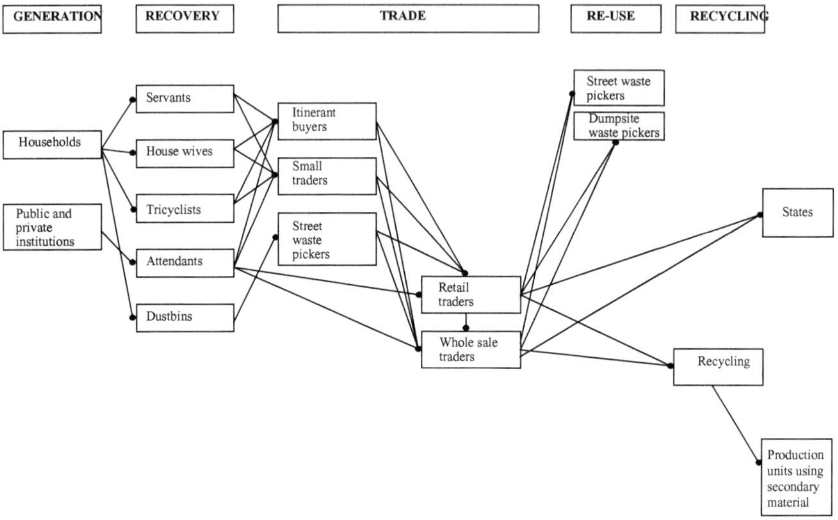

Figure 7.3. Plastic waste flow

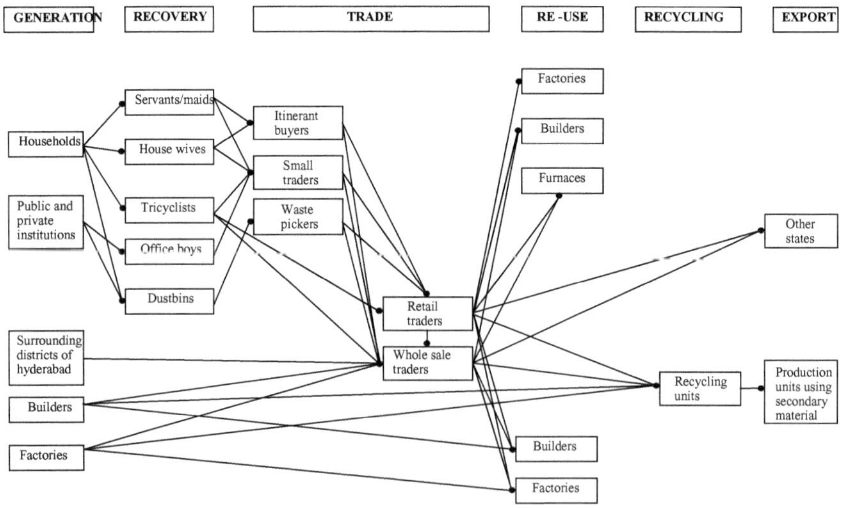

Figure 7.4. Metal waste flow

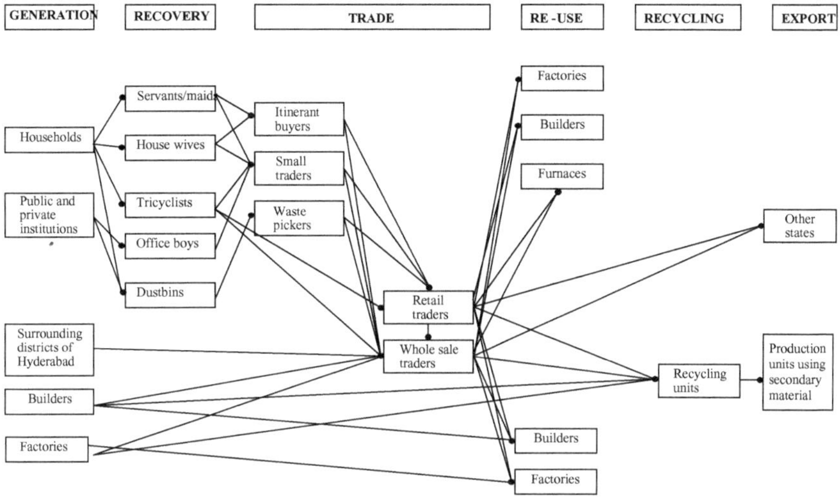

Figure 7.5. Glass waste flow

7.2. RECYCLING ENTERPRISES

The recycling enterprises are the final units in the commodity chain of recycling. Their economic viability determines the viability of the whole chain of suppliers attached to them, as indicated in the previous diagrams. They determine the prices for their products and the prices paid for the secondary materials used as raw materials in such production, and price changes are passed down the chain. Therefore, the analysis starts from these private sector recycling units as end units in the value-added chain. The entrepreneurs generally have an 'other caste' status, i.e. are not from backward or scheduled castes. They are well educated, with half of them having a bachelors degree in engineering or commerce, and the other half an MA degree in similar subject areas.

Recycling units have become more popular in the 1990s. In the 1970s only one was set up, whereas six were started up in the 1990s. The recent entry of new units is an indication of the viability of the recycling units. The presence of recycling units with both small and large capital investment indicates that units of varied size can operate successfully in recycling activities (Table 7.1). Larger units have the equipment to carry out all stages of processing and production. Smaller units produce intermediate goods, as the investment for further equipment is too high. The units are able to obtain reasonable to large margins of profits, with an average of almost 10 percent (based on turnover/costs). The plastic companies are able to obtain higher profit levels than units recycling other types of materials (see tables in the Appendix to this chapter)[6].

6. This is similar to results obtained in earlier studies of recycling units in Bangalore (Van Beukering, 1994).

Employment in the recycling units is on a contract basis as well as casual work (daily recruitment). However, workers tend to remain with the same unit over a longer period of time. On average, women formed almost 40 percent of the workers employed (Table 7.2). The units employed on average fifty workers, with the smallest unit employing only 19, and the largest unit 120 workers Wage levels for men varied between Rs1,400 – 3,000 per month, whereas for women it remained between Rs 900 to a maximum of 1,900. This indicates that women earn some 66-75 percent of what men earn in this sector. Their wages are generally lower per hour, and men work overtime whereas women generally do not. The only types of fringe benefits generally given to the workers are a yearly bonus and credit against their wages. A few units also provide compensation for costs of illnesses. Many workers in the units complain of pains in back and chest, and respiratory infections.

Currently, these units are viable according to the profit margins found. This viability is based on several ways of reducing costs. To begin with, the workers are not engaged in formal wage labor, and receive few social security benefits. Secondly, the units are located in industrial areas, so they face little protest from neighbourhood residents or environmental activists. Some units are located in the rural areas surrounding Hyderabad, as the government has given tax reductions and other concessions to promote non-agricultural units there. Finally, the absence of trade unions and lack of awareness about environmental effects of these recycling units in the rural areas maintain their profit margins at the moment. However, these units may become economically non-viable, if the costs of environmental externalities caused by these units are internalized, or when they start to comply with the rules of labour legislation.

Table 7.1. Characteristics of the recycling enterprises in Hyderabad

Unit No.	*Waste recycled*	*Location of the enterprise*	*Years of set up of the unit*	*Age of the unit (in Years)*	*Capital size (fixed capital in Rs)*
1	Waste paper	Patan Cheru	1985	15 years	2,000,000
2	Iron scrap	Jeedimetla	1980	20 years	30,000,000
3	Plastic waste	Jeedimetla	1995	5 years	700,000
4	Waste paper	Medchal	1992	8 years	2,500,000
5	Plastic, waste paper	Patan Cheru	1991	9 years	950,000
6	Iron scrap	Nacharam	1972	28 years	4,000,000
7	Plastic waste pipes	Jeedimetla	1999	1 year	1,000,000
8	Plastic waste	Kattedan	1994	6 years	1,500,000
9	Plastic waste	Jeedimetla	1997	3 years	1,000,000
10	Waste paper	Gandam guda	1997	3 years	600,000

Source: field survey

Table 7.2. Wages and nature of employment in different types of recycling units

Units	*Number of males employed*	*Wages of male worker*	*Number of females employed*	*Wages of women workers*	*Total workers (M+F)*	*Nature of employment*
1	32	1,500 to 1,700	23	1,100 to 1,900	55	Contract
2	38	1,500 to 1,700	28	9,00 to 1,150	66	Contract
3	22	1,600 to 2,300	20	1,100 to 1,450	42	Contract
4	18	1,500 to 2,200	12	1,100 to 1,500	30	Contract
5	16	1,400 to 3,000	3	1,300	19	Contract
6	20	2,000 to 2,200	40	1,000 to 1,200	60	Casual + Contract
7	25	1,700 to 2,600	15	1,200 to 1,500	40	Contract
8	22	1,800 to 2,500	18	1,200 to 1,500	40	Contract
9	110	1,800 to 3,000	10	1,100 to 1,500	120	Casual + contract
10	20	1,800 to 2,600	10	1,250 to 1,650	30	Contract

Source: field survey among workers (not the same units as in previous tables)

7.3. WHOLESALE TRADERS

Wholesale traders are the dominant suppliers of secondary materials to the recycling units. Their background is still mainly that of forward castes, although 20 percent of them have a scheduled or backward caste background. Almost half has a tertiary education degree. Another 20 percent of the traders have only primary education, and the remainder a secondary education. The wholesale units are generally well-established, with 60 percent being established in or before the 1970s. In the 1990s, a second group started up activities (24 percent).

The wholesale traders usually specialize in a single waste since they deal in larger quantities of mainly uncontaminated waste. However, 40 percent also deal in a second material, which forms up to 10 percent of their turnover. There are 9 paper wholesalers, 9 in metal, 4 dealers in bottles, and 2 in plastics. They work in business premises; 60 percent invest in their own premises, and forty percent of the wholesalers rent it. Their main activity is sorting out waste, storing it and transporting it to the recycling units, which buy from them.

They sell the waste either to other local wholesale traders and recycling units, or directly to recycling units located elsewhere in Maharastra, Tamil Nadu and Karnataka. Recycling units in other states pay relatively higher prices. The wholesale traders are less affected than retailers by the fluctuations in the prices of waste materials since they handle large quantities of materials. However, they run risks from unreliable suppliers and buyers Wholesale traders need to pay their suppliers immediately and also give them loans in times of needs. The recycling enterprises, which buy the materials, expect them to supply materials on credit for periods ranging from 15 up to 90 days. Therefore, the working capital invested in the business can be quite high.

The average value of waste bought weekly is Rs 133,125. The wholesalers earn an average net income of Rs 30,269 per week, after deducting costs of transportation, labour, non-wage benefits, land and infrastructure (Table 7.3). It is evident from the data that income varies widely between traders As profit margins are rather low, it is the size of the turnover which determines the size of net income of the units.

The units set up in the last 20 years employ about 5 people on average, with the older units employing seven people on average (Table 7.3). Women and children mainly carry out sorting, whereas men are involved in the carrying and transporting of waste. There is no shift system in these units.

Table 7.3. Distribution of wholesale trader units, according to age, size of the unit, and average net income.

Cat.	*Age of the unit (In years)*	*Number of units*	*Size of the unit (Number of workers)*	*Net income (rs per week)*
1	1-8	6	5	19,535
2	9-16	4	5	10,760
3	17-25	4	4	7,538
4	26-32	11	7	55,879
All		25		30,269

Source: field survey

Wholesalers purchase secondary materials from a variety of sources. These include waste pickers, itinerant buyers, retail trades, government offices and others For the four waste fractions covered in the field survey, the major proportion of the materials purchased comes from retail traders (ranging from 63 percent for metal and paper, 68 percent for bottles, to 95 percent for plastic). The wholesalers clearly prefer buying secondary materials that have already undergone the sorting and cleaning process carried out by retail traders (see tables Appendix).

The extent to which wholesalers buy materials from different sources of waste suppliers varies with the characteristics of their units. The following tables indicate differences by size of units. Small units buying paper tend to buy exclusively from retail traders, whereas large units buy half from retailers, 35 percent from itinerant buyers and 15 percent from waste pickers Plastic waste units buy almost exclusively from retail traders, with only 4 percent being purchased from itinerant buyers and waste pickers combined (they are mainly small units). Wholesalers buying metal scrap buy 60-70 percent from retail traders, between 15-20 percent from itinerant buyers, and less than 5 percent from pickers For larger units, government offices are also an important source (15 percent). Among wholesalers purchasing bottles, retail traders are the main source for small units (85 percent), going down to 65 percent among larger units. The latter also purchase 27 percent from itinerant buyers The figures indicate that itinerant buyers are a more important source of secondary materials than usually acknowledged, confirming earlier studies by Furedy (1992) on this type of trader. It also makes clear that waste pickers have little direct access to wholesalers in selling waste, with amounts of waste ranging between 1-3 percent of materials sold to wholesalers

7.4. RETAIL TRADERS

Retail traders in Hyderabad are largely from 'backward' castes (71 percent). Another 24 percent are from 'other caste' backgrounds. This suggests that the lower in the value chain, the lower the social background in the Hindu caste hierarchy. Their education levels are lower than those of wholesalers: 55 percent have a secondary education, and only a small number have a tertiary education. The vast majority are men: only 4 percent women retailers were found. The traders already had some business experience before starting these units, and form part of a larger family and caste community network of people working in this sector. This provides them with financial and moral support.

Units of retail traders were started in the late 1980s and 1990s. More than half were started after 1995 (32 out of 55). They are mainly located in their own residential premises (rented). They operate with a license from the MCH, but are nevertheless often harassed by the premise owners They do not invest capital in buying premises or transportation, but hire this when necessary. They also do not hire outside wage labor, but utilize unpaid family labor of female and male members in order to reduce costs. The units do manage to expand their activities and the number of people working there over time.

Retail traders mainly carry out sorting and cleaning, and stock the waste into sufficient quantities before selling it. Their main mode of transportation to sell the waste is by car or rickshaw. They sell the waste to wholesale traders but transportation costs are paid by retail traders themselves.

Retail traders purchase a combination of paper, bottles, plastic, broken glass, metal scrap, aluminum, copper and other wastes, and do not specialize in one type of material. 64 percent buy all the waste fractions covered in the study. Another 25 percent buy three out of four types of materials. Only 11 percent specialize to a greater extent.

Traders purchase from a variety of sources as well: street and dumpsite pickers, itinerant buyers, government offices and households. The availability of waste materials as well as the quality determine from whom the retail traders buy (see tables Appendix). Retail traders buy 60-80 percent of their materials from itinerant buyers The exception is metal, which they also purchase from other retail buyers This suggests that for retail traders, itinerant buyers are the most important category of suppliers It also suggests that metal scrap is not available in large quantities, so retail traders have to purchase it from colleagues to obtain enough to sell profitably. Retail traders do buy from pickers: between 12 – 33 percent of their materials come from them. This is particularly true for paper waste, and least true for plastic. Traders do try and bind waste pickers and itinerant buyers as suppliers to them by providing them with loans and benefits in kind.

There are clear seasonal variations in the business for retail traders During the rainy season, business output drops by 25-50 percent as the different types of pickers and itinerant buyers supply less waste. The retail traders are aware of the import policy of the government. They also have waste licenses, which are useful to them when buying and selling the waste and to avoid any problems from the Municipal Corporation of Hyderabad.

The retail traders make an average net income of Rs 817 per week, after costs of renting a storage space, transportation, electricity, and water charges. For this group as well as the wholesalers, the level of turnover is more important than levels of profit margin in determining income. Average weekly value (for the last week during the field survey in 1999) purchased by the units is Rs 4,045. The composition of the waste purchased also influences the level of income. On average, paper provides 34 percent of the total value of waste materials purchased, plastic 12 percent, metal 28 percent and bottles 26 percent.

7.5. ITINERANT BUYERS

Itinerant waste buyers go from door-to-door purchasing 'dry' waste items from households and domestic workers throughout the day. From the point of view of itinerant waste buyers, high income areas yield a better quality of waste materials, but domestic workers normally collect and sell them. They say that they have better luck in middle-income areas where women of the households are more likely to sell themselves than in the high-income areas where waste is given to maidservants and domestic workers The itinerant waste buyers also cover residential areas, shops and offices. They work individually rather than in groups and remain in the same areas overtime. This ensures continuity and a greater degree of trust in the relationship with their clientele.

Itinerant buyers are mainly from a scheduled caste or scheduled tribe background (73 percent). The remaining group is from 'backward' castes. They are overwhelmingly men, with only 3 percent women among them. They are largely illiterate, and have come from a rural background in Andhra Pradesh, migrating to Hyderabad in search of work. The seasonal itinerant buyers migrate back and forth to the rural areas, and only work in 'buying waste' when there is no work in the rural areas.

The itinerant waste buyers usually own bicycles, which are their means of transport when canvassing areas to purchase and transport the wastes. They generally work for 10 to 11 hours per day on average, within a 10 km radius from their place of residence. They collect waste for six days in a week and rest on Sunday. During the rainy season, they only work for 3 to 4 hours, and their income declines drastically (see table appendix). In order to smooth out the fluctuations in income, they resort to borrowing and repay the loans during the peak season. Itinerant buyers run health risks during their work. They complain of headaches, cuts by sharp objects, skin ailments, and aches and pains.

The itinerant waste buyer purchases all types of waste and does not specialize, as none of the waste is available in adequate quantities. They sell the waste on the day it is collected. They sort out the waste at the trader's place before selling it to him. Hence, no storage and sorting costs are involved. Therefore, the only cost the itinerant buyer has is the bicycle he uses, and the rest of the value of the waste is income.

Itinerant buyers working all year round in the business generally have enduring ties with their retail traders (77 percent of the traders interviewed). There are two ways in which these ties are reflected. In the first, the trader gives interest-free loans and provides working capital on a daily basis. The working capital provided in the morning is deducted daily in the evening from the payment made to the itinerant buyers for the waste sold to the trader. The loans provided to them can go up to Rs 5,000 at a time. That amount has to be repaid in two installments, with accounts settled on Diwali (annual festival). The itinerant buyers also receive annual gifts from their traders on Diwali. The second way of promoting enduring ties is to provide only interest free loans, with similar repayment schedules as in the first method. These methods require the itinerant waste buyer to sell their waste to one trader only, for a long time.

Seasonal itinerant buyers generally take neither interest-free loans nor daily working capital (23 percent of the traders). They can sell their waste to any trader they like. The tied method is beneficial to the itinerant waste buyers for obtaining interest free loans and working capital, reducing costs and providing a minimum of financial security against sudden calamities. It is beneficial to the traders in ensuring continuity of supply of materials for selling.

The average level of income per week for all itinerant traders was around Rs 1,223. in 1999. It varied between traders with enduring ties with their retail traders, who earned Rs 1,152 and those with no such ties, who earned Rs 1,456. The variations in income between slack and peak season indicate that itinerant traders with strong ties allow a greater fluctuation than traders with no enduring ties with retail traders Because they have no access to loans from such traders, they try and keep their income up by buying as much as possible (see table Appendix).

7.6. STREET AND DUMP SITE WASTE PICKERS

Social background

The main types of non-household and non-institutional collectors of waste for trade and recycling are the street- and dumpsite waste pickers They include younger men, women and children. In Hyderabad, waste pickers are mainly recent rural migrants to the city, moving there as family for mainly economic reasons. Both groups come mainly from scheduled caste backgrounds, and to some extent from scheduled tribe backgrounds. More women are found among them than in other categories of working people in the waste chain: 28 percent of the street pickers and 42 percent of the dump pickers Children work mainly among the dump pickers: they are 15 percent of the people interviewed there, compared to only 4 percent among street pickers In both groups, 80 percent of the people are illiterate, and the majority of their children do not go to school either.

The pickers live in dire circumstances. They have little access to public services. The vast majority does not possess ration cards (88-90 percent do not), and they make use of public water taps or bore wells to obtain water. Among dump pickers, 65 percent have lived less than five years in Hyderabad, whereas among street pickers 53 percent have lived there shorter than five years The waste pickers at the Mansurabad site come from the same rural area, are related, and live together in one neighbourhood in Hyderabad. Waste pickers at Golconda and Gandamguda have varied social and geographic backgrounds. This suggests that migrants may move up in the value chain as they live longer in the area, gaining access to waste with a higher unit value.

Economic activities

Street pickers concentrate their picking activities mainly on slums, industrial areas and residential areas and institutions. Three-quarters of them work seven days a week, the others six days a week. There is a clear seasonal difference in the length of time they work, as the rainy season makes picking difficult and the materials less valuable. They carry their materials to a retail trader daily for sorting and selling.

Dump pickers concentrate on their one site, to which they gain access by paying Rs 10 daily to the municipal officials at the dumpsite. They pick a variety of waste materials and sort and sell it on the same day to retail traders Transportation is another important cost, as they are not allowed on public transportation with their materials. Some hire rickshaws together, others cadge a ride from the municipal waste collection crews, and others walk.

Both groups of pickers sell through 'tied' and 'untied' methods to their retail traders Untied selling allows them to sell to any trader they like, whereas tied selling means they have access to a sizeable loan (up to Rs 5,000) but are required to sell to the retail traders providing this loan[7]. In both groups, the vast majority remains untied to any one trader. However, more than 20 percent of the dump pickers make use of loans from traders, compared to only 10 percent among street pickers Within both groups, some are able to take loans through the connections in the rural areas where they came from, a preferred option.

The waste collected by both groups of waste pickers includes paper, plastic, polythene bags, metal scrap, pads, broken glass, and bottles[8]. The amount of materials in the slack season is between 70-75 percent of the amounts collected in the peak season. Fuel wood and bottles (and some clothing and household items collected) are reused by the pickers themselves. Dumpsite pickers collect similar types of waste as street pickers in larger amounts. The largest quantity consists of paper, followed by polythene, plastic, and metal scrap[9]. For most materials affected by rain, the amounts come down to 60 percent of the peak season.

Average incomes for pickers vary between the two groups of pickers Street pickers earn an average of Rs 274 per week, when they work in a 'tied' relation with a trader, and Rs 247 when they sell on the open market. In the slack season, they earn an average 68 percent of their earnings in the peak season. Untied pickers earn 73 percent of their peak wages, and tied pickers 63 percent. This suggests that for street pickers, working for a variety of buyers actually increases income; tied pickers have a lower income, but some access to loans to help them through the slack season (Table 7.4).

Dump pickers earn an average of Rs 391 when they work in a 'tied' relation with a trader, and Rs 388 when they sell on the open market. In the slack season, tied dump pickers earn about 68 percent of the peak season income, whereas untied pickers earn only 60 percent. This suggests that for dump pickers, having a fixed relation with a

---

7. Pickers indicate that they can take out loans up to Rs 1,000 without being tied to any retail trader.
8. The quantity of waste collected by street pickers came to 24 kg paper, 25 kg plastic, 8 kg polythene, 19 kg metal scrap, 14 kg pads, 21 bottles and 2 kg broken glass in a week during peak season. But it came down to 17 kg, 18 kg, 6 kg, 14 kg, 10 kg, 4 kg, 15 (no) and 2 kg respectively in the slack season.
9. The average weekly quantity of waste collected comes to 49 kg paper, 22 kg plastic, 35 kg polythene, 21 kg metal scrap, 15 kg pads, 20 kg broken glass, 2 bottles and 3 kg of other materials during peak season, but it comes to 29 kg, 13 kg, 20 kg, 13 kg, 9 kg, 12 kg, 1 (no) and 2 kg respectively in a week in the slack season.

trader reduces fluctuations in income, and provides access to a basic form of social security. However, only 22 percent of the pickers have such a relation (Table 7.4).

Table 7.4. Working methods and incomes of street and dump pickers in Hyderabad, India

Sl.	*Type of description*	*Street waste pickers*	*Dump site waste pickers*
1	Tied method	12 (10 percent)	16 (22 percent)
	Untied method	108 (90 percent)	56 (78 percent)
	Total	120 (100 percent)	72 (100 percent)
2	To whom waste is sold	Retail traders, wholesalers	Retail traders, wholesalers
3	working days/ week	7	6
4	Tied method	Value (Rs)	Value (Rs)
	Peak season	336	466
	Slack season	211	316
	All seasons	274	391
5	Untied method	Value (Rs)	Value (Rs)
	Peak season	286	496
	Slack season	208	280
	All seasons	247	388
6	All methods	Value (Rs)	Value (Rs)
	Peak season	311	481
	Slack season	210	298
	All seasons	261	390

Source: field survey

Working hazards

Almost all the street pickers complained of health problems: mainly headaches, injuries by sharp objects and eye problems. Dump pickers had similar problems, but also mentioned respiratory ailments, suggesting that the air conditions at the dump are an added problem. Street pickers spend Rs 21 per week on health treatment, and dump pickers Rs 25. Both groups use primarily private health care facilities.

Street pickers are also regularly harassed by both police and public, who prevent them from going to particular areas. Unmarried girls and women picking waste also face sexual harassment from the municipal staff, and sometimes their goods are stolen. The public transport system does not allow them on the bus, and thus raises the costs of transport for them.

7.7. ASSESSING SUSTAINABLE DEVELOPMENT IN THE INORGANIC RECYCLING SYSTEM

The final section takes up the issue of assessing the contribution of the inorganic waste recycling system to various aspects of sustainable development in the urban area of Hyderabad. The indicators presented in chapter 1 are used, albeit flexibly, according to the extent of information available in the study. It focuses on three areas of contribution: socio-economic aspects, public health and environmental aspects.

Contributions to socio-economic aspects of sustainable development

Economic efficiency and viability

The recycling sector is characterized by its totally private nature: there is no government subsidy to any actor/enterprise in the value chain. This means that currently the allocative efficiency of the system is still high enough to make it profitable for the various actors in the commodity chain to carry out their sorting, trading and recycling activities. The profit margins of the recycling units have been seen to lie between 5 – 22 percent with an average of 10 percent. Net income among wholesalers is around Rs30,000 per week. The average weekly incomes of retail traders and itinerant buyers were found to be Rs817 and Rs1,222 respectively. The weekly average incomes of street and dump pickers were Rs250 and Rs389 respectively. These figures show that the majority of groups in the commodity chain can make a reasonable living from this economic activity.

Although various actors indicate that there is more competition now within their groups, there were no clear-cut indications that price levels were going down within the commodity chain in the Hyderabad area. Materials were drawn largely from the city and region, and sold in the same area. There is little to no connection made as yet to waste from foreign sources outside India.

The operational performance of the enterprises, traders, and pickers in the chain is well-adapted to the fine-meshed network necessary to retrieve mixed waste effectively from households, streets and dumpsites. The lowest-paid pickers and itinerant buyers are distributed all over the city, selling to retail traders at neighbourhood level. Wholesalers buy mainly from retail traders, and recycling units mainly from wholesalers At the level of wholesalers, 550 tons of paper, 39 tons of plastic, 63 tons of metal scrap, and 14,000 bottles are recovered per week. The MCH collects an average 6,800 tons per week; the recycling chain adds some 10 percent to that figure through its activities[10].

---

10. The figure on the MCH comes from chapter 3, Table 3.3. Obviously, these figures are indicative only. However, they are based on an extensive field survey, and their validity is at least as good as other estimates found in the literature.

The commodity chain is not very interested in increasing its productive efficiency through technological innovations or investment in new machinery. This is shown among recycling units, where small ones make only intermediate products and large ones final products for which more machinery is necessary. Wholesalers and retailers try to invest as little as possible, indicated by the preference among wholesalers to hire storage premises and among retail traders to hire both premises and transport as and when necessary. Increasing productive efficiency is mainly done in a 'low road' manner by reducing costs throughout the commodity chain. The major way of reducing costs is not to conform to legal requirements in terms of taxes (for the wholesalers and recycling units working to some extent informally). Neither do the units or traders and pickers conform in any way to labour legislation requirements, as I will indicate further in the next section.

Employment and labour conditions

Levels of employment created in the commodity chain are quite substantial. The ten recycling units covered created employment for 500 people, of whom almost 40 percent women. The 25 wholesalers covered employed 143 people totally. The 55 retailers employ on average two family members as unpaid labor, coming to 110 people. Itinerant buyers are an important category in the commodity chain in Hyderabad, with 60 people found in only one sample. This suggests that this category may well be larger than usually suggested in the literature (with the exception of Furedy, who has indicated its importance). Pickers covered were almost 200 people, not including the family members working with the respondents. However, these numbers are samples from populations of unknown sizes, which do not allow one to extrapolate the numbers to the city as a whole.

Security of employment is low among all workers in the sector. Only in the recycling units are employees found who have contracts for a longer period of time. Nevertheless, the workers in the wholesale units and recycling units with no official contracts do tend to work there for longer periods of time. Fringe benefits are not given according to official rules in the whole commodity chain, but yearly bonuses, gifts in kind and usually some form of loan is possible from the employer (or buyer of goods) to cover large expenses or calamities. The loans made by retail traders to their pickers provide an informal type of social security to them. However, only 10-20 percent of the pickers have access to this kind of social security.

Incomes have been indicated above for a number of categories of entrepreneurs and self-employed traders and pickers The wage labourers in the recycling units earn between Rs 900 – 2,600 per month. Women earn about 60 percent of the wage levels of men, and form 40 percent of the workers Again, the figures suggest that wage levels in the sector are not lower than comparable jobs for the different groups of people involved. However, the social status of the work is much lower than comparable jobs in the construction industry, for instance.

The gender composition of the workforce shows that women are concentrated in wage work in recycling and wholesaler units with less security than men workers, and in the vulnerable street picking and dump picking. They are not found among the category of traders, where medium-level incomes are made. This skewed pattern has to do with gendered constructions of public space, which do not allow women of certain castes to move from household to household to buy materials. Those women who do street picking are from scheduled castes and tribes, and are not expected to conform to the norms of higher caste groups.

Working conditions are not good and the majority of street and dump pickers complain of health problems, which they associate with their work. They spend a fair amount of money on private health care, which is more expensive but also more easily accessible than public sector health care (although it is free, the hours conflict with the working hours of the pickers).

Regulation of the sector

The legality of the actors/enterprises in the commodity chain goes down from recycling units to pickers The recycling units are most often registered enterprises with single ownership or proprietorship. They pay taxes, and have legal connections to electricity and land tenure. However, they usually do not conform to the labor legislation pertaining to permanent workers

Wholesalers in the sample preferred to be located in industrial estates or outside the city, in order to avoid strict environmental regulations.

Retail traders work mainly from their home premises, which the majority rent. Both aspects make them vulnerable to harassment from the property owner and neighbors, because of the environmental problems associated with waste. It also means that they are not formally regulated as businesses, and are part of the 'informal economy' in Hyderabad. They are open to harassment by the police and municipal officers because of this, and have to pay bribes regularly in order to prevent closure or confiscation of their goods.

Itinerant buyers are not recognized as official collectors and are also part of the 'informal economy'. They have to create their own legitimacy by building up trust with the households from whom they obtain their materials over a longer period of time. Pickers are considered completely illegal and mistrusted even more by residents, municipal crews, and police. They are harassed by the various groups mentioned, and have to bribe police and municipal officers for access to the waste (especially at the dumpsite).

This aura of illegality that hangs over the small-scale and 'informal' trading activities has prevented local and national authorities from recognizing the economic and environmental contributions the recycling sector makes to recovering inorganic waste materials. Oddly enough, this also applies to international agencies promoting safer and more effective solid waste management (such as the World Bank). The thrust of their policy interventions is in the direction of more effective collection and safer disposal (see WB SWM Roadshow in India in 2000; new Rules for SWM by GOI), and prevents them from developing new initiatives including recycling and recovery activities.

System viability
Currently the waste recycling system is viable, because of its 'informality' and reduced costs of labour and non-paid taxes. If it were to be formalized and included as part of the urban SWM system with a recognized contribution to reducing volumes of wastes for disposal and the environmental contribution of reduction of use of virgin resources, the system may become less viable, because increased taxes and labour costs will reduce profits in recycling units. These units pass on costs down the commodity chain, making waste picking less viable or completely unviable financially.

Contributions to public health and environmental aspects of sustainable development

The environmental impact of reuse and recycling activities has been assessed in terms of two parameters: reduction in volume of waste disposed of at the dumpsite, and the perceived impact on environmental health conditions (in India called public health).

Cleanliness of neighbourhoods
The environmental or public health impact of waste recovery and recycling at the neighbourhood level is perceived to be negative. This applies more to the street pickers than to itinerant buyers and is related to the fact that waste materials are taken out of the rubbish bins to be sorted by the pickers They are also considered to be potential thieves, and residents are very hesitant about their presence in their neighbourhood. Itinerant buyers do not pose an environmental health problem, as they take unmixed waste out of the neighbourhood before trading it further. Retail traders are perceived as a negative environmental health hazard, as they store materials in their home premises in residential areas, and have pickers coming and going with the waste they sell. Wholesalers are usually located outside residential areas, and not perceived by residents as a hazard.

Health of workers
The recycling commodity chain relies on labor-intensive recovery and transformation activities, and exposes most workers to continuous threat of injuries and possible infections from contaminated waste. The workers most at risk include the women and children sorting waste in recycling and trading units, and the street and dumpsite

pickers searching for waste in their respective locations. This is reflected in the expenditure on health that almost everyone in these groups makes weekly to counteract the negative effects of their occupational hazards. They do not use protective equipment, their only tool being a long stick to rummage through waste piles.

Environmental impact – reduction of waste volumes
In the previous section, the reduction of waste volumes realized in the trade and recycling sector was calculated at 10 percent of the waste collected by the MCH weekly. This is a substantial contribution to more sustainable SWM. This applies especially in the context in which existing dumpsites are becoming full, and new dumpsites are yet to be taken into use. In fact, this would argue for introducing further and more stringent recycling measures, so that existing dumpsites can remain in use for a longer period of time.

However, the opposite trend is likely to occur. The privatization of SWC is reducing the access of waste pickers and itinerant buyers to waste at the neighbourhood level (see chapter 3), and the question remains whether the secondary collection points are able to offer a substitution at the same level of recovery.

7.8. SYSTEM WIDE CONCERNS

A major worry is that neither authorities nor residents in Hyderabad recognise the recycling sector as an integral part of a SWM system in which concerns over environmental health and environmental impacts are integrated. Rather, the public health perspective is still the leading concern in efforts to improve the system, a concern which remains predominant in the new national Rules for SWM laid down in 2001. The implications are that matters will get worse before they can get better in terms of contributions to environmental aspects of sustainable development in SWM, as access to waste is closed off more effectively and the amount of waste recovered is reduced.

The alternative utilized in European countries is to shift the burden of recovery from private sector initiative to households, who are required to provide their labour for free. They are required to sort waste materials and to offer them in specific containers to the public authorities, or to take sorted materials to secondary collection points. This is an option which would be highly resisted in India for several reasons. Currently, households sort materials that are saleable to itinerant buyers or other traders Providing free labour would be anathema to middle-class and high-income households because of the 'unclean' association of waste, and would therefore be left to servants, who would have little interest in performing such tasks well if there was no extra income to be derived from waste.

Alternatives suggested by engineers from ERM in Chennai to allow waste pickers legal access to secondary collection stations, and organize waste unloading in such a fashion that they could do this efficiently, were not accepted by international donors, who feared the necessity of dealing with small-scale cooperatives in a larger SWM project. National and local authorities in India could well introduce such measures to ensure that a maximum amount of waste is recovered before being removed to the dumpsite. This is even more important in the context of new dumpsites likely to be located well away from Hyderabad, increasing transportation costs substantially.

Finally, the lack of recognition is a lost opportunity to integrate an existing completely private market recovery system that is financially and economically viable, into the citywide system. Recognizing existing efforts offers the opportunity to introduce new, locally adapted low-cost technologies into the system to reduce the current health hazards run by the people working in waste recovery, and to increase the productivity of their efforts.

APPENDIX TO CHAPTER 7

Table App. 7.1. Production process and output in recycling waste by type of waste

Waste recycled	*Output manufactured*	*Description of the processes in manufacturing*		
		Process – I	*Process – II*	*Process – III*
Waste paper	Machine glassed craft, paper and packing material	Pulping (sorting)	Chemical mixing	Drying
Iron scrap	M.S. Inguard, Tor steel (8mm-20mm)	Melting of scrap	Induction F/C route	Rolling mills
Plastic waste	Polythene bags, plastic components	Grinding	Granules	Final production
Waste paper	Packing materials	Sorting/pulping	Chemical mixing	Drying
Plastic, waste paper	EGG containers	Pulping/sorting	Chemical mixing	Drying
Iron scrap	M.S. Inguard	Melting	Oxidation, de-slagging	Tapping
Plastic waste pipes	Kisan PVC (Agri) pipes	Grinding	Drawing	Finished pipes
Plastic waste	Poly bags, plastic components	Pulping, Melting	Chemical mixing	Final production
Plastic waste	Plastic components	Melting	Chemical mixing	Final production
Waste paper	Paper & packing material	Pulping	Chemical mixing	Final production, sheets folding

Source: Field Survey

Table App. 7.2. Costs and returns of recycling enterprises (per month in Rs)

Description of the items	*Description of the unit*									
	Unit 1	*Unit 2*	*Unit 3*	*Unit 4*	*Unit 5*	*Unit 6*	*Unit 7*	*Unit 8*	*Unit 9*	*Unit 10*
Costs										
Raw-materials	2,246,000	18,000,000	291,000	1,662,000	1,974,830	12,416,000	337,500	475,000	175,000	40,000
Labour Cost										
A Wages	44,400	829,000	12,220	20,000	256,500	251,000	34,400	49,200	30,900	26,000
B Non-wages	11,500	50,000	-	-	80,000	50,000	-	-	5000	-
Other expenditure	60,000	675,000	34,000	23,000	100,000	550,000	25,000	100,000	31,000	4000
Total cost	2,340,500	19,554,000	337,000	1,705,000	2,411,330	13,267,000	396,900	624,200	241,900	70,000
Grass value of output (monthly)	2,500,000	20,400,000	385,000	1,800,000	2,916,000	14,280,000	434,000	700,000	255,750	90,000
Net income (B-A=c)	159,500	846,000	47,000	95,000	504,670	1,013,000	37,100	75,800	13,850	20,000
C: B = D	6%	4%	12%	5%	17%	7%	9%	11%	5%	22%

Source: Field Survey among recycling units

Table App. 7.3. Percentage of paper waste purchased by wholesale tradersfrom different agents (last week) by size of unit (Quantity in Kgs)

Size of the unit (No. of workers)	*Street/Dump site waste pickers*	*Itinerant buyers*	*Retail traders*	*Government offices*	*Others*	*Total*
1 – 5	455 (0.6)	1,400 (2.2)	65,450 (96.7)	-	350 (0.5)	67,655 (100)
6 – 10	700 (0.2)	7,000 (1.8)	230,300 (60.2)	1,400 (0.4)	143,500 (37.4)	382,900 (100)
11 – 15	-	-	-	-	-	-
16 – 21	15,000 (15.0)	35,000 (35.0)	50,000 (50.0)	-	-	100,000 (100)
Total = 11	16,155 (2.9)	43,400 (7.9)	345,750 (62.8)	1,400 (0.3)	143,850 (26.1)	550,55 (100)

Source: Field Survey

Table App. 7.4. Percentage of plastic waste purchased by wholesale Traders from different agents (last week) by size of unit (quantity in Kgs)

Size of the unit (No. of workers)	*Street/Dump site waste pickers*	*Itinerant buyers*	*Retail traders*	*Government offices*	*Others*	*Total*
1 – 5	357 (1.9)	350 (1.9)	16,205 (90.3)	-	1,050 (5.9)	17,962 (100)
6 – 10	-	-	21,000 (100)	-	-	21,000 (100)
11 – 15	-	-	-	-	-	-
16 – 21	-	-	-	-	-	-
Total = 7	357 (0.9)	350 (0.8)	37,205 (95.5)	-	1,050 (2.8)	38,962 (100)

Source: Field Survey

Table App. 7.5. Percentage of metal waste purchased by wholesale traders from different agents (last week) by size of unit (quantity in Kgs)

Size of the unit (No. of workers)	*Street/Dump site waste pickers*	*Itinerat buyers*	*Retail traders*	*Governmet offices*	*Others*	*Total*
1 – 5	840	8,995	23,135	700	4,200	37,870
	(2.2)	(23.7)	(61.1)	(1.8)	(11.2)	(100)
6 – 10	350	3,010	12,600	-	3,850	19,810
	(1.8)	(15.2)	(63.6)		(19.4)	(100)
11 – 15	350	700	4,900	1,050	-	7,000
	(5.0)	(10.0)	(70.0)	(15.0)		(100)
16 – 21	-	-	-	-	-	-
Total = 12	1,540	12,705	40,635	1,750	8,050	64,680
	(2.4)	(19.6)	(62.8)	(2.7)	(12.5)	(100)

Source: Field Survey

Table App. 7.6. Percentage of bottle waste purchased by wholesale traders from different agents (last week) by size of unit (quantity in Nos)

Size of the unit (No. of workers)	*Street/Dump site waste pickers*	*Itinerant buyers*	*Retail traders*	*Government offices*	*Others*	*Total*
1 – 5	3,640	525	24,640	-	-	28,805
	(12.6)	(1.8)	(85.6)			(100)
6 – 10	-	3,500	3,500	-	-	7,000
		(50.0)	(50.0)			(100)
11 – 15	-	49,000	119,000	-	14,000	182,000
		(26.9)	(65.4)		(7.7)	(100)
16 – 21	-	-	-	-	-	-
Total = 8	3,640	53,025	147,140	-	14,000	217,805
	(1.7)	(24.3)	(67.5)		(6.5)	(100)

Source: Field Survey

Table App. 7.7. Percentage of paper waste purchased by retail traders by size of unit (last week) (quantity in Kgs)

Size of the unit (No. of workers)	*Number of units*	*Street/Dump site waste pickers*	*Itinerant buyers*	*Retail traders*	*Others*	*Total*
1 – 3	46	4,149 (27.4)	954 (63.1)	-	1,435 (9.5)	15,127 (100)
4 – 6	8	595 (32.9)	752 (41.6)	-	462 (25.5)	1,809 (100)
7 – 9	1	2,730 (48.8)	2,870 (51.2)	-	-	5,600 (100)
Total	55	7,474 (33.2)	13,165 (58.4)	-	1,897 (8.4)	22,536 (100)

Source: Field Survey

Table App. 7.8. Percentage of plastic waste purchased by retail traders by size of unit (last week) (quantity in Kgs)

Size of the unit (No. of workers)	*Number of units*	*Street/Dump site waste pickers*	*Itinerant buyers*	*Retail traders*	*Others*	*Total*
1 – 3	46	283 (11.5)	1,801 (72.9)	-	385 (15.6)	2,469 (100)
4 – 6	8	105 (11.7)	791 (88.3)	-	-	896 (100)
7 – 9	1	70 (14.2)	420 (85.7)	-	-	490 (100)
Total	55	458 (11.9)	3,012 (78.2)	-	385 (9.9)	3,855 (100)

Source: Field Survey

Table App. 7.9. Percentage of metal purchased by retail traders by size of unit (last week) (quantity in Kgs)

Size of the unit (No. of workers)	*Number of units*	*Street/Dump site waste pickers*	*Itinerant buyers*	*Retail traders*	*Others*	*Total*
1 – 3	46	2,605 (14.3)	7,305 (39.9)	7,490 (409)	890 (4.1)	18,290 (100)
4 – 6	8	500 (18.2)	1,949 (709)	-	301 (10.9)	2,750 (100)
7 – 9	1	-	-	-	-	-
Total	55	3,105 (14.8)	9,254 (43.9)	7,490(35.6)	1,191 (3.7)	21,040 (100)

Source: Field Survey

Table App. 7.10. Percentage of bottles purchased by retail traders by size of unit (last week) (quantity in Nos)

Size of the unit (No. of workers)	*Number of units*	*Street/Dump site waste pickers*	*Itinerant buyers*	*Retail traders*	*Others*	*Total*
1 – 3	46	4,595	13,780	350	2,135	20,860
		(22.1)	(66.1)	(1.6)	(10.2)	(100)
4 – 6	8	380	2,888	-	350	3,618
		(10.5)	(79.8)	-	(9.7)	(100)
7 – 9	1	-	-		-	-
Total	55	4,975	16,668	350	2,485	2,4478
		(20.3)	(68.1)	(1.4)	(10.2)	(100)

Source: Field Survey

Table App. 7.11. Amounts/values of waste collected according to method of organization and type of waste

Description of waste collected	Tied method (46)						Untied method (14)						All methods (60)					
	Peak season		Slack season		All seasons		Peak season		Slack season		All seasons		Peak season		Slack season		All seasons	
	Quantity (kg)	Value (Rs)	Quantity (kg)	Value (Rs)	Quantity (kg)	Value (Rs)	Quantity (kg)	Value (Rs)	Quantity (kg)	Value (Rs)	Quantity (kg)	Value (Rs)	Quantity (kg)	Value (Rs.)	Quantity (kg)	Value (Rs.)	Quantity (kg)	Value (Rs.)
Waste paper	206.23	618.69	123.12	369.36	164.63	494.02	179.71	539.13	130.65	391.95	155.18	465.54	200.10	600.12	124.87	374.63	162.48	487.37
Plastic	54.67	437.36	32.64	261.12	43.65	349.24	77.50	620.00	56.35	450.80	66.92	535.40	59.99	479.97	38.17	305.37	49.08	392.67
Iron scrap	49.13	196.52	29.34	117.36	39.23	156.94	54.28	217.12	39.47	157.88	46.87	187.50	50.33	201.32	31.71	126.81	41.02	164.06
Pads	7.17	14.34	4.29	8.58	5.73	11.47	13.57	27.14	9.87	19.74	11.72	23.44	8.66	17.32	5.59	11.18	7.13	14.25
Bottles	77.54	155.08	46.30	92.60	61.92	123.84	88.00	176.00	63.98	127.96	75.99	151.98	79.98	159.96	50.42	100.85	65.20	130.41
Steel	6.82	20.46	4.08	12.24	5.45	16.35	35.42	106.26	25.76	77.28	30.59	91.77	13.49	40.48	9.13	27.41	11.31	33.95
All	401.56	1442.45	239.77	861.26	320.61	1151.86	448.00	1685.65	326.08	1225.61	387.27	1455.63	412.55	1499.17	259.89	946.25	336.22	1222.71

ANNE M. KARANJA, MOSES M. IKIARA, THEO C. DAVIES

CHAPTER 8
REUSE, RECOVERY AND RECYCLING OF URBAN INORGANIC SOLID WASTE IN NAIROBI

8.1. INTRODUCTION

Traditionally, local governments centred their urban solid waste management strategies on effective collection, transportation and disposal services. The primary objective was to protect and improve public health standards.

This approach came under increasing pressure and scrutiny in the industrialised world in the 1960s and 1970s, as changing consumption patterns resulted in rising waste generation rates further increasing the already overstretched waste sinks. The increasing amounts of waste became socially unacceptable. Therefore, a new perspective on urban solid waste management was devised in developed countries. This perspective, entailing the reduction of waste through prevention, reuse, recycling and energy recovery, seeks to promote a more environmentally sustainable form of solid waste management. Through recovery, reuse, recycling and composting, waste traders and recyclers contribute to more sustainable development by promoting cleaner urban neighbourhoods, financial viability and the reduction of the volumes of waste destined for disposal, as well as the creation of employment for predominantly poor people (Baud *et al.*, 2001).

Urban SWM strategies in developing countries are still largely focused on public health goals (see chapter 6). However, the increase in waste flows is compounding the already serious problems of SWM in these countries as well. Therefore, developing countries also need new forms of regulatory and incentive-yielding structures to encourage the recovery, recycling and reuse of inorganic waste materials, in order to come to more 'integrated sustainable waste management systems'. In Kenya, this is particularly important, as the public sector collection of SW has now almost completely disappeared (chapter 4).

Urban governance currently stresses the importance of alliances and partnerships among various actors providing urban services – including SWM – as instruments that offer greater probability of reaching socio-economically viable systems of service. It

I. Baud et al. (eds.), Solid Waste Management and Recycling, 161-194.

is therefore important to include a discussion on the nature of such alliances existing in SWM activities, as well as their contributions to urban sustainable development.

This chapter deals with the arrangements for inorganic resource recovery and recycling in Nairobi, focusing on the actors, processes and activities of waste recovery, reuse and recycling. The chapter identifies the actors and activities involved, describing the level, organisation and nature of their operations, and the institutional framework under which they operate. Secondly, the kinds of alliances that actors have formed are examined, including the types of co-operation and conflicts involved. Finally, the extent to which these activities contribute to environmental and socio-economic aspects of sustainable development is analysed.

Recovery, reuse and recycling activities can reduce environmental stresses through the reduction of waste destined for disposal, or left to accumulate in homes and neighbourhoods. Such activities also provide increased opportunities for employment and income for those directly involved in such activities. However, when not appropriately regulated, they can produce environmental stresses themselves, by using environmentally threatening production methods, unsafe disposal of waste generated during production, and by threatening the health of actors and nearby areas.

The questions guiding the discussion are:

- Who are the actors involved in the recovery, reuse, trade and recycling of inorganic waste materials in Nairobi, and within what regulatory framework do they operate?
- How are activities and processes in this area organised, and what are their contributions to sustainable development?
- What are the main alliances between the actors, and how do these contribute to sustainable development?

Data collection entailed examination of the institutional arrangements involved in the management of SW in the city and the commodity chains under which the activities are organised[1]. Actors studied included households, waste pickers, and itinerant buyers, waste dealers, and small and large-scale recycling enterprises. From each of these actors, information was collected on social and demographic characteristics, operational patterns and their dynamics, the rules and regulations (formal or informal), and where they fit into the commodity chain. Collection, trading, reuse and recycling arrangements, the materials preferred, pricing and price fluctuations, net incomes,

1. The discussion is based on analyses of field data collected between 1997 and 1998 in Nairobi and also on secondary materials on the subject. Various methods of data collection were employed. The survey method both through structured and unstructured interview schedules served as the main medium of primary data collection. Other methods included focus group discussions, key informant interviews as well as direct observation. Review of secondary materials was also used.

economic mobility and motivations of the entrepreneurs were studied. Close examination of the types of partnerships, types of co-operation, and conflicts and tensions, helped assess their impacts on effective SWM and sustainable development.

8.2. INORGANIC WASTE DEFINED

Inorganic waste refers to waste composed of material other than plant or animal matter. This includes sand, dust, glass and many other synthetics. Research on SWM however normally considers waste materials such as paper, sand, dust, glass, plastic, rubber, metal, bones, textiles and leather as inorganic although they are arguably organic. The unique aspect of inorganic waste is its non- or slow biodegradability, as a result of which its disposal presents particularly ominous implications on the capacity of local and global waste sinks.

Regulatory framework and policies related to inorganic waste recovery and recycling

In making SWM more environmentally friendly, waste minimisation, recovery, reuse and recycling of waste materials should be incorporated in existing waste management strategies. In its regulations on solid waste management, Kenya's policy, like those of many other developing countries, does not mention the recovery, reuse and recycling of urban inorganic solid waste. However, with the increase in waste flows and the problems faced by the municipal authority in managing the waste efficiently and effectively, more attention will have to be paid to such activities. The Environmental Management and Co-ordination Act of 1999 (EMCA, as discussed in chapter 4) promises to do so but has yet to be implemented.

Policies influencing waste recovery and recycling

The recovery, reuse and recycling of inorganic solid waste in Nairobi is not perceived as a municipal concern or of any relevance to NCC's mandate for SWM. The city neither sponsors nor espouses the recycling, recovery or separation of inorganic waste in policy or practice (Lamba, 1994). Existing activities relating to these processes are openly obstructed and legally undermined, particularly the lower levels. Waste picking activities in Nairobi remain officially unrecognised, socially unaccepted and highly stigmatised. To the NCC and city dwellers, waste pickers are a nuisance group of criminals (Hake, 1977: 179-83, Syagga 1997).

Small-scale waste recycling is similarly not recognised as an activity relevant to SWM and reducing environmental stresses. Even in academic evaluations of small-scale recycling enterprises, (see for instance McCormick (1991) issues of industrialisation and growth take precedence in theoretical considerations. The studies in this area do not examine environmental or public health issues in relation to *jua kali* activities in which inorganic waste materials are used.

The Factory Act, Cap 514, has a slight relevance to large-scale recycling companies using waste materials in production. This Act requires that factories are kept clean and free from effluvia resulting from the drainage of waste, sanitary convenience or nuisance. It also requires that accumulations of waste from the various production processes be cleared daily from the factory premises by a suitable method. It does not extend to other related manufacturing activities taking place outside the factory premises, such as the sourcing or transportation of raw materials inputs. The Act refers to waste only as 'dirt and refuse' and does not make specific reference to hazardous wastes. Such hazardous waste often ends up in the *Dandora* dumpsite and also in the undesignated waste dumps within the city. The formal SWM policy does not classify waste on the basis of health and environmental risks. Generators of such waste are left to dispose the waste without control.

The Scrap Metal Act, Cap 503, gives marginal relevance to the recovery and recycling of scrap metal. This law requires that an intending scrap metal dealer acquire a license from local government. It also prohibits the storage of scrap metal in a place other than that specified in the license, or otherwise authorised in writing by the licensing officer. The act does not cover other waste materials or even other waste management processes with regard to scrap metal, which is in the first instance only implicitly defined as waste. It has nothing on waste recycling or reduction of the rate of generation of scrap metal. Like other NCC SWM policies, it faces serious problems of enforcement.

8.3. ACTORS IN RECOVERY, REUSE, TRADE AND RECYCLING

The recovery and recycling activities in Nairobi are organised in several commodity chains, consisting of a combination of waste pickers, itinerant buyers, dealers or traders and wholesalers[2], as well as small-scale and large-scale recycling units. These activities take place in a private market and involve a number of activities. Waste pickers and itinerant buyers operate at the lower-income end of the chain and large recycling enterprises at the highest-income end. Dealers and wholesalers operating at the intermediate income levels of the commodity chain provide the link between the pickers and recyclers

Waste pickers obtain waste materials from streets and dumps to sell to traders and wholesalers The materials are then sold to small- and large-scale recycling enterprises. However, these actors sometimes sell materials directly to reusers, e.g. bottles and newspapers Similarly some of the actors, especially those at the higher income-end of the chain, occasionally source waste material inputs from outside the local chain. Within the commodity chain, large recyclers using waste materials in production, normally determine prices, which influence those down the line.

2. There are few wholesalers in Nairobi, one of them specializes in waste paper.

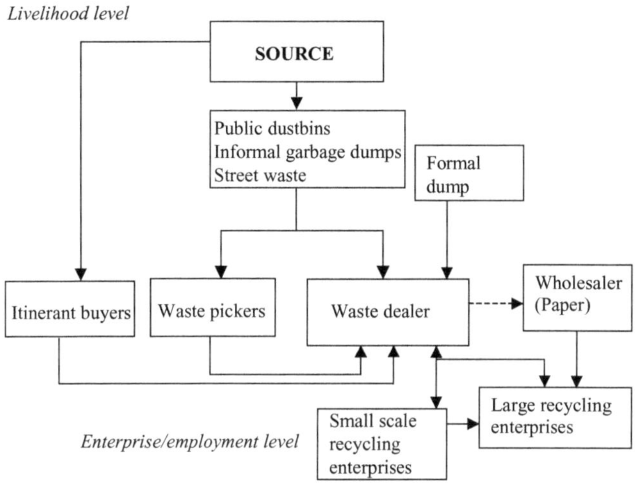

Figure 8.1. The recycling chain for inorganic waste materials in Nairobi

The social and operational relationships that exist in the chain of recovery, reuse and recycling processes find expression in the way prices are fixed, the exchange of vital information regarding sales and material preferences, as well as the arrangements made to transport waste. Forms of social security include advancing interest free loans and sometimes non-material support in times of adversity between traders and pickers Assistance is also provided through the supply of waste material on credit. These relationships are fundamental in analysing the effects of these activities on sustainable development as they enable actors to survive. Relationships are found:

- Amongst the waste pickers (intra-group)
- Between waste pickers and buyers
- Between itinerant byers and dealers
- Between dealers and wholesalers,
- Between wholesalers and recycling factories
- Between dealers and small and large-scale recycling units
- Between these categories and the producers of solid waste

The lower-income actors have a livelihood approach (see Figure 8.1). Waste picking is a survival strategy for the poor and unemployed of the city. People engaged in it generally lack access to basic necessities of life. Their lives are also characterised by insecurity and vulnerability. Waste picking is in their view the only legitimate means of eking out a living, through actual generation of income as well as access to non-cash benefits. For the larger more formal actors, it is an investment from which they endeavour to generate returns. The two are however intricately intertwined with larger scale, more formal production enterprises providing the impetus for the activities at lower income levels.

The capacity of different actors to contribute to aspects of sustainable development is a function of the social, financial and legal realm in which they operate, and which determines their position in the chain. For the lower levels the activities are illegal and their working conditions anything but favourable.

8.4. STREET AND DUMP WASTE PICKING

According to Hake (1977: 35-6) the phenomenon of homeless people living off streets was a feature of Nairobi as early as 1901, and is responsible for the enactment of vagrancy laws in 1902. In his survey of garbage pickers in Nairobi, Odegi-Awuondo found a 60-year-old garbage picker who had been collecting and trading in bones since 1944 (Odegi-Awuondo, 1994: 61-2). Waste picking therefore is an old activity in Nairobi, but has changed in scale and arrangements in accordance with the demographic and socio-economic transformations of the city (Karanja, thesis in preparation).

The first known recovery of waste materials for sale in Nairobi can be traced to 1963, when people recovered the copper-holed coins used by the colonial government, which were discarded by the new government at a dumping ground in *Mathare* North[3]. A kilogram of these coins sold for Ksh. 2.50 to scrap metal dealers Other types of materials retrieved for sale and reuse in the 1960s included bottles, scrap metal, rubber, tin cans, packing cases and cartons as well as building materials (Hake 1977: 186). Waste picking continued throughout the 1970s but was rather insignificant and largely associated with the *parking boy*[4] phenomenon. The main materials picked included steel fragments, bones, paper, aluminium, copper and tins.

The market for these was limited and not many people were involved. Waste picking became much more noticeable in the city from the mid-1980s onwards. An overwhelming majority of 63 (94 percent) amongst street waste pickers and 54 (73 percent) of dump waste pickers in our samples affirmed having started waste picking at around this time. This was further corroborated through the case histories of long-time pickers who emphasised that the numbers of waste pickers increased and competition for waste materials became one of the major problems in waste picking during this period. It is also at this time, according to the pickers, that serious fluctuation of prices began. Ironically, there was also an increase in the waste materials available for recovery in the streets, as indiscriminate dumping became rampant. Waste generation rates at this time continued to increase steadily while the collection levels

3. This is one of the oldest and largest slums in the city. It is not unusual for waste to be deposited in low-income areas.
4. Now dubbed 'parking family' as a proliferation of more people, of all ages including women, move into the city to eke out subsistence from activities hitherto the preserve of *parking boys*. These range from begging, stealing, assisting car parking, cleaning as well as guarding for token payments.

by the then main service provider, the NCC, were going downhill. The waste management problem in Nairobi had reached crisis proportions (chapter 4). This suggests that several factors contributed to the increased numbers of people engaging in waste picking and related activities.

8.5. SOCIAL ASPECTS OF WASTE PICKING AND THE COMMODITY CHAIN

There are two main types of waste pickers in Nairobi, depending on the locality of activity, namely street waste pickers and dump waste pickers Street waste pickers separate waste in small open city waste sites, mainly in the streets and public dustbins. Dump waste pickers on the other hand operate in the large formal or informal disposal[5] sites where final disposal of waste takes place. These factors significantly differentiate these categories in terms of organisation, networks as well as the types of materials preferred, their marketability and incomes. They also have different implications for occupational and health effects. However, both categories predominantly consist of young (average age of 25 to 36 years)[6] men, with low educational and skills levels, who have resided in the city for about 15 years They are commonly referred to as *chokora,* a derogatory Kiswahili term describing people without fixed abode whose main 'occupation' is to rummage through dustbins in search of food and other 'valuables'. They are also viewed as petty criminals. As mentioned earlier, their identity is closely tied with that of *parking boys.*

There are few women pickers in the streets, which is indicative of the social and cultural definitions of the types of activities especially for women (see Huysman, 1994). Female pickers are much more prevalent in large waste dumps, away from the public eye. Because of their inability to participate in the regular waste picking arrangements found at the dump, their incomes are lower than those of male pickers but they rely on the *Dandora* dump not only for income but also for household essentials like food, cooking fuel, soap etc. Most of these are single female heads of households characterised by extreme poverty.

20 percent of the dump waste pickers at *Dandora* reside within the dump itself and like their counterparts in the streets, they obtain other basic essentials like food and clothing from the garbage heaps. The streets are also home to a significant number of street waste pickers Garbage is a source of cash as well as non-cash income, the latter

---

5. The data on dump waste picker(s) was obtained from surveys amongst waste pickers at the *Dandora* dumpsite and also from two of the informal ones, *Kawangware* and *Embakasi.*
6. These data do not include street children who are far younger than this and who engage in recovery of waste materials, mainly food, on an intermittent basis. Though they may sell some materials to dealers, they are not regular participants in the commodity chain. It is estimated that there are now about 60,000 of such children (Daily Nation, March 14th, 1998).

of which is not easily quantifiable but in this case seems to be a significant source of motivation or survival tactic.

These actors have devised several methods of coping with the occupational constraints typical of their work. Certain 'tricks' are applied to try and increase gains from waste picking. These include the wetting of paper to increase its weight and the `fixing' of weighing machines by traders in order to lower the weight. All players in the chain are knowledgeable on these and other important aspects including the quality and the demand patterns of the differentiated types of materials. The most striking of these are the informal and highly recognised territorial, picking and trading arrangements found amongst street waste pickers as well as dump waste pickers These are most sophisticated at the *Dandora* dumpsite and comprise of arrangements through which competitive accessibility to the 'high quality' waste disposed of by private companies and household utility manufacturers is ensured.

8.6. MATERIALS, SALES AND INCOMES

Most (80 percent) of the waste materials recovered at the *Dandora* dumpsite are sold to *Mukuru* recycling project, a CBO initiated by a local church in one of the low-income residential areas neighbouring the dump. From the project the materials are transferred directly to recycling factories after a period of accumulation. This arrangement as conceived by the project leaders has successfully helped circumvent the 'exploitative' dealers operating around the dumpsite and has resulted in relatively higher prices and incomes for the waste pickers at the dump. Other initiatives meant to uplift the economic viability of these activities include the recycling of plastic within *Mukuru* itself. With technical and financial assistance from UNEP, an agreement for the transfer of technology and skills has been reached between *Mukuru* and *Skyplast*, one of the city's large-scale plastics recycling companies[7].

Although street waste picker(s) operate outside of such arrangements, they tend to earn more from sales on average than dump waste picker(s). This is attributed to the variations in the types and amounts of materials collected in terms of quality and quantity. The table below provides data on the materials, prices, and quantities collected.

7. This project was started by a priest in one of the local Catholic churches and also comprises of a rehabilitation program for drug addicted dump waste pickers An arrangement has also been made for pickers injured at the dump to access medical attention from a Catholic run hospital at reduced charges (interview with Mukuru recycling project manager).

Table 8.1. Types of materials average quantities (kg.) and prices (Ksh.) for street and dump pickers

Material	*Street pickers, no. (percent)*	*Prices in Ksh./kg/pcs*	*Amount p/ day in kg*	*Dump pickers, no. (percent)*	*Price-in Ksh./kg/pcs*	*Amount per/day*
Paper	50 (73)	2	6	44 (60)	2	5
Plastic	23 (34)	2.50	4	60 (81)	3	4
W/bottles*	29 (43)	1-10 p/pc	9 pcs	26 (35)	1-8/pc	0-30
B/glass**	17 (25)	2.50	5	37 (50)	2	5
Bones	25 (37)	3	3	53 (72)	3	3
Steel	23 (34)	2.50	3	41 (55)	1	3
Scrap iron	30 (44)	2.50	4	46 (62)	2	4

* W/bottles – whole bottles
** B/glass – broken glass (*vunjika*)

It is apparent from the table above that dump pickers retrieve more waste on average than street pickers Waste at the dump, is concentrated in one area and less time is spent on segregation. On the other hand, materials recovered by street pickers fetch higher prices. The reasons are:

- Street pickers retrieve relatively cleaner waste materials compared to dump pickers Waste materials obtained from the streets, often proximate to the point of generation, are much cleaner and immediately saleable while waste reclaimed from the dump is dirtier and fetches lower prices. Plastics and bottles for instance need more thorough cleaning while tin cans are set on fire to remove the paint and residues. The costs of re-using or recycling materials obtained at the dump is thus made higher by the process of cleaning, leaving a lower profit margin.
- The dealers that street pickers sell pickings to offer better prices than those at the dump. Traders at the dump control prices much more effectively while street pickers on the other hand sell to dealers located further apart from one another, making this more difficult.

Specialisation or monoculture garbage collection, as defined by Odegi-Awuondo (1994), apparently is non-existent among the two groups of pickers Any item perceived to be valuable either in terms of marketability or re-usability is salvaged. The main differentiating factor between the materials preferences between street and dump pickers is the capacity of the material to withstand damage and corrosion during collection, transport and disposal, and the demand by large waste recyclers Odegi-Awuondo (1994) found that the most popular items for street waste pickers are paper, scrap metal and bottles. Dump waste pickers prefer plastic, bones and scrap metals. Fairly extensive segregation of waste therefore is undertaken at various stages prior to final disposal. In addition, proportions that survive this separation are prone to much spoilage as they get mixed with other wet and dirty waste materials. This may explain the preponderance of waste plastics amongst dump pickers and that of paper amongst street pickers Others that withstand damage are scrap metal, bones, steel, bottles. The diagram below represents the materials preferred by street and dump pickers.

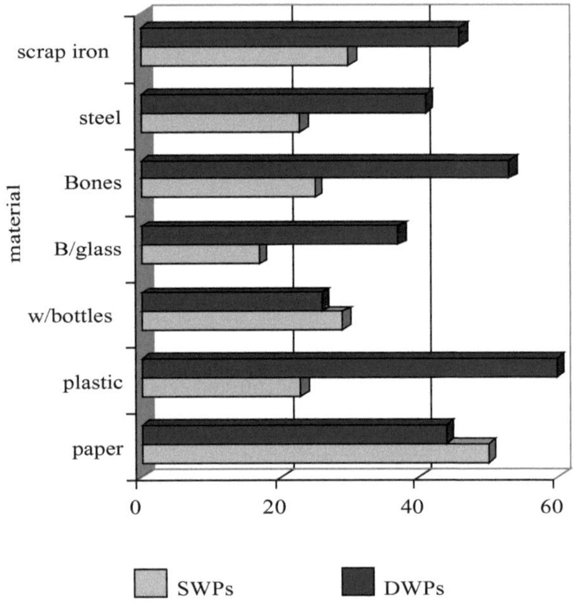

Figure 8.2. Materials preferences for street waste pickers and dump waste pickers

The recovery of paper at the dump on this graph is overstated due to the collection of clean paper from offices and other institutions in the CBD by one of the groups of the *Mukuru* Recycling Project.

The price of paper depends on its type and condition. Clean white paper fetches about Ksh. 4 per kg while brown paper sells for Ksh. 2 per kg. Soiled, i.e. wet and/or muddy, paper is often rejected or sold for a marginal price. Seasonally, 'good' paper is abundantly available during the sunny dry season, as not too much damage is inflicted. The prices of paper have declined over the last few years from Ksh. 8-12 per kg in the mid and late 1980s, to the current Ksh. 2-4 per kg.

In our surveys, waste pickers associated the price declines to the fall in factory demand in turn resulting from the outsourcing of certain raw material inputs. The local market for waste paper according to the waste pickers has declined because of the importation of waste paper by the LSREs. Many LSREs that previously consumed most of the local waste paper gathered by waste pickers took to importing waste paper from neighbouring countries (Uganda, South Africa) and from as far as France. Although more concrete data on this aspect was not obtained especially from public sector sources (e.g. Ministry of Trade and Industry), those involved for substantial lengths of time in waste picking and trade were consistent in tracing the sharp price fluctuations to the late 1980s followed by much more steady declines in the 1990s. The trends were largely blamed on the SAPs-led liberalisation reforms especially those on trade and finance. These actors saw the imposition of taxes on imported waste paper as a mechanism that could help correct this anomaly but the counter explanations i.e. the *prevailing short supply of high quality waste paper* and how this had affected the LSREs production processes seemed plausible and made economic sense. Moreover, owing to the lack of clear policy on the importation of waste paper, the government was reluctant to interfere with the sector in the direction of tax impositions or adjustments because of the other benefits accruing.

The demand for scrap-metals on the other hand is higher because these are extensively recycled by both small- and large-scale enterprises. There also is less reuse at source and along the commodity chain.

The preponderance of plastic materials among dump waste pickers stands out clearly. Dump waste pickers continued to gather plastics regardless of the very low prices, because demand at the recycling factories had at the time fallen. Large heaps of plastic materials could be observed at the traders' yards as purchases continued albeit at very low prices, in the hope of selling it '*when the factories re-open*'. The low demand and poor prices however may have caused less recovery of plastics upstream (i.e. by street waste pickers and itinerant byers), resulting in its plentiful presence at the dumps and inadvertent preference by dump waste pickers.

Bottles are salvaged extensively before the collection and transportation of waste from source and are therefore not found in any significant amount at the dump. There is also a buy-back or bottle-returns arrangement for soft drink bottles by the manufacturers via the *Central Glass Industries (CGI)*, a large-scale glass reprocessing company. Broken glass (*vunjika*) is too heavy to be carried about and is therefore avoided by street pickers, and is instead popular at the dumps. However, CGI had at the time of our filedwork ceased the purchase of local broken glass resulting in excessive and conspicuous output at the dumps and at traders' premises.

Aluminium, copper, brass, lead and zinc (non-ferrous metals), which are easily differentiated, manually sorted and cleaned fetch higher and stable prices of up to Ksh. 30 per kg. They yield considerable environmental and economic benefit as they require relatively less energy in recycling (see Powell, 1983) and have more applications than the ferrous metals (iron and steel). These are however very rarely available in the city especially in the streets. Bones also fetch better prices but are similarly difficult to find and require specialisation and are thus deemed unfeasible especially by street waste pickers Hence a relatively higher presence at the dumpsites.

Although not indicated on the diagram, food waste is also extensively collected largely for direct consumption. Other materials salvaged for direct reuse include clothes, food, household items, and fuel wood. Amongst waste pickers only 3 percent reported not reusing any of the waste materials recovered. Unless one has a supplementary source of income (which is not the case with these groups), it is unlikely that a monthly income averaging Ksh. 3,000 (US$ 40), can support an average HH size of 4 for street waste picker(s) and 6 for dump waste picker(s). Both had an average share contribution of household expenses of Ksh. 1,600 (US$ 22) per month. 75 percent and 57 percent of street and dump pickers respectively are married with children.

8.7. ITINERANT BUYERS

Mali kwa mali[8]
(Kiswahili for 'itinerant buyers') were amongst the earliest type of hawkers to be recognised in Nairobi. In 1967, the City Council issued up to 50 licenses for barter hawking (Hake, 1977: 180). Itinerant buyers in current Nairobi are also largely men, older (average age was 49 years) than waste pickers and long-time residents in the city. Their main mode of waste segregation is on a house-to-house basis and is motivated by the need to access 'uncontaminated' waste. This activity is concentrated in the middle and high-income residential areas but is on the decline. Like waste pickers, Itinerant buyers have low educational and skill levels but have spent a relatively longer time in waste work. They have more experience and also involve family in the sorting, cleaning or packing of the materials gathered, which is normally carried out at home.

Itinerant buying of waste materials in Nairobi became noticeable in the early 1970s and was initially via barter. Itinerant buyers provided new household utility items like buckets and kitchenware obtained from wholesale shops in the city. In exchange itinerant buyers obtained bottles, tin cans, newspapers as well as items of clothing, which they sold to traders at '*gikomba*', the largest open-air second-hand clothes market in the city. From here the materials were sold to individual consumers and probably found their way back to households. Exchange deals involved a lot of haggling with each party trying to maximise their gains. The itinerant buyers interviewed in this study reminisced over the diminished economic viability[9]. The activity was also more respectable and largely done out of choice.

Though the main motivation and types of materials bought have remained the same, new players have been added to the commodity chain. In the upper income areas,

8. Literally meaning, 'goods for goods', denoting the earlier mode of reclaiming waste via barter between itinerant buyers and households.

where householders place little importance on waste once it is removed from the house for collection, domestic workers and extended family members now play a key role in these activities. In these areas itinerant buyers operate through links with domestic workers Most of the transactions are secretly conducted during the day when householders are away at work. This way, house workers share in the proceeds of such dealings and thus have developed an interest in the household's waste management arrangements. The materials obtained are sold to neighbourhood-based traders Itinerant buyers therefore tend to work in areas close to both the selling sites and to their place of residence, transporting waste materials home for sorting and packing with the use of family labour. In terms of socio-economic viability, the work of itinerant buyers has become more difficult and less profitable. Their monthly incomes, averaging Ksh. 3,500 (US$ 46) per month are only slightly higher than those of waste pickers Nevertheless Itinerant buyers have side occupations through which they generate supplementary income. These include farming and small-scale real estate investments. To become an itinerant buyer, one requires soft loans from relatives and/or friends ranging from Ksh. 1,000 (US$ 14) to Ksh. 6,300 (US$ 90). Such amounts are out of reach for the lowest income operators in the chain.

8.8. THE DEALER: CENTRALITY IN THE CHAIN

Dealers are the main trade intermediaries in the channels through which accumulated waste materials recovered by pickers and buyers are passed on to the small and large-scale waste reprocessors The position of dealers or small buyers as a category in the recycling chain is explained by the existence of demand for recyclable waste materials. Scrap metal dealing for instance existed in Nairobi in the 1960s, 'serving the useful purpose of retrieving waste material for reuse'. Licensed dealers purchased waste scrap metal from a network of collectors who made their living by sourcing these from 'odd corners' and selling them to dealers These fetched 60 cents for copper, aluminium 40 cents and 5 cents for other materials (see Hake 1977: 186).

Dealers' activities consist of aggregation of waste materials by buying small quantities from waste pickers and itinerant buyers and delivering large amounts to factories on demand. Because of the relatively high volumes required by reprocessors and thus high costs, it is not possible for waste pickers to deliver waste materials directly to recycling enterprises.

---

9. Directly translated from the Kikuyu language. Most of the pickers, traders and itinerant buyers are from the Kikuyu ethnic group (JICA 1998). Beside the proximity of Central Province their place of origin to the city, it has been claimed that they had a tradition for successful trading or 'peddling' (Hake 1977: 179). However, their preponderance in waste picking is also indicative of the erosion of their erstwhile relatively stronger social and economic base resulting in involvement in activities hitherto ethnically shunned (JICA, 1998).

Types of materials traded

Trade and recycling of waste materials in Nairobi does not seem to be sufficiently developed to enable specialisation. Dealers do not specialise, as such specialisation may result in business failure if/when the demand for any particular material diminishes. Thus diversification in waste materials is aimed at avoiding or reducing such risks. This contrasts with the situation in India, where specialisation especially by large dealers is clearly observed, though waste trade operations there are larger than in Nairobi. (Baud and Schenk, 1994).

Paper is the most preferred waste material. Other important materials include scrap iron, plastic and whole bottles. Data on material preferences and prices by waste dealers are given in the table below.

Table 8.2. Materials traded: types, purchase and sale prices

Type	*Number (percent)*	*Average purchase prices in Ksh.*	*Average selling prices in Ksh.*
Paper	24 (77)	2	3
Broken glass	12 (39)	1	3
Steel	13 (42)	3	5
Scrap iron	22 (71)	3	5
Plastic	17 (55)	3	5
Whole bottles	17 (55)	1-8 per piece	1-15 p/kg
Bones	9 (29)	2.50	4
Aluminium	8 (26)	7	12
Copper	6 (19)	7	10

Note: N=36. Totals add up to more 100 percent due to multiple responses.

At the time of fieldwork, most of the dealers, especially around the *Dandora* waste dump, had accumulated large amounts of plastic. Surprisingly, they continued to purchase it from waste pickers but offered much lower prices to them. However, they made extensive use of unpaid family labour and indebted pickers, in order to lower their own operating costs.

The physical location of a dealer's business is determined by the availability of good materials in the surrounding areas, the presence of waste pickers as well as the proximity of the site to the dealer's home. Many dealer activities in Nairobi are concentrated around dumpsites. Those who purchase materials from street waste pickers are more widely scattered in the residential estates. Like waste pickers, dealers decrease in number outwards from the city centre. They are rare in high-income residential areas where large private companies have taken over household waste collection.

Dealers operate in open yards measuring about ½ acre, rented from private owners for a small monthly fee or on vacant municipality land, e.g. road reserves. Rent for the latter is paid in the form of bribes to policemen or the NCC *askaris*.

Operational costs and incomes

Operational costs for dealers are higher than those of other actors The high costs of transporting waste materials to the reprocessing factories is largely covered by the dealer. A typical waste dealer in Nairobi invests about Ksh. 6,000 (US$ 78) to start trading. Capital is raised from personal savings, soft loans from friends and relatives as well as group savings (merry-go-round)[10]. The need to earn income is the main reason dealers give for their involvement in this work. They also state that the activity was previously more profitable ('before factories closed').

Unlike waste pickers, dealers are required to have trade licenses from the Nairobi City Council. There is no standard format and fee for this. Bribery is rampant and somewhat accepted as a necessary component of the application and processing procedures. The amount paid for the license can range from Ksh. 300 to Ksh. 7,000 (US$ 4 – 92) per year. As a result, some dealers opt to operate without it but have to constantly bribe the council *askaris* arguing that one has to give bribes anyway whether they have the license or not. This is considered a necessary expenditure for the survival and smooth running of the business. Failure to do this often results in the confiscation of one's stock of materials and expulsion from the trading site.

Their incomes average Ksh. 13,000 (US$ 171) per month and range from Ksh. 2,500 to 40,000 (US$ 33 – 526) per month. Most dealers also earn supplementary incomes through other activities, which are co-managed with spouses and other relatives. 60 percent of the dealers in this study were also trading in '*mitumba*' (second-hand clothes from abroad which are preferred to local manufactures) and 48 percent in charcoal. Though these seem to be recent initiatives (perhaps prompted by decline in waste dealing), trade in second-hand clothes is particularly profitable though complicated. This group showed attempts at diversification or, as some put it, preparation to exit should 'business continue to decline'. Dealers also have a slightly larger household, averaging 9 persons, which often includes extended family members They help with the work in exchange for accommodation and assistance in the search for employment.

---

10. Rotating group saving schemes are a common phenomenon in the city and can operate on neighbourhood, work, professional or gender basis. From the small amounts contributed by each member, an informal form of financing at low/zero interest rates on a rotational basis is conducted.

8.9. INORGANIC WASTE RECYCLING: THE UPPER END OF THE CHAIN

Waste reprocessing in Nairobi is mainly carried out in large companies. The exception is scrap metal, which is also reprocessed by small-scale enterprises, at the *jua kali* level. This is central in assessing the importance of waste recovery, reuse and recycling to aspects of sustainable development. Through the buyers, we identified the most marketable materials, *viz.* paper, scrap iron, plastics and whole bottles, in that order of importance. Four large-scale companies were interviewed concerning their trade and recycling activities. These were *Chandaria (*which includes *Madhupaper), Central Glass Industries (CGI), Roll Mill* and *Skyplast.* Further information on large-scale reprocessing of paper and glass was also obtained from *Chandaria* and *CGI* through their representatives attending the EU SWM project stakeholder workshop. In the rest of the discussion we shall use paper and glass (well established large-scale recycling) and scrap metal (small-scale recycling) to access the contributions to urban sustainable development by small and large-scale waste recovery and recycling activities.

Waste paper recycling: Chandaria and Madhupaper

Although waste paper recycling in Nairobi is the most important sub-sector in waste recovery, reuse and recycling and is carried out on a large-scale, there are only a few large enterprises involved in the activity. These are Chandaria and Madhupaper, both owned and managed by the Chandaria group of companies[11]. The central player in waste paper trade is a single wholesaler, *Kamongo,* and not the dealers or small buyers, as is the case in scrap metal, glass and plastics. *Kamongo* is also under the same ownership, but poses as the main broker for waste paper for the giant Pan African Paper Mills located in *Webuye* in Western Kenya, also owned by *Chandaria.*

Chandaria acquired Madhupaper in 1987[12]. In highly politicised circumstances. This made *Chandaria* the only paper recycling company in Nairobi. Below is a diagram[13] of the paper recovery and recycling chain.

11. This is a group of Asian business magnates, with interests ranging from high technology equipment and pharmaceuticals to waste paper recycling, not only in Kenya but also in the East African region.
12. This was during the single-party political rule, generally characterized by wide-scale abuse of political power against those perceived to be anti-establishment.
13. There are other large recycling companies outside Nairobi These however still obtain a high proportion of their waste paper from Nairobi through *Kamongo* wholesalers and are moreover still associated with *Chandaria* in ownership, organisation and operations. We did not find small-scale paper recycling activities in the city.

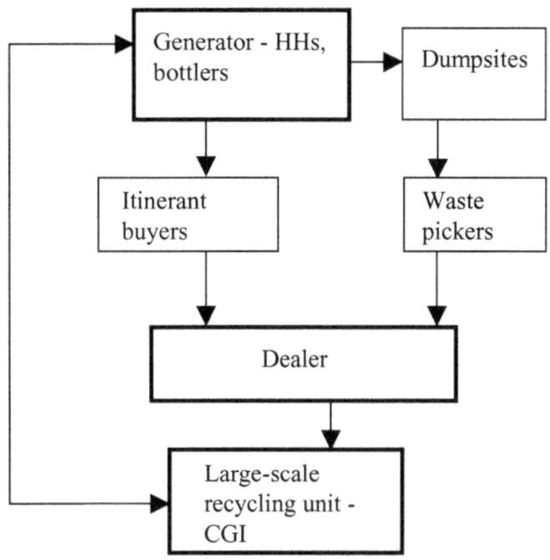

Figure 8.3. Chain for the recovery and recycling of paper in Nairobi

Recycling processes in Chandaria

Chandaria produces items like toilet paper, serviettes, table napkins and kitchen rolls from waste paper. The factory uses about 24 tons of paper per day. Its sister company *Madhupaper* uses 20 tons per day. White virgin paper pulp, which is imported at Ksh. 25 per kg, is used at a daily rate of 24 tons, and waste paper is used at the rate of 20 tons daily. The factory ensures minimal generation of wastewater by not using fresh water for pulping. The production process also requires the use of some chemical and organic dyes. Caustic soda is the main cleaning agent used. Stagnation of water is avoided in order to keep germicides to a minimum. A large bale of tissue paper containing about 40 rolls wholesales at Ksh. 356.50 (US$ 5) inclusive of VAT. Retail prices range between Ksh. 9-12 per roll depending on the type of outlet with supermarkets offering lower prices than *dukas* and *kiosks*.

Employees, payment and working conditions

Chandaria has about 700 employees, a high proportion of which are casual. The lowest wage paid is Ksh. 180 (US$ 2.3) per day for casuals and Ksh. 2,000 (US$ 25) per month for permanent staff. Permanent employees are provided with other benefits like medical cover. Manual labourers and machine operators are provided with protective gear like dust coats, masks, rubber boots and milk for those coming in close contact with chemicals to help neutralise the chemical fumes that may have been inhaled. The company also has health and compensation insurance in case of injuries suffered in the course of work. Permanent workers have a more regular medical cover in accordance with their work and salary grades.

Capital and current expenditures for *Chandaria* are high. Machines and other capital equipment are imported and require careful maintenance in order to ensure optimal daily capacity. *Chandaria* recently took to the importation of waste materials from neighbouring countries. This affected the economic viability of the activities of the people lower down the chain.

Large companies maximise or maintain the productivity of resources by using cheaper raw material sources where legally admissible. On local outsourcing behaviour of large car assemblers in Kenya, Masinde (1996) found that assemblies and franchise holders did not voluntarily procure their requirements from local sources, except those components and parts whose local sourcing was legally mandated. Even then, local procurement of these was still lower than 100 percent. Moreover, citing lower quality of local sources, they regularly petitioned the government about local acquisition requirements and several exemptions were awarded. None of the components were then acquired locally even when available.

Other than cleanliness, easy, timely and sufficient deliveries, there does not seem to be exceptionally high standards required for waste paper as raw material input in reprocessing. That means that the use of local sources would be relatively easy, compared to more technologically complex, high quality raw material requirements in other manufacturing activities.

Chandaria argues that substituting raw material sources is part of a larger strategy aimed at dealing with the escalating costs of production and harsh economic conditions, which were made worse by high levels of contamination in local waste paper. Moreover, as the manager hastened to add during the workshop, the company had had to make several other adjustments, including the retrenchment of a proportion (35 percent) of the permanent staff, and price increases of their products.

It is difficult for us to pinpoint the specific factor that induced this behaviour given the prevailing mix of negative political and economic conditions. At the time of the workshop (2000), *Chandaria's* demand for local waste materials had reportedly fallen even further due to the shortage of power and water. The search for alternatives had increased production costs remarkably. The manager contended that price increases were not considered tenable in the face of increasing competition from cheap imports.

Glass reprocessing

The reuse and recycling of glass waste in Nairobi begins with the household. Householders recover bottles and other items of glass for reuse within the house. Bottles may be returned directly to CGI though this is not a significant channel. Returnable bottles are normally exchanged with the purchase of drinks at retail outlets. A significant proportion of bottles and *vunjika* is sold or bartered with itinerant byers who in turn trade these to small buyers or dealers A small proportion is cleaned and sold by itinerant byers to individual reusers like small clinics and dispensaries. A lot of bottles and *vunjika* nevertheless still find their way to dumps (formal and informal). Like paper, glass recycling is only done by large-scale reprocessors.

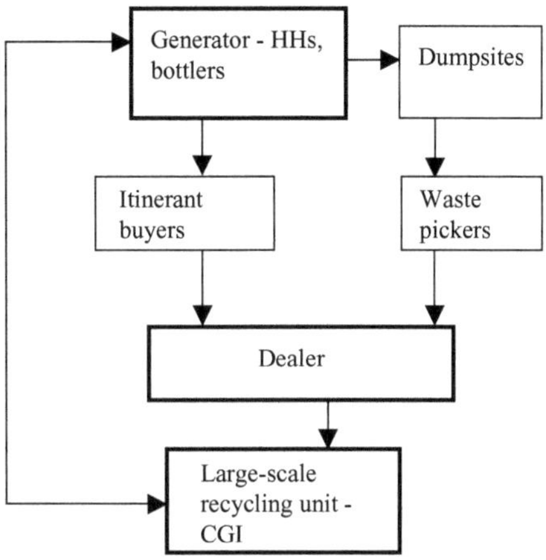

Figure 8.4. Chain for the recovery, reuse and recycling of glass

Glass is reprocessed by CGI, a subsidiary of *Kenya Breweries Ltd.* (KBL). CGI started recycling waste glass in 1987, using waste glass obtained from its parent company KBL and other large bottling companies like *Coca-Cola*. Reprocessing at this time simply entailed the sanitization of bottles for reuse. They also purchased broken glass (*vunjika*) from waste dealers for reprocessing into drinking glasses, water jugs, tumblers as well as bottles. Like *Chandaria*, CGI started buying bottles and broken glass from Tanzania, Uganda and Rwanda. These were obtained at varying prices according to type; green and clear glass at Ksh. 2,500 (US$ 33) per ton and brown glass at Ksh. 1,800 (US$ 24) per ton.

CGI is not explicitly recognised as a *waste glass-recycling factory*, but as a formal enterprise undertaking bottle recycling for its parent company KBL. At its formation in 1987, capital finance largely came from KBL. However, cleaning bottles was soon found to be insufficient to generate and sustain the company's recurrent expenditure and CGI began cleaning bottles for other multinational bottling companies. Additionally, its survival was to be enhanced by the reprocessing of cullet into glassware. This move made the recovery and trade of *vunjika* economically feasible for waste pickers and dealers However, because of the high maintenance costs of the precision equipment used, the company soon found broken glass recycling to be unprofitable. Operations were discontinued in the early 1990s and the full production line closed completely in 1999. The waste glass collectors and dealers were left with large heaps of broken glass with no alternative outlet.

Because there was no recognition in policy of the socio-economic and environmental relevance of these activities, there was no legal recourse for the small local waste collectors and dealers By the time of the CGI's 'factory closure', most dealers had spent an average of Ksh. 20,000 (US$ 263), a substantial amount given the prevailing business uncertainty and the constant fall of prices of other materials. The company now concentrates on recycling its own (KBL's) bottles, obtained from the numerous brewing subsidiaries as well as distributorships in the East African region. This has affected the economic viability of bottle and vunjika recovery locally[14].

With the energy crisis in 2000, which necessitated a power-rationing program, the company decreased both permanent and casual staff by 25 percent[15]. The raw materials acquisition budget increased by 20 percent, strengthening the need to seek alternative and cheaper sources. According to one of the managers, at the height of the power crisis the factory was operating at about 80 percent of normal production capacity and expected a further decline unless the economic situation in the country improved.

Large-scale metal recycling: the case of Roll Mill

There are about 8 metal rolling mills in Nairobi, of which three were studied[16]. During the fieldwork (1997/98), most were closed due to 'poor business'. One of these, Roll Mill, was started in 1986, and is a large-scale metal recycling company. The factory consumes about 30 tons of scrap metal per day, the bulk of which is obtained from local dealers and brokers like *KIOI* and *ROMA*, at about Ksh. 4 per kg. These materials are in turn obtained from waste pickers at the cost of Ksh. 2 per kg. Other important inputs like chemicals, virgin steel bars and hard coal are imported from outside the country (mainly from India). The company's main products are twisted metals and ground bars. These are consumed locally but also exported to neighbouring Tanzania and Uganda.

Employment and working conditions in Roll Mill

Roll Mill provides protective gear like helmets, boots, overalls and gloves to its factory workers, a majority (90 percent) of whom are casual workers, earning about Ksh. 140 (US$ 2) p/day. The company has no staff medical insurance cover except for serious industrial accidents. Other occupational health hazards emanating from routine exposure to chemicals and other harmful solvents are not covered. Some of the workers at *Roll Mills* said that attempts at legal action against illegal dismissal and

14. It was however alleged during the stakeholders workshop that CGI did not stop recycling of vunjika completely but continued, albeit at a lower scale, with raw material inputs largely coming from outside the country.
15. was no information available on working conditions in the factory.
16. They were Steel Structures Ltd., Premier Rollers, City Engineering, Kikuyu Steel millers, Insteel and the three that we studied EMCO, Morris and Co. ltd. and Roll Mill.

compensation for factory injuries have been foiled in the past through bribery and deals between government officers from the Ministry of Labour and the factory owners Factory regulations concerning the disposal of factory waste and sanitation are also flouted, due to poor enforcement mechanisms and corruption of government inspectors

Small-scale metal recycling

In his study on garbage collectors, Odegi-Awuondo (1994) found that scrap metal was one of the most popular items collected, second only to paper. Earlier on we have observed the popularity of scrap iron amongst the collection and trade actors and also that it is one of the waste materials recycled both at large and small-scale levels.

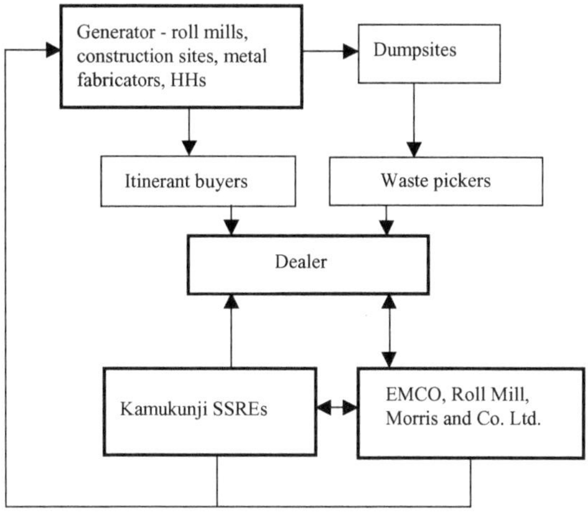

Figure 8.5. Scrap metal recovery and recycling chain in Nairobi

Small-scale metal recycling
We selected a sample of small-scale metal recyclers from a cluster of scrap metal workers at *Kamukunji,* a place about two kilometres from the city centre.

Kamukunji metal recycling enterprises came into being in 1978 when some artisans were evicted from their site on *Grogan* road to pave way for the construction of a housing estate. Others came in from the nearby *Burma* market and from other parts of the country to start small-scale businesses. These artisans later requested land from the government and obtained tentative allocation in 1985 after the District Commissioner (DC) quashed an eviction order by the NCC. Kamukunji gained government recognition in 1986 when the president ordered the construction of sheds to protect the workers from the harsh sun (*jua kali*). He also advised artisans to form associations through which government could channel assistance[17]. It was then that the *Kamukunji*

Jua Kali Association (KJKA) came into being. Its objectives ranged from improving artisan welfare with regard to the working conditions, to providing infrastructure, technology, and training in marketing and finance. The total population of metal workers was estimated at more than 2,000, settled on 1.5 hectares of land (Frijns *et al.*, 1997).

Although government conferred a degree of legitimacy on the entire informal small-scale enterprise sector, its development remained on the fringes of Kenya's development agenda. Most of the policies emanating from this recognition were never implemented. Government failed therefore to offer practical support to the sector's development (Government of Kenya (GoK), 1986, 1988, 1992, 1996; Helmsing 1993; King 1996; Frijns *et al.*, 1997).

Organisational aspects at Kamukunji

Kamukunji[18] is strategically located within easy access to raw materials from the nearby large factories in *Industrial Area* and scrap metal yards in the surrounding areas. There is also a ready market for finished goods from the neighbouring population and travellers at the nearby *Country Bus Station*. Goods destined for sale outside Nairobi or elsewhere are easily transported to the relevant destinations through this bus park.

The enterprises were established recently, and are small with an average of 2 workers inclusive of the proprietor[19]. The typical owner of the small-scale metal recycling enterprises is a man, ages between 20-60 years and averaging 40 years in age. Employees are mainly hired to assist in the production and finishing of goods and put in long hours, depending on the workload.

Start up capital/credit

Only small amounts of money were used to start the businesses and are normally raised from own savings, relatives and friends as well as rotating savings groups. The modal amount is Ksh. 10,000 (US$ 132) and the mean Ksh. 7,000 (US$ 92). Because of the strong kinship and ethnic based arrangements of production in *Kamukunji,* an artisan can enter recycling with as little as Ksh. 3,000 (US$ 40). However, finance remains one of the most serious barriers to entry and to enterprise growth.

---

17. This information obtained through interview with the then *Kamukunji Jua Kali* Association secretary in 1998 and updated in 2000 during the EU SWM project stakeholder workshop.
18. A total of 33 enterprises and their workers were purposively selected and interviewed in order to understand the production processes of the enterprises.
19. In their classification of Small Sized Enterprises (SSEs) in Kenya according to size, Parker and Torres 1994, show that about 80 percent are in the 1-2 non-growing category while the 11-50 is absent or 'missing'.

Small-scale metal recyclers largely depend on familial and informal sources of credit. The only institutional sources of credit are NGOs like KREP, which loans to groups based on the Grameen model. Many entrepreneurs consider the amounts loaned too small especially because they are not sufficient to procure capital goods and in their view have only helped them increase their already excessive stock[20]. This suggests that the entrepreneurs are aware of the investments necessary for growth but lack the means to make these. Some of the machines used require as much as Ksh. 100,000 (US$ 1,500), way out of the reach of these enterprises.

Waste materials usage

Small metal recycling enterprises use waste materials because they are cheaper[21], as 86 percent of the entrepreneurs said. Materials are obtained from foundries and rolling mills in the industrial area via dealers They are also obtained from retail shops located near *Kamukunji*. A small number of the enterprises obtain some of the waste materials from waste pickers[22]. Steel sheets, iron and aluminium metal obtained from metal fabrication factories in the industrial areas are the most commonly used materials. The virgin materials consist of metal sheets, galvanized iron, tin plates and aluminium. These are more expensive and only small quantities are used. Over 50 percent of the enterprises do not use any virgin materials. The types of waste materials and the proportion of enterprises using them are detailed on the table below.

Table 8.4. Types and number of enterprises using waste materials

Type of waste material	*No. enterprises using*	*Percent*
Metal bars	4	12
Steel sheets	20	61
Scrap Iron	13	39
Aluminium	13	39
Used tyres	3	9
Metal drums	4	12
Plastic drums	2	6
Shoes and soles	1	3
Leather	1	3

Note: responses totalling over 100 percent due to multiple responses (N=33)

20. It is clearly acknowledged among the entrepreneurs that large stock does not necessarily ensure higher sales and thus has little impact on earnings.
21. Elsewhere in this discussion it is shown that this was previously the case but has now changed and has become one of the major problems faced by these enterprises.
22. Tin cans used to make paraffin lamps obtained from fuel stations are purchased from waste pickers

The individual quantities of materials used were not obtained but the reported total monthly quantities ranged from 20 –800 kg, averaging 242 kg per enterprise per month. The table below indicates that enterprises consume relatively low quantities, perhaps also indicating low production levels.

Table 8.5. Total monthly consumption of waste material by small-scale enterprises

Quantity in Kg.	*No. enterprises using*	*Percent*
< 1	1	3
1-100	11	33
101-200	6	18
201- 300	2	6
301 – 400	5	15
–500	2	6
>500	3	9
Not known	3	9
Total	33	99

Sources and prices of waste materials

Waste materials (factory rejects like metal sheets, metal slugs) produced in metal foundries at the *Industrial Area* are sold cheaply to 'workers[23]' and in turn sold to the small-scale enterprises. This system evolved with the small-scale-recycling sub-sector. Previously, these factories sold metal waste directly to artisans at *Kamukunji* but as demand increased, stimulated by the expansion of the *jua kali* activities, 'new' intermediary positions emerged. Large-scale recycling units established scrap metal dealerships around the *Kamukunji* area. Two of these, *KIOI*[24] and *ROMA,* established in the early 1990s, are now the largest scrap metal dealers in the area. They obtain metal waste in bulk from large-scale recycling units for sale to small-scale metal recyclers in the city and from waste pickers at cheap prices for sale to LSRCs. Other less important sources include building sites, car garages, factories and dumps. The scrap metal dealerships in the areas surrounding *Kamukunji* [25] are now owned by Asian businessmen closely associated with the large-scale metal foundries in the *Industrial Area.*

23. These, as we later found out, are actually Asian scrap metal dealers in close business association with the large-scale recycling units.
24. '*Kioi'* in Kikuyu means collector or gatherer. The name is in this instance used by an Asian businessman essentially to camouflage the underhand business relations and operations with the large-scale metal foundries in the city.
25. *Industrial Area* and the nearby *Burma market* and *City Stadium* and even those further away in *Kariobangi*

The larger Asian waste traders are replacing smaller dealers whose main source of waste materials was the municipal waste stream. This process has implications on the socio-economic viability of the small-scale recycling enterprises at *Kamukunji*, particularly with regard to the production costs. *Kamukunji* entrepreneurs indicate that buying waste materials is one of the major problems in SS metal recycling, second only to the lack of markets owing to the intricate dealer-dominated waste materials market arrangements at the site.

Use of virgin materials

The virgin materials used comprised of metal sheets, galvanized iron, tin plates and aluminium. These are more expensive and only small quantities are used. 17 (52 percent) of the enterprises hardly use any virgin materials. The quantities ranged from 0-300 kg per month and averaged 82 kg. The table below provides information of virgin material usage by the 17 enterprises.

Table 8.7. Types of virgin materials and number of enterprises using them

Type of material	*Number enterprises using*	*Percent*
Paints	11	65
Clay	2	12
Galvanised iron sheets	11	65
Tin plates	5	29
Steel	2	12
Aluminium	1	6
Black sheets	1	6

Table 8.8. Quantities of virgin materials use per month and number of enterprises using them

Quantity in Kg/p.m.	*No. enterprises using*	*Percent*
< 1	13	76
1-50	8	47
51-100	2	12
101-150	3	18
151 – 200	4	23
>200	3	18

Note: (N=17)

Source: Karanja, fieldwork[a]

a. The data indicate the total consumption of virgin materials per month. Disaggregation at enterprise level of consumption of specific material types proved difficult.

Technology in small-scale metal recycling
Sheet metal fabrication requires the least skills and resources. It is mainly small-scale in size and informal in operations and uses simple tools and equipment (Frijns *et al.* 1997). The enterprises studied mainly used simple technology to perform functions such as bending, riveting, straightening, folding, rolling, shaping, moulding etc. Other simple tools like hammers, chisels, scissors, pliers, axles, brushes, anvil, and saw are commonly used. Some of the enterprises own a bending/punching machine, riveting machine, roller and welding machine, drill and grinder. Services are also hired from neighbouring enterprises at a fee. Electricity for use with these machines is tapped from the nearby *Undugu Society* and sub-divided amongst a few enterprises that also share the bill. Shortage of power sometimes dictates that some production functions like heavy welding have to be performed during low power peaks. Few of the entrepreneurs have formal training for the work they are engaged in. Skills are mainly picked up through informal apprenticeships.

Products and markets
A great variety of household utility items are made at *Kamukunji*. They include sheet metal boxes, charcoal *jiko* (cooking stoves), bicycle carriers, farm implements like hoes and *jembes*, water basics like tanks, gutters, and watering cans, wheelbarrows, poultry-keeping essentials, cooking pots as well as griddles. Charcoal *jikos*, sheet metal boxes, buckets and cooking pots are the most common products, with 40 percent of the enterprises making *jikos*, boxes, drums[26], buckets and cooking pots.

Goods manufactured at *Kamukunji* are largely sold to individuals, retailers and wholesalers, small-scale enterprises in and outside Nairobi. Six enterprises (18.2 percent) also sell some of their products to Zaire, Rwanda, Burundi, Central Africa, Uganda, Tanzania and Germany. However, this is on a one-off basis and largely results from the piecemeal government sponsored regional integration in trade fares under COMESA[27] or such other regional trade bodies. These do not translate into regular customised orders Moreover, without access to institutional credit, it is unlikely that these enterprises have the capacity to service big orders even if they were available. As a result, products are mainly made for display and sale at the site, aiming at the immediate local market. Most entrepreneurs depend on 'chance' sales from the immediate local neighbourhood (93 percent of the entrepreneurs).

An outstanding feature at *Kamukunji* are the huge, dusty, neatly arranged stacks and rows of buckets, *jikos*, metal sheet boxes and wheelbarrows. Some of the items were produced 'almost a year ago'. Aggressive marketing by intermediaries for commission

26. Drum here refers to recycled oil or chemical containers that are cut into half or simply cleaned and sold for reuse as cooking pots for institutions like schools or as feeding troughs for livestock. They are also used for brewing *busaa* or *chang'aa (*local illicit drinks) or for water storage.

27. Common market for East and Southern African.

payments elicits varying views from the entrepreneurs, with some contending that it enhances sales and others that it lowers prices and discourages buyers The area generally has a criminalized image. Efforts to promote products outside the *Kamukunji* precincts are hampered by the bureaucracy and high costs of fulfilling government requirements for retail permits. Entrepreneurs lament that business in *Kamukunji* is now very low in comparison to 10-15 years ago. The economic decline in the country has also had a negative impact on these activities.

There does not seem to be significant seasonality in the production or marketing of goods, with the exception of sheet metal boxes, which peak in October and November and between December and February. These are popular with boarding school students because of their capacity to withstand long distance haulage. Incomes earned during such infrequent 'booms' reach Ksh. 10,000 (US$ 32) per month. At the lowest season, incomes fall to as low as Ksh. 1,600 (US$ 21) per month. *Kamukunji* entrepreneurs concede to sometimes selling products below cost out of desperation ('at least to continue producing').

Although regular sales are desired, there is hardly any production re-organisation in terms of specialisation and/or innovation. Entrepreneurs continue to produce even when sales are very low. Simplicity, cost and availability of tools or machines rather than product marketability seem to determine product choice. Despite the apparent stiff competition for an unstable and shrinking market, entrepreneurs avoid risky investments or experiments for fear of losing what they consider a 'steady income', even if low. Large stocks are kept for the simple reason that 'one never knows when a big buyer may come by'.

Prices

Prices depend on the production costs and sale outlets available for the product. They also depend on whether the item is marketed by the entrepreneur or by a broker. A small *jiko* costs Ksh. 120 (US$ 1.6) at *Kamukunji,* but may cost between Ksh. 150-200 (US$ 2-3) at the supermarket. A medium sized one may cost Ksh. 180 (US$ 2) locally and Ksh. 300 at the supermarket. However, wheelbarrows are rather contentiously priced, selling for as much as Ksh. 2,200 (US$ 30) at *Kamukunji* but much less (Ksh. 1,700; US$ 22) in some retail outlets.

Earnings

About 40 percent of the entrepreneurs in our survey reported an average monthly income of Ksh. 4,200 (US$ 56) and a working capital of Ksh. 6,300 (US$ 84) per month. Due to poor record-keeping as well as incomplete cost calculations characteristic of *Jua kali* enterprises, it was difficult to confirm these data.

Kinship and labour ties in small-scale metal recycling

Social and ethnic relations in *Kamukunji* are strong and are the main channels through which training is obtained. An entrepreneur may give free apprenticeships to his relatives, only getting involved in production at the management level. Owners also rent out their workshop and tools to relatives and friends. Someone working for piece wages or a casual employee can under certain conditions produce his own goods. When he makes 'substantial' savings, he may buy his own materials and use a colleague's (relative or friend) workshop to make and market his own goods. However, for subsistence such metal workers largely rely on irregular and unreliable earnings from casual employment or piecework in the workshops. Employees are hired when needed and are paid a daily wage of about Ksh. 120 – 200 (US$ 2 – 3).

According to Kinyanjui (1996), due to diversification, poor bookkeeping and the use of family labour and credit, SSEs are often seen as lacking in entrepreneurship and growth. These factors, however, are often those that allow people to enter, conduct business and survive in an otherwise hostile business environment. The use of (un)paid family labour helps keep production costs down and improves profit margins, as the full costs of production are not taken into account. The work translates into a form of apprenticeship or training for the worker, whose labour is given at no immediate cost.

There are two angles to diversification, namely the involvement of the entrepreneur in other income generating activities and the investment of high proportions of the earnings in areas other than the enterprise. According to McCormick (1991), rather than invest in the city, small-scale entrepreneurs avoid risk by investing in land, rental property or children's education. 'When business is good', entrepreneurs at *Kamukunji* prefer to invest in land and other assets rather than in the business, because these are considered more stable sources of security in old age and in the event of collapse of business ventures. 60 percent of the small-scale metal entrepreneurs in this study reported that they had purchased a piece of land and/or built a permanent house in their rural homes. Although the land purchased is very small (averaging 0.5 acres and sometimes consisting of group-owned, unsurveyed plots in government sponsored settlement schemes), this is considered a significant achievement. Whether it is precipitated by the fear of collapse of businesses given the deteriorating business environment, or simply a cultural or traditional norm or pre-occupation of entrepreneurs, is not clear. Respondents in our surveys seemed pre-occupied with the 'imminent collapse' of business, due to the poor economic and political conditions prevailing.

In many parts of Africa, residence in the city for rural-urban migrants is mainly to earn money. It is believed that cities have more and better opportunities for this. The hope and expectation to return 'home' eventually is nurtured through the maintenance of strong ties[28] (Huntington 1977, Oucho 1986, Gugler 1997). Even those born in the city and committed to an urban life know of another 'real home' and the possibility of

28. This includes, remittances, visits to and fro as well as involvement and contributions to family, clan and community activities both in town and 'at home'.

returning there. Thus, despite the relatively lower number of immediate returns, buying land and/or building a house 'at home' is a formalised means of claiming their position in the rural community. Social allegiance through family, clan or ethnic obligations is thus demanding on income. The stated levels of income or profits may therefore not be the most useful means of determining the socio-economic feasibility of small-scale waste metal recycling.

8.10. CONTRIBUTIONS TO SUSTAINABLE DEVELOPMENT

Changing relations between actors in recovery and recycling

Although traders are friendly with waste pickers, they are at the same time authoritative and stern, especially during sale transactions. Most maintain meticulous records of the waste amounts bought and money owed to the waste pickers. There seems to be a high degree of trust in their transactions. Other relations include the provision of interest free loans to pickers which are repaid with waste or by performing tasks for the dealer. Whether or not this phenomenon entails 'tying' or 'bonding' of the picker to the dealer is not clear but it is apparent that some sort of over extraction may result. The agreements through which indebted pickers repay either through waste materials or labour (performing such tasks as cleaning, sorting, packing or transportation of materials for sale) seem to be capriciously defined.

Dealers maintain strong connections with other actors in the commodity chain. The need to keep themselves abreast with information on the purchase and sale prices as well as the demand for various waste materials is underscored. This information is used to determine sale and purchase prices of various materials and may be selectively passed on to the pickers who in turn adjust their picking preferences accordingly. Cash tips and other inducements have sometimes to be given to factory employees in charge of sourcing raw materials, in return for which the regular purchase of high volumes of materials from the particular dealer is enabled. The extent to which this can ensure the dealer's survival however has been constrained by decreases in the consumption of local waste materials by large-scale waste reprocessors. There are also too many dealers who are all similarly skilled and eager to buy these rights. This means that the amounts required for the numerous trade transactions are rising and cannot be justified by the profits earned. Moreover, factories now prefer clean, sorted materials unlike earlier when factories undertook the cleaning and sorting processes themselves. A lot of dealers have now found it necessary to undertake these in order to make their deliveries much more attractive to the factories. This has further increased operational costs for the dealers.

The relations between small-scale metal recyclers and dealers seem to have changed markedly causing distortions in the commodity chains to the advantage of large-scale reprocessors These changes seem to have been orchestrated by the latter in an effort to enhance competitiveness in the face of a harsh business environment in turn resulting from the harsh economic conditions prevailing in the country. Both the small-scale recyclers as well as the in-house waste recovery arrangements at the *Kamukunji* grounds have been affected negatively.[29] The latter's production costs have increased while the in-house women's group waste collection group's main waste materials outlet is now unreliable. These factors have weakened the economic viability of the lower-income end actors in the commodity chain. Co-ordination at this end of the chain is lacking or at best informal.

The NCC has no policy towards informal waste picking and its role in the waste materials recycling industry. JICA 1998 surmised that waste picking (or scavenging as they called it)[30] was an economic 'good' that needed to be managed and controlled in a 'socially acceptable' manner. Their survey of *Dandora* waste pickers found 'evidence of disease, poverty, malnutrition and hunger' and a meagre income of Ksh. 120 (US$ 2) per day per picker. It appears that waste picking is a response to poverty and also to the failure of the NCC to provide effective waste collection services. However, in addition to income earning (cash and non-cash) waste picking also provides a *de facto* waste collection and disposal service.

NGOs have no role in waste picking and trade activities. These activities are typically perceived as private businesses without any role in community development or urban waste management. At the community level, accusations that the activities cause environmental and health problems, including the bad odour and vermin (flies, cockroaches, rodents) are common. Dealers have also been known to purchase items stolen by domestic workers from households causing suspicion by householders and sometimes disapproval on the location of waste buying shops or transactions within the residential estate. Because of their type of work, presentation and conditions of living, waste picker(s) are shunned in most residential areas in the city.

Socio-economic issues

Most of the actors in the commodity chain seem to earn only a subsistence income. However, both itinerant byers and dealers who are more entrepreneurial earn relatively higher incomes from their work. Whereas waste pickers abhor the involvement

---

29. This recovery of recyclable scrap metal waste is carried out by a women's group, the *Kamukunji Women's Jua Kali Group*. This group purchases most of the scrap metal produced by enterprises at the grounds for sale to larger scale factories. The group has been allocated an area within the grounds for sorting, puching and storage of the materials.
30. Their reproachful reference to it inadvertently buttressed the NCC's dismissive perception. This having been a government-initiated project is bound to influence policy. JICA only recognised the significance of waste picking in an argument that its eradication would possibly increase the level of crime in the city and not as an aspect of SWM.

of family in waste work, dealers and itinerant byers frequently enlist family labour in sorting, packing and cleaning. Materials collected by itinerant byers are sometimes taken home and family labour enlisted in these processes. Dealers use unpaid family labour for the transportation of waste and the manning of premises in exchange for pseudo-apprenticeship. Extended family members are also provided with accommodation and subsistence in exchange for involvement and labour contributions to the dealership. Although indicative of longer hours of involvement in work especially for itinerant byers, the use of unpaid family labour results in lower operational costs and thus higher incomes.

Despite the relatively higher entrepreneurial capacity, dealers do not have any advantaging skills in waste trading. However, they are longer-time residents in the city and have more experience in waste-related work. Although only 11 (30 percent) of those interviewed had been involved in waste picking or itinerant purchasing before engaging in waste trade, it is still a notable finding especially with respect to the formation of relevant trade relationships. An experienced dealer is better placed to handle the sometimes tranquillized, rowdy and hostile waste pickers and to broker the trade deals with large-scale waste recyclers Skills in marketing and in the formation of networks are necessary for the maximisation of earnings, business survival and stability. Dealers possess knowledge on the types and demand of differentiated waste materials, the available outlets, their requirements and methods of short-cutting some of the processes in order to gain some advantage.

Although most of the actors express their dissatisfaction with the activities they are engaged in, their capacity to move from one level to the next or to a more 'pleasant' activity seems to be low. There is no evidence of mobility from one level to another at this lower income- end of the commodity chain with the exception of the few noted before. Hopes or ambitions for formal sector employment are now even more illusionary given the on-going retrenchment drive[31]. Similarly, joining 'a better kind of business' requires the accumulation of capital, which most actors (with the exception of dealers) are presently incapable of. The need to 'change to other types of work' is articulated across the whole spectrum of activities but with knowledge and evidence of the absence of affordable, legal and socially accepted options. Minimal planning or strategizing towards this is observed, except amongst dealers.

For waste pickers operating at the lowest income-level, theirs is to fulfil 'a pressing demand of daily life', i.e. survival (Gutkind 1967: 37). In targeting more highly re-saleable waste itinerant byers fair slightly better in terms of income. Dealers are the focal point of this sub-group of activities. Theirs is entrepreneurial, notwithstanding their use of unpaid family labour and dependence on higher-level actors in the chain.

31. Formal private sector employment in Kenya has always been much smaller. Until recently public sector was the largest provider of formal employment.

Assessment of the socio-economic, public health and environmental dimensions of this requires recognition of the wider social, economic and political milieu in which the chain is based. Issues related to sustainable development with regard to these actors are constrained not only by the actors' positions in the chain and nature of activities but also by the wider social, economic and political environment in which these activities are embedded and, more importantly, by their non-inclusion in waste management policies.

Environmental and health issues

Waste pickers live and dwell in dilapidated surroundings without many of the basic facilities for basic hygiene. Their working conditions are equally deplorable, without use of any safety equipment or attire. The exposure to contaminants is highly prevalent. Most of the actors in this part of the chain complain of numerous health problems attributable in their judgement to involvement in waste work. An overwhelming 89 percent of the waste pickers reported frequent affliction from ailments they considered to be resulting from handling waste. The most frequently named included headaches and stomach and diarrhoea problems, cuts and bruises, respiratory and skin ailments, as well as pains and aches in the body. Although epidemiological studies have not been conducted to establish a relationship between waste work and disease, these seem inevitable given the nature and conditions of work and habitation. These diseases may result from walking long distances in unfavourable weather conditions, consuming waste food from dumps or coming into contact with corrosive and injurious materials as well as disease causing vectors, thriving in the waste.

Although waste pickers cause littering in the streets and surrounding areas, the removal of waste materials from the waste stream that would otherwise end up in disposal sites contributes to environmental benefits of waste management. Due to the absence of data it is not possible to quantify the amount of waste diverted from the municipal waste stream by the activities of these actors

Owing to the negative attitude, frivolity and disregard with which these activities are viewed by the authorities and the public, it is not possible to compute the volume of waste diverted from the municipal waste stream. Regarding the *Dandora* dump, not only is the total population living in and off the dump not known but also that of dump waste picker(s) is largely based on estimates by NGOs and CBOs involved. The Catholic Priest Father Alex of *Mukuru* recycling project and the NCC dumpsite office put the total population as varying from 500 to 2,000. Moreover, not all those found at the dump are engaged in waste picking. NCC officers working in the disposal-fee collection office, who interact fairly closely with pickers, acknowledge that the dump is a popular hide-out for many of the city's criminals.

Using a crude estimate of 2,000 pickers and average amounts separated per day as those interviewed provided, the average amount of plastics separated per day at the dump would be 9,000 kg (approximately 1 ton) while that of paper would be 4,000 kg (approximately ½ ton). Informal dumps are not only smaller in surface area but also harbour a much smaller and more scattered population of waste pickers The volumes separated are as such even more difficult to estimate.

In the small-scale enterprise sector, the myriad of machinery and equipment used makes it difficult to design standard devices for use in all the sub-sectors There are no laws, regulations or even advice requiring the use of protective devices. Artisans are left to discern for themselves what is dangerous or what is not and what measures to institute.

Kamukunji is in common day language in Nairobi known as '*clang clang*' in reference to the constant loud sharp sounds resonating from the persistent hammering of metallic objects into desired shapes and sizes. The artisans are also tightly packed in the muddy/dusty[32] uneven space available, which is also used as the storage and display area. Each metal worker occupies an area of about 6.2 m^2 on which to perform all the necessary activities. The artisans do not use any protective gear like earmuffs, gloves or eye protection to avoid 'welding eyes' and have to purchase medical care privately in times of illness. Some accidents have occurred for instance during the cleaning of drums previously used for storage of 'poison' (toxic chemicals) resulting in 'deep hard- to- heal' chemical burns. The only recourse for meeting the usually high medical bills in such instances is 'harambee', where metal workers pool together the funds required, to access treatment. The jua kali recycling sector is therefore a 'work at your own risk' trade.

This notwithstanding, health issues are not a big concern according to the KJKA secretary, as the artisans 'are used' to the noise and all other harmful occupational conditions under which they work. What was of concern and thus recently resolved were the environmental and poor sanitary conditions especially in relation to human conveniences in the grounds. The association recently petitioned the NCC for a more constant and regular supply of water and also privatised the only toilet in the area. They also requested the government to confirm their occupation and use of the area. Approval was received through a title deed issued to KJKA

For large-scale waste reprocessing companies, it is required by industrial law in Kenya that factory workers, especially those handling harmful or dangerous equipment, be provided with the necessary protective attire, and that other safety measures are put in place. These are stipulated under the in the Ministry of Labour's labour standards, health and safety regulations, which are implemented by the Inspectorate department. However, they have not been adequately enforced. In large-scale recycling units, job

32. Depending on season.

insecurity especially in this era of mass retrenchment hinders any sort of petition or reporting to the ministry of any violations of the regulations because of fear of losing employment. The large reservoir of labour available makes it very easy to replace 'errant' workers. Many of the workers in the factories are sufficiently aware of the health risks but have little alternative for employment. The rare monitory visits by the Ministry of Labour are primarily used to obtain 'bribes' from owners of the large-scale recycling units. Although managers in the large-scale recycling units claimed to be adhering to the government requirements, fieldwork visits to some of the factories proved otherwise. Workers could be observed going about work without basic protective gear like masks and gloves.

8.11. CONCLUSION

The arrangements for resource recovery and recycling discussed in this chapter have potential to contribute to SWM and sustainable development but some serious contradictions continue to impede its growth and development. It remains severely neglected in government regulation. Government policy both at national and local levels does not consider the modality's significance to SWM and sustainable development. There has been official recognition of these activities especially at the lower-income levels, while some of the actors higher up the commodity chain have propitiously gained from recent government policy in a manner threatening to the socio-economic viability of lower level. For environmental as well as socio-economic purposes, there is need to stimulate and facilitate the recovery and recycling of local waste materials through incentives.

It is contended on the one hand that failure on the part of the government to recognise and support these activities and the partnerships therein precludes the contribution of this modality to SWM and the opportunity to design sustainable alternatives[33] or solutions to the problem. These activities have the potential to increase the quantities of inorganic waste materials diverted from the municipal waste stream especially through recovery and recycling opportunities. It can also generate new categories of work and thus produce or increase the employment and income earning opportunities of the city dwellers (Lardinois and Furedy, 1999).

33. Sustainable alternatives here as outlined by Schubeler (1996) refer to strategies that aim at waste minimisation, increased waste recovery and reuse and the safe and environmentally sound disposal of waste. These activities in addition increase opportunities for employment or to earn income.

PART III: REUSE OF URBAN ORGANIC SOLID WASTE

CHRISTINE FUREDY

CHAPTER 9
URBAN ORGANIC SOLID WASTE: REUSE PRACTICES AND ISSUES FOR SOLID WASTE MANAGEMENT IN DEVELOPING COUNTRIES

9.1. INTRODUCTION

This chapter explains the interest in urban organic solid wastes, the relevance of these wastes to municipal solid waste management, the main ways in which organics are generated and reused, and the issues that arise from the wish to ensure safe and effective reuse as part of sustainable development in cities in Africa and Asia.

Urban administrations nowadays are seeking ways to divert organic wastes from municipal solid waste streams for a variety of reasons, as noted below. Recommendations are made for separation at source so that safe composting can be carried out. Private companies are being encouraged to undertake composting, often via forms of public-private partnerships. More attention is being paid to the role of non-governmental organisations in promoting citizen awareness of organic waste issues, and co-operation with separation at source. This project draws attention to the many informal ways that organic wastes are currently reused, which are rarely taken into account in official plans for managing organic wastes.

The nature and uses of urban organic solid waste[1] have been discussed and researched from a number of disciplinary and policy perspectives. *Food, Fuel and Fertilizer from Organic Wastes* (NRC, 1981), a pioneering book with a technological orientation, reported on research in the context of the late 1970's concern about limited world resources. The World Bank Water Supply and Sanitation section set up the Integrated Resource Recovery Programme in the early 1980s. Environmentalists' interest in urban waste recycling developed at the same time. An interest in low-cost techniques prompted the documentation of organic waste reuse by WASTE, a Dutch consulting foundation (Lardinois and van der Klundert, 1994). Aquaculture research and projects incorporating human excreta and solid wastes became established (see, e.g., Edwards, 1992). Epidemiologists and water and sanitation experts initiated health risk studies

1. The focus here is on what is generally known as 'garbage' or 'refuse,' and not on human and animal excreta, although municipal solid wastes contain varying amounts of both.

I. Baud et al. (eds.), Solid Waste Management and Recycling, 197-211.

(see Blumenthal *et al.*, 1989, 2001; Khouri *et al.*, 1994). This approach has not yet been applied, however, to the reuse of the organic wastes from municipal refuse.

In the late 1980s, a 'food security' perspective on development emerged. The International Institute for Environment and Development was the first group to emphasise recycling and reuse as a basic principle for sustainable development, and this theme, integrated into the Brundtland Report (World Commission on Environment and Development, 1987), is particularly emphasised in relation to food production (Mitlin and Satterthwaite, 1994; van der Bliek, 1992; Koc, McRae, Mougeot and Welsh, 1999; Hardoy *et al.*, 2001).

Geographers have been interested in urban agriculture for several decades (Yeung, 1985). This field is now interdisciplinary and has broad bilateral and international agency support, as evidenced by the establishment of the Cities Feeding People programme at the International Development Research Centre (Canada) and the collaborative Centre for Research on Urban Agriculture and Forestry (Netherlands and Canada), as well as networks such as City Farmer and The Urban Agriculture Network (Mougeot, 1999).

While early writing about urban agriculture made little reference to the use of urban organic wastes as inputs, in the past ten years, more scholars have recognised links between urban solid waste management and the reuse of organic wastes in urban and periurban agriculture (Allison and Harriss, 1996). Two aspects are stressed: that urban organic wastes largely derive from resources removed from near-urban and rural areas which could be returned to the production cycle, and that reuse can benefit urban waste management by reducing some disposal costs and curtailing toxic, infectious and unaesthetic residues (van der Bliek, 1992; Furedy, 1995; Smit *et al.*, 1996).

From the mid 1990s there have been initiatives aimed at bringing together the interest in food production and urban solid waste management. This was one aim of the Urban Waste Expertise Programme funded by the Dutch government, which has supported a number of projects in composting (see 't Hart and Pluijmers, 1996; van der Klundert *et al.*, 2001). The International Water Management Institute of the UN Food and Agriculture Organisation is supporting research on waste composting for urban and peri-urban agriculture (Drechsel and Kunze, 2001). Recently the British DFID program on natural resource systems has integrated the perspective of peri-urban stakeholders in Hubli-Dharwad, India and Kumasi, Ghana (Nunan, 2000; Brook and Davila, 2000). Nutrient flow analysis has emerged as a sub-topic in urban and peri-urban resource systems (Belevi *et al.*, 2000; Drechsel and Kunze, 2001), and waste-to-energy in the urban context is again being explored (Grover *et al*, 2002; Dunnet, 1998).

It is usual now to find that policy recommendations for municipal solid waste management make reference to the need to address the treatment and reuse of organics wastes specifically (Rosenberg and Furedy, 1996; India, 2000).

In short, a base is being laid for a broad framework for understanding urban organic solid waste in developing countries. The current emphasis in municipal solid waste management is upon public-private partnerships (the co-operation of governments, companies, NGOs and international agencies) (see Baud *et al.*, 2001, and the concluding chapter of this book). This approach is being applied to organic wastes with reference to composting.

Nevertheless, this attention to urban organic solid waste has not, so far, produced systematic or adequate research on the nature and quantities of organic wastes in cities of developing countries. The positive and problematic aspects of their exploitation in any one city or region have not been explored and the implications of reuse for solid waste management are not well understood. Every piece of research that directly addresses the reuse of urban organic wastes and its system-wide implications helps to lay the foundation for the needed comprehensive approach. This is the importance of the project work reported on in this section.

9.2. URBAN ORGANIC SOLID WASTES AND MUNICIPAL SOLID WASTE MANAGEMENT

Urban organic solid wastes include not only the organic material in municipal waste streams, but wastes generated by gardening, urban agriculture, park and road maintenance, livestock keeping, food processing, tanning, and the like. (Although human excreta are also organic wastes, they are not usually covered in discussion of waste reuse in solid waste management and are not included in this discussion.) The generators can be classed as bulk generators of raw wastes (such as green markets, parks, stables, slaughterhouses), bulk generators of processed wastes (such as food processing industries, large hotel/institutional kitchens), and small generators of raw and processed wastes (such as households). Most of the organics in waste streams are generated by kitchens in the course of daily living.

It is the organics that are put out for general collection and so are mixed in the solid wastes that most concern municipal managers. Interest in controlling the organic fraction of waste streams (which typically comprises from 35-70 percent of total municipal waste generated in large cities of developing countries) has a long history. Composting and reuse techniques (including use for animal feed, fuel and construction) have been documented in Africa and Asia, going back hundreds of years The interest in urban organic solid waste has become more general, however, in the context of environmental thinking about waste reduction, strategic planning for solid waste management (see Rosenberg and Furedy, 1996), and greenhouse gas emissions. In addition, those interested in aiding small farmers and livestock keepers view urban organics as recoverable resources.

The changing nature of solid waste in cities of developing countries complicates the already difficult task of dealing with such organic wastes. Although there is little understanding of precisely how the composition of solid waste is changing, due to lack of baseline data and reliable research (see Furedy, 1998) the general trend noted is: the great increase of plastic film (small plastic bags), pieces of hard plastic, broken glass, medical wastes, and industrial residues from unregulated industries. The non-biodegradable and toxic elements present problems for the reuse of urban organics, especially for practices of conveying city garbage to peri-urban farms or of farming on old dumpsites.

Organic wastes represent a great challenge for solid waste management because of their decomposability, their seasonal variation in nature and quantity, their admixture with non-biodegradable wastes, and the practical difficulties of marketing compost products.

In addition to the complex nature of these wastes, management must take into account the varied 'actors' who generate, handle, and use the wastes and their products. The most numerous generators are household residents but hotels, restaurants, food stalls, shops, markets, butchers, food processing plants and some industries also contribute, along with parks and roads departments—a mix of official, formal and informal actors. Organic wastes are handled by generators, official collection crews, dump workers, and those who transport organics to end users, principally near-urban farmers and livestock keepers.

9.3. CUSTOMARY AND INFORMAL PRACTICES IN RECOVERY AND REUSE OF ORGANICS

Throughout the developing world, organic wastes are generally in high demand for animal feed, fuel, fertilizer, as soil amenders and even in construction. Any large city and its hinterland support many practices by which significant amounts of the organics generated are reused. The figure below depicts some of the main reuse practices, and the table provides further details.

Type and origin of wastes, or site	Materials included	Reuse and diversion practices
Residential kitchen and yard wastes	Kitchen wastes with some garden trimmings, leaves, grass cuttings	Backyard composting for home gardening; domestic animal feeding; neighbourhood composting and vermicomposting
Restaurant and canteen food wastes	Raw peelings and stems, rotten fruits and vegetables and leftover cooked foods, bones	Sold to poultry, pig, goat farms; bones to feed and fertilizer factories
Shop/ institutional wastes	Organic and food waste. May contain other wastes	Households, shops, institutions may separate organics for community composting
Market wastes	Organic wastes of fruit and vegetable markets	Sold or given to farmers; collected for centralized compost plants
Processed industrial wastes (canning industries, breweries, etc.)	Food wastes, bagasse, organic residues	Sold to fertilizer companies; sold for domestic fuel
Parks and road verges	Grass clippings, branches, leaves	Composted by Parks departments; scavenged for fuel and construction; used as animal fodder
Mixed municipal solid wastes	Full range of local solid wastes, incl. small industries' wastes, biomedical wastes, human and animal excreta	Diverted to peri-urban farms, 'central' compost plants, neighbourhood compost schemes
Organic material 'mined' from garbage dumps	Decomposed mixed municipal wastes with non-biodegradable residues	Nearby farmers collect waste from current and old dumps; municipality may auction waste pits; waste may be sieved at the site
Old garbage dumps	Decomposed garbage	Old dumps are often cultivated in Asia, and to some extent in Africa
Animal excreta	Cattle, poultry, pig dung from urban and peri-urban farms and stables	Used for fuel, for construction, for fertilizer
Slaughterhouse, butcher, tannery wastes	Bones, skin, intestines, horns, scrapings, etc.	Sold to fertilizer and feed factories: rendered, composted; sometimes applied to farm fields with minimal processing
Racecourses	Horse dung	Sold for mushroom growing, horti-culture

Table 9.1. Main practices of urban organic waste reuse in developing countries

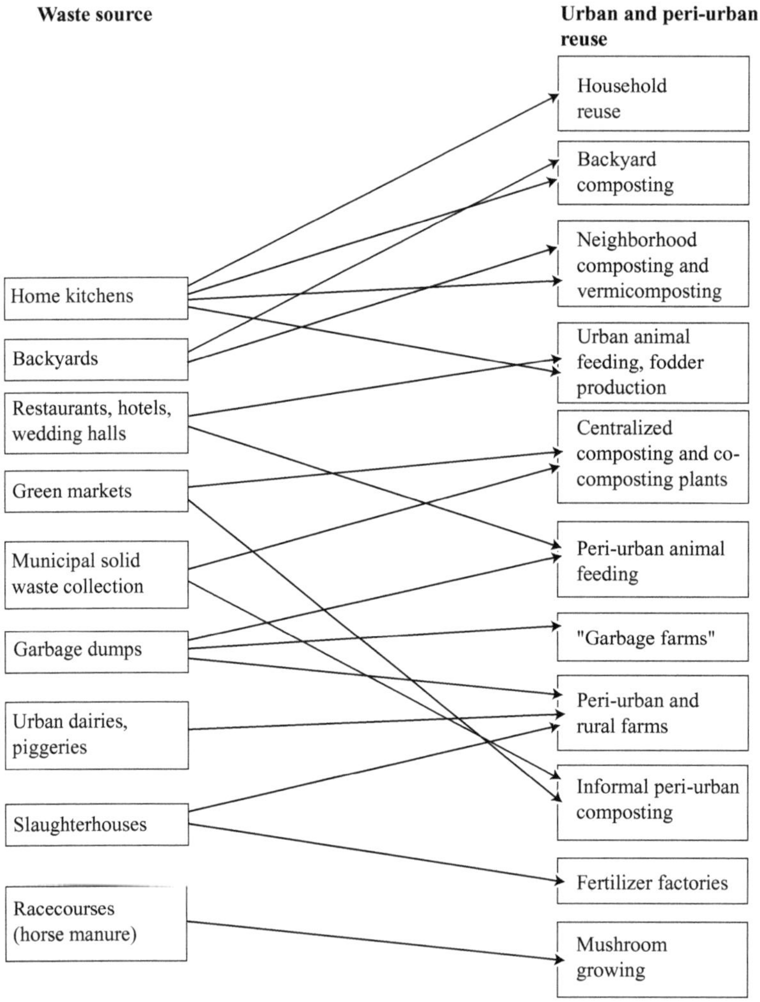

Figure 9.1. Reuse of urban organic solid wastes in developing countries
Source: adapted from Furedy, Maclaren and Whitney 1997

Most, but not all, recovery and reuse is 'informal' and long-standing, or 'customary.' (Fertilizer and feed production are carried out by the corporate sector as well). The quantities of wastes involved and the costs and benefits of reuse practices cannot be estimated at the present level of knowledge. The informal and undocumented nature of most practices poses problems for ensuring safe and sustainable waste reuse.

One type of reuse that has received particular attention in the Hyderabad research of this project is farm reuse of organics. The main practices associated with farm reuse are:

- the cultivation of dumpsites (closed or operating) (Furedy and Ghosh, 1984; Drescher, 1994; Furedy, 2002);
- the 'mining' of dumpsites for organic matter (see Rosenberg and Furedy, 1996: 72; Scheu and Bhattacharya, 1995; Nunan, 2000; Brook and Davila, 2000; Djabatey, 1996; Asomani-Boateng, 1999; Harris *et al.*, 2001; Kiango and Amend, 2001);
- the transportation of mixed solid waste to farms, where inorganics are largely removed before the waste is applied to soils (Lewcock, 1995; WASTE, 1998; IBSRAM, 2000; Nunan, 2001; Dulac, 2001);
- on-farm co-composting of urban animal and farm wastes (which also applies to 'backyard farms') (Birley and Lock,1999:139; Nunan, 2001).

In many cities of India, farmers have access to municipal dumps to collect decomposed material. In some cases the municipality auctions pits at dumps (Nunan, 2000). Municipal solid waste may be transported to peri-urban farms under private contracts and agreements with municipal collection crews (who deliver to the farms instead of disposing at the garbage dump). In addition, farmers delivering produce to green markets may carry back market wastes in their empty vehicles. The cultivation of old dumpsites by small farmers has been observed around Accra, Ghana (Drescher, 1994; Asomani-Boateng, 1999; Osborn, 2000).

There is evidence for a steady decline in farm reuse near large cities in Asia, mainly due to contamination of the waste with inorganics and high transportation costs (Nunan, 2001; Brook and Davila, 2000). Around West African cities, where chemical and other commercial fertilizers are scarce and expensive, on the other hand, farm reuse appears to be increasing (e.g. in Kano, Nigeria, cf. Lewcock, 1994).

The use of food wastes and green wastes for animal feed and fodder is a widespread practice. This includes food and peelings fed to household animals, as well as food and canning wastes used in commercial animal husbandry (e.g. piggeries, goat farms). In Khartoum, it has been estimated that goats consume about 30 percent of urban solid waste (Richardson and Whitney, 1995).

Sieved, decomposed municipal waste may be used as a cover material at dumpsites (Scheu and Battacharyya, 1995). Urban and peri-urban cattle dung and crop residues are widely used as fuel in parts of Asia.

To the extent that reuse is of materials that do not enter the municipal waste management system, these practices reduce the burden of waste collection for a city. Furthermore, less disposal space is needed when wastes are diverted, or removed, from dumps. Even with high levels of informal reuse, however, substantial amounts of organics remain in the municipal waste flow, material that solid waste managers wish to treat and to divert from disposal.

9.4. ISSUES IN COMPOSTING OF URBAN ORGANICS

Although organic wastes can be exploited for energy – formally, by anaerobic digestion to produce gases, pelletisation and briquetting (Furedy and Doig, 2002), and informally for domestic fuel – the use of decomposed biomass in food and plant production (in agriculture, horticulture, forestry and aquaculture) remains the means by which the greatest amount of urban organic wastes are reused. Composting (controlled decomposition of organics by numerous micro-organisms) is the preferred method of processing. The experience with composting of urban organics, however, has been fraught with multiple problems: of feedstock; of plant operation; of the quality and price of the product; of marketing; of consumer understanding; and of institutional support.

Motivations for composting

From the point of view of solid waste management composting is a means of both treating and reducing the amount of waste requiring final disposal. Backyard composters are concerned with recovering resources, but most households cannot reuse their organics and must put them out for collection. They thus have no incentive for keeping organics separate from other wastes. Pilot projects in composting often try to integrate general environmental, health and specific social concerns (Lardinois and Furedy, 1999). Private fertilizer companies may undertake composting from urban wastes to enhance their 'green' image rather than to make a profit on that line (Patel, 2000). A major issue for solid waste managers is what will motivate household and institutional waste generators to undertake and keep up the difficult task of carefully separating wet wastes. The long run success of urban organic solid waste composting would seem to depend on such separation.

City experiences with mechanized compost plants

The failures, inadequacies, and high cost of large, mechanized compost plants have been noted in many countries (see Rosenberg and Furedy, 1996:124, 317; Lardinois and Marchand, 2000; Etuah-Jackson *et al.*, 2001). As well as problems with inappropriate machinery, lack of expertise, high operating costs, difficulties in developing markets and so on, such plants have produced poor quality compost because they usually used mixed municipal wastes as feedstock. The problem of contamination is increasing as municipal wastes contain more glass, biomedical wastes, and plastics. City governments find it hard to acquire convenient sites for large plants. A few large-to-medium plants are still operating with simpler technology, and with government subsidies, e.g. the autonomous Karnataka Compost Development Corporation plant in Bangalore (Lardinois and Furedy, 1999; Lardinois and Marchand, 2000) and the Accra municipal plant (Asomani-Boateng and Haight, 1999). These problems have led to consideration of privatisation and small-scale units.

Private enterprise in composting urban organic solid waste

The interest of private companies in composting is reviving in South and South-east Asia. This has been encouraged by the move for public-private partnerships in solid waste management (cf. chapter 5). The units are often set up by agro-chemical companies and may receive assistance from municipalities in the form of access to free municipal wastes, and a rent-free site, as is the case with the Excel Industries plant in Bombay and some of the company's franchises (Chakraborty, 2000).

Private companies that invest in research and development are in a better position to make a profit than any other operation. An example is Terra Firma Bio-Technologies that is doing vermicomposting in Bangalore (Lardinois and Marchand, 2000).

Neighbourhood-scale composting

International agencies, bilateral aid programmes and NGOs are promoting neighbourhood-scale compost plants, which draw their feedstock from area residences, and are usually run by NGOs or CBOs, with assistance (such as access to land) from municipal councils. There have been numerous pilot projects of this kind in the past decade. For instance, UNCHS has supported small-scale composting, particularly in Nairobi (Harrison & Paumard, 2000); the World Bank continues to encourage composting of solid waste in Indonesia (Mockler, 1998; Hoornweg *et al.*, 1999; Zurbrugg and Aristanti, 1999); the Dutch government via the foundation WASTE has supported projects and research in India, the Philippines and Mali (WASTE 1996-; van der Klundert *et al.*, 2001).

The most recent assessments of experiments in small-scale composting suggest that it is of limited relevance for the solid waste reduction of larger cities. Lardinois and Marchand analysed the financial and technical feasibility of three such plants (in Bangalore, Kathmandu, and Manila) and concluded small-scale plants are financially unfeasible and usually lack the necessary technical knowledge; they produce little compost and, consequently, the selling price is very high. For instance, the cost of compost produced by the Centre for Environmental Education pilot plant in Bangalore was $US 1,514 per ton, with a hidden cost of another $US 724 (Lardinois and Marchand, 2000). The problem of small plants being located far from agriculturalists is noted for African examples (Birley and Lock, 1999). These plants can, however, play a role in environmental education and employment of disadvantaged persons.

Vermicomposting

Compost composed largely of worm castings from worms fed organic wastes has a greater fertilizer value than aerobic compost from organics and can usually be sold at a higher price. A number of pilot projects have been undertaken with solid wastes (Spiaggi *et al.*, 2000; Yasmeen, 2001), and private fertilizer companies have adopted this technique (for example, Terra Firma Bio-Technologies in Bangalore). The NGO undertakings, however, are too small to be significant in city waste treatment, and there are unresolved health issues (see section 9.5 below). It seems likely that, leaving aside its incorporation in some commercial operations, vermiculture using urban organics will be confined to demonstration projects to serve mainly educational purposes or to small undertakings where there are niche markets with customers willing to pay a higher price than that of compost. For instance, squatter families lacking solid waste collection might persist with vermicomposting after pilot work, if they can also sell the worms for fish bait or poultry feed (cf. Spiaggi *et al.*, 2000). The pilot project in Hyderabad is an example of a subsidised project that has yet to develop a secure market for its product.

Waste generator co-operation with organic waste management

To produce compost acceptable in quality and price to buyers, on the scale that can significantly reduce urban organic solid waste, would seem to require large-scale 'separation at source'. That is, the organic wastes are kept separate or segregated by waste generators and are separately collected for processing. Source-separated organics originate from two main sources in a city: residential consumers (mainly households) and bulk or single-source generators (food processing plants, wholesale market terminals, green markets, large hotels, large restaurants, large institutions, parks). In many cities of developing countries, the available separated organic wastes are in high demand and are extensively exploited through informal networks (Furedy, 1995). To date, cities have had limited success in gaining residential waste generator co-operation in separating organics that have hitherto been put out for collection mixed with other wastes. Reports on pilot schemes have highlighted many barriers to compliance, although householders may express willingness to separate when initially questioned (Lardinois and Furedy, 1999; Asomani-Boateng, 1999; Pitot, 2001; Le, 1995). It is important to note that in addition to waste-generator cooperation, the separate collection of residential organics on a large-scale requires radical changes in existing solid waste management systems and the unit costs are very high (Lardinois and Furedy, 1999: 187,195). An approach that does not demand complete waste generator compliance with separation is doorstep sorting by waste collectors. This can be implemented where collectors receive wastes from door-to-door and immediately sort organics from inorganics. It is being applied in projects for small-scale neighbourhood composting (Dulac, 2001).

'Compost credits': International agency support for composting

The Global Environmental Facility (GEF) has become a source of international funds to subsidise compost making from urban organic wastes. The GEF offers funds for processes that reduce greenhouse gases (GHGs). While research has not established the extent to which anaerobic composting reduces GHGs, GEF has accepted the assumption that composting is preferable to sanitary landfilling or open dumping. Regions and cities have begun applying for funds under the rubric of 'compost credits'. The West Java and Jakarta Environmental Management Program has secured a large GEF grant to support a community-based organic waste compost scheme (Hoornweg, 2000) and India is preparing an application to GEF.

Persisting issues

It should not be assumed that composting is the best way to reuse all urban organic wastes. As Harriss and his colleagues point out, the 'purer' wastes (i.e. not mixed with other garbage) will have a higher value as animal feed (Harriss *et al.*, 2001). However, with respect to the organic wastes generated in urban residences and institutions (the wastes that comprise the bulk of the responsibility of solid waste managers), composting remains the process by which a large quantity could be processed and diverted from disposal. But composting of urban organic solid waste is not financially sustainable under the current methods of cost accounting by solid waste management departments in developing countries. As long as most wastes are disposed of by low-cost open dumping, composting is an expensive option for cities and towns in these areas.

The composting units in Asia that are the most successful and that manage to recover their costs from their sales are private firms with a wide range of products, sold to a variety of customers (nationally and even abroad). Their compost, however, is out of the reach of small peri-urban farmers who greatly need compost and who would be willing to buy a low-priced waste-derived but pure product (see chapter 10; Nunan, 2000).

Neighbourhood-level composting projects can play a role, however, in public education about solid waste management. Good compliance with source separation or doorstep sorting is essential to these projects. The most progress in source separation has been in cities where the municipal government has strongly endorsed separation, and community groups and NGOs implement this or doorstep sorting. Support of regional and national governments for separation further strengthens waste generator co-operation (Lardinois and Furedy, 1999: 192).

Apart from the debates over what sort of composting is viable, there remains a broader question of understanding the current uses of urban organic wastes, who 'owns' the wastes, and how the competing claims to these resources should be articulated and resolved (Harriss *et al.*, 2001).

9.5. HEALTH ISSUES IN ORGANIC WASTE REUSE

Because the reuse of urban organic wastes is so varied and is largely conducted informally, the potential health implications are very complex.

Health risks may be associated with every aspect of reuse, including: transportation and handling of organic wastes; processing; the application of organics from mixed municipal wastes to soils; cultivation on old dumps; and feeding animals with wastes. Many actors are involved, and there are risks of human-animal disease links.

A thorough general overview of potential problems related to this topic can be found in the report *Health Impacts of Peri-urban Resource Development* (Birley and Lock, 1999) which covers organic waste reuse in urban and peri-urban agriculture. A summary of problems and possible control measures can be found in the March 2001 issue of *Urban Agriculture Magazine*. De Zeeuw and Lock classify the main problems as:

- survival of pathogenic organisms in residues
- zoonosis associated with animal wastes
- increase of disease vectors
- respiratory problems from dust and gases
- injuries from sharp fragments
- contamination of crops from heavy metal take-up and agrochemical residues via wastes and their leachates.

It is often asserted that composting of urban organic wastes eliminates or substantially reduces any health risks to farmers and consumers The reality is not so straightforward. Decomposition must take place at a temperature of at least 60° Centigrade to destroy most pathogens and the product should be matured for several weeks (Brunt *et al.*, 1985). A question mark hangs over vermicomposting, which takes place at lower temperatures. In Europe, anaerobic processing of organic wastes is required prior to vermicomposting, but this is not done in vermicomposting in developing countries (Furedy, 2001).

At the present time, few organic wastes are being safely composted in Africa and Asia. Far more organic wastes enter food production via 'garbage farming' (i.e. use of municipal solid wastes) on farms, cultivation of old dumps, and animal feeding than via composting, and these activities are rarely controlled. Hence, the numbers of persons at risk must be very large.

There has been, however, little specific research on these health risks. If testing is done of soils or compost, it is usually for the occurrence of heavy metals and even then, there is no agreement on appropriate standards for safe heavy metal bioaccumulation for the wide variety of crops grown (Furedy, 2001). An encouraging result of pathogenic tests is cited by Brook and Davila (2000). Tests were done of compost from urban wastes in the U.K. Natural Resources Systems Programme in Kumasi, Ghana and the levels of pathogens in the soil were not high and were not linked to the pattern of local diseases. It is likely that higher risks apply to workers who handle and process organic wastes, especially those coming from mixed municipal wastes than to consumers of crops grown on waste-fed soils.

Drawing upon the considerable work done on the use of human excreta and wastewaters in agriculture, international discussion has begun on the possible control measures for waste reuse in urban and periurban agriculture (for instance crop selection, worker education and protection, controlled disposal of hazardous and biomedical wastes) (FAO, 2000; de Zeeuw and Lock, 2001; Blumenthal *et al.*, 2001). The informality of most urban organic solid waste reuse in Africa and Asia is the major challenge to progress in reducing health risks (Furedy, 2001; Cointreau-Levine *et al.*, 1998). If cities moved to control the health risks from organic waste reuse, a substantial part of the current informal, and even official, reuse practices would be curtailed. As the editors of *Urban Agriculture Magazine* have emphasised, what is needed is a multi-level approach that seeks to balance health risks against the benefits of waste reuse (Lock and van Veenhuizen, 2001).

9.6. INSTITUTIONAL CONTEXT

The institutional context or 'framework' for the management of urban organic solid waste is usually seen as encompassing aspects such as who undertakes waste processing (the city, private companies, NGOs) and whether there are specific policies and regulations pertaining to the management of organic wastes. Brook and Davila go further to suggest the inclusion of factors affecting farmers' ability to buy compost or use recommended control measures. So, land tenure systems and networks for disseminating information to farmers and other actors should be examined (Brook and Davila, 2000: 71, 136).

A broader issue is that of who legally owns the wastes generated in a city and how competing claims for accessing wastes can be resolved. Most city governments will assert that they own these resources but none have rules for considering competing actual or potential uses. Whether a country or city has policies and programs that undermine socially responsible reuse might also be considered under this rubric. The preference given to impractical waste-to-energy schemes or pelletisation plants to the neglect of assistance for composting is a barrier in some countries (Furedy and Doig, 2002).

Except where there are municipal compost plants and slaughterhouses supplying fertilizer factories, there are few specific regulations for organic waste reuse in developing countries. Urban and peri-urban farmers are not seen as stakeholders In other words, the institutional context is not developed. This is particularly true for African countries.

The most important development in the institutional context to support sustainable reuse of organic wastes would be a move towards strategic planning in solid waste management, so that the management system would take account of negative and positive environmental externalities. If compost production was recognised as diverting wastes from disposal and improving soils, and some of the financial savings of composting and other safe uses of organic wastes were used to improve and promote these activities, then these forms of waste processing would become more feasible for cities (Dulac, 2001). Cities that are implementing door-to-door waste collection for residences have the option of integrating into their collection systems door-step sorting to recover organics, and, later, separation at source.

Two recent developments mentioned above suggest that changes are pending in the institutional context, at least in Asia. The first is the Indian Ministry of Environment and Forests' adoption of solid waste management rules for large cities that include requirements for source separation and composting of organic wastes (India, 2000). The second is the possibility of getting substantial funds from the GEF for initiating composting of urban wastes. These grants are to facilitate cross-sector support for composting. In all probability some African countries will apply for such funding soon.

9.7. CONCLUSION

Sustainable development requires the reduction of waste loading on local and global 'sinks' (Hardoy *et al.*, 2001: 354). Activities that enhance productivity and employment come under the rubric of sustainable development, as long as they do not endanger the environment and hence the health of people and animals. The safe reuse of organic wastes from cities is thus a cornerstone of sustainable development.

Solid waste management would be greatly helped if organic wastes could be treated and reused effectively. The unresolved issues in implementing these goals arise largely from safety, practicality, and equity in meeting the needs of the different social entities that are stakeholders in organic waste reuse.

As noted above, there are many obstacles to truly safe reuse of organics from multiple sources in cities. The application of high standards to processes and products without considerable institutional support would result in the reduction of reuse rather than its increase. Little consideration has been given, as yet, to striking a balance in the tension between waste reduction and public health.

Markets clearly exist in large Asian and African cities for uncontaminated organics. In Asia, private and informal actors capture large quantities of these organics, just as the recyclable inorganics are dealt with by waste trading and recycling enterprises. The problematic organics for municipal solid waste departments are those that are mixed in household, institutional and shop wastes. To render these less of a burden on solid waste management, composting with separation at source is promoted as the ideal. Achieving this on an adequate scale in large cities seems unattainable soon since waste generators have no strong incentives for keeping their organic wastes fit for safe composting. Doorstep sorting, however, offers an easier approach to capturing a good deal of residential organic waste.

Subsidising the entrance of private firms into composting by supplying free wastes and even land meets the need for dealing with some of the un-used urban organics. The products of profit-making firms are generally too expensive, however, for small and basic-crop farmers Local farmers will be the losers if municipal or private undertakings divert the organics that the farmers accessed previously.

Policy dilemmas are grounded in the fact that the interests of the actors or stakeholders who wish to access or to manage urban organic solid waste differ: small and marginal farmers seek a low-cost and easily available input for their fields; private compost-making firms wish to capture the available, uncontaminated organics; solid waste managers see subsidies to such firms as the easiest way to reduce some of their responsibility for organic wastes; NGOs may see compost-making as serving social goals, such as employment for the disadvantaged. A sound policy for urban organic waste management will seek to accommodate all these interests. The partnership approach may help to achieve this goal (see concluding chapter).

In the meantime, changes are taking place nationally and internationally. The effects of the requirement in India that 'class one' cities undertake composting are yet to be gauged, while the interest of the GEF in reducing green house gases through composting is linking local practices to global concerns.

S. GALAB, S. SUDHAKAR REDDYAND AND ISA BAUD

CHAPTER 10
URBAN ORGANIC SOLID WASTE: PRACTICES IN HYDERABAD[1]

10.1. INTRODUCTION

This chapter focuses specifically on the way that organic waste moves through the solid waste management system in Hyderabad. The activities are examined in the light of the possible contributions that can be made to a more sustainable development of the system. The management of organic solid waste in terms of collection, transportation, treatment and ultimate disposal is posing a serious challenge to local governments in India, where a large proportion of waste remains organic (see chapter 3). Although dumping at disposal sites is still done in Hyderabad, urban administrations in India are increasingly looking for other ways to dispose of organic waste, as land is scarce near cities and the costs of transporting waste over long distances are prohibitive. This is reflected, for instance, in the yearly meetings organised by the Urban Think Tank in India, for local authorities in India and the region[2], of which the 1999 meeting was on SWM.

Recovering organic waste for composting is recognized as an important strategy for reducing waste flows, although this premise is not yet widely reflected in the activities of local authorities in Hyderabad (UNEP, 2001). Source separation of organic materials would enhance the quality of organic waste for such recovery activities and make them economically feasible, which is presently not the case in India (UNEP, 2001). In the literature, waste reduction activities refer to diverting post-consumption residues from final disposal at the local level by separation and composting of uncontaminated organic material. For municipal waste, households would have to segregate the waste at source. This also means that urban administrations would have to encourage source separation of 'wet' and 'dry' wastes by households for the reuse of organic wastes. This demands greater community awareness and participation in solid waste manage-

1. We gratefully acknowledge the assistance of Anil Yadav and Umamaheswar Rao, CESS research students, in the collection of field data. In a later stage Ms. R. Dhanalahshmi helped us in corss-checking some of our earlier findings and in completing the data set.

2. This is part of a UNDP-World Bank Water and Sanitation Programme, financially supported by DFID, (Department for International Development, UK).

I. Baud et al. (eds.), Solid Waste Management and Recycling, 213-227.

ment. Non-governmental organisations could provide the necessary education regarding people's participation in solid waste management. The services of waste pickers and itinerant buyers can be utilized, especially if they can handle uncontaminated source separated solid waste (Baud and Schenk, 1994; and Furedy, 1997 a).

This chapter examines the extent to which the activities recognized in the literature as being important for organic reuse, actually occur in Hyderabad.

10.2. ORGANIC WASTE

In chapter 3, the waste generated in Hyderabad and its changing composition is discussed. Very little information on the composition of the waste exists, and what there is, is usually based on estimates of a general nature. A study of waste characteristics in 1997 suggests that about 55 percent of the waste at three dumpsites in Hyderabad was compostable matter (Save Systems, 1997)[3].

There are two categories of main generators of organic waste in Hyderabad. The first are the bulk waste generators of organic waste (both mixed and one type of organic waste). These include stables, dairy farms, hotels, restaurants, hostels, function halls, markets, and slaughterhouses. The second are the continuous generators of small amounts of organic waste mixed with other waste, mainly consisting of households.

The market for organic waste is quite varied, and is shown in Figure 10.1. Waste from bulk waste generators goes into the municipal waste stream as well as into private sector channels for reuse. Dung from dairy farms and stables is collected by private trucks and is transported directly to farmers Waste from hotels and restaurants is collected by herdsmen, employees and Municipal Corporation of Hyderabad (MCH) truck crews (see also chapter 3). The food waste collected by herdsmen is used as fodder for cattle, and is also used to feed pigs, poultry, goats and sheep. Employees and municipal crews throw the waste into dustbins from which it goes into the transportation and disposal channels by MCH and private contractors Market waste is transported mainly by MCH workers; a small part of it goes on for vermicomposting[4]. Organic waste from slaughterhouses is collected by MCH workers, and goes through the usual municipal channels to the dumpsites.

---

3. The garbage examined was collected from Gandamguda, Golconda and Autonagar dumpsites. It was collected every day (5 samples) for one week in the period of August (rainy season). The figures were on wet basis, with moisture levels of 55-60 percent.
4. The project originally started as a pilot, but in 2001 was working on a larger scale, after recovering from a fire which stopped work for several months. According to recent information it had 15 beds for composting, able to handle 90 trucks of waste. Organic waste is combined with neem oil cake and cow dung to make it more useful (Dhanalakshmi, 2002).

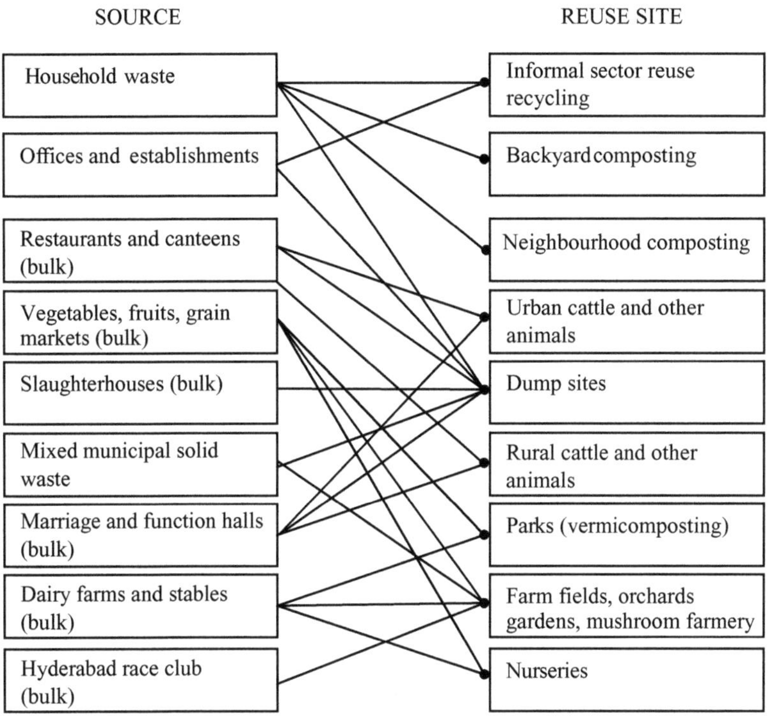

Figure 10.1. The agents of organic waste supply and demand in Hyderabad

Note: This figure is based on one produced by Furedy, Maclaren, and Whitney (Koc *et al.*, 1999).[5]

Hotels/restaurants generate an estimated 80 tons of waste daily. Edible waste is given to beggars, and cattle farms take vegetable waste. There are 34 markets and 5 slaughterhouses, which generate organic waste. Markets generate between 20 tons to 200 kgs of mainly organic waste daily (for an extensive description, see chapter 3). On the premises of the Kothapet fruit market, one acre of land is allocated to a vermicomposting unit run by an NGO, *viz.* Society for Preservation of Environment and Quality of Life (SPEQL). All the markets face problems with the accumulated waste, which leads to unhygienic conditions on the premises of the markets. The five slaughterhouses are run by the MCH. The sheep and goat slaughterhouses generate approxi-

5. The full reference is: Furedy, C., MacLaren, V. and Whitney, J, 'Waste Reuse for Food Production in Asian cities: Health and economic perspectives', in Koc, M. et al. (1999), Hunger-Proof Cities, Ottawa, IDRC.

mately 10 tons of organic waste per day and the cattle slaughterhouses about 20 tons of solid waste. Organic waste from these slaughterhouses is collected by MCH. These slaughterhouses face the problem of water clogging because of the huge accumulation of waste.

Waste from households, the second type of generator, is collected by private as well as MCH workers Servants take edible leftovers, and the rest of the organic waste is put in the dustbins. In the collection system of the Municipality there is no separation of wet and dry waste. In some areas where the tricycle collection scheme exists, residential welfare associations and NGOs/CBOs use the waste collected for vermicomposting, neighbourhood composting and backyard composting[6]. The main bulk of the household waste collected by MCH and private workers goes to the dumpsites.

10.3. RE-USING AND RECYCLING ORGANIC WASTE

There have been many and varied initiatives to reuse and recycle organic waste materials, but their outputs have been mixed. Among the bulk generators, there is a sizeable number of generators who promote reuse through private sector channels (fee figure above). These initiatives remain outside the purview of the municipal waste stream, and are based on economic feasibility of the re-sale of waste materials.
There have also been several attempts to carry out large-scale public sector composting over the years More recently, several initiatives have been undertaken to promote decentralised composting through alliances between local government and NGOs, based on the use of both household and bulk generator organic waste (such as the Fruit Market where SPEQL undertook activities; see later sections and chapter 11). The mixed organic waste generated by households has been incorporated in both types of composting processes.

Large-scale public sector composting

Composting of municipal waste has recently become topical through the new rules issued by the Ministry of Environment and Forests, which mandate composting as a way to reduce waste levels at dumpsites (India, 2000). The Municipal Corporation of Hyderabad was involved in initiatives to use solid waste for composting since the late 1950s. Early attempts foundered by the end of 1960s, due to high prices of the compost (Snel, 1997). Again in 1977, a large-scale mechanical compost plant was established in cooperation with NEERI, which was later leased to a private company. It stopped in 1986, as the plant failed to maintain demand for its product in comparison with cow dung being sold, which also had a higher nitrogen content than the compost.

6. These initiatives will be discussed later in this chapter.

Currently, the MCH has undertaken initiatives for both large-scale and small-scale private garbage-processing units. The municipal corporation of Hyderabad has entered into an agreement with a private sector company for anaerobic composting units to generate power and pellets (Selco International Limited). The company has established a plant at Gandamguda dumpsite producing fuel pellets from mixed waste[7].

Decentralised Composting

Decentralised composting can be broadly divided into two categories, *viz.* single source waste and multi-source mixed waste composting. As part of its 'voluntary garbage disposal scheme' (VGDS: see chapter 3 for details), the MCH also included decentralised composting at the neighbourhood level on an experimental basis. Originally, the Reddy Foundation undertook to compost waste in a neighbourhood park, where it was to be converted into manure through a vermiculture process. The municipal corporation of Hyderabad allotted a plot there for the NGO to undertake decentralised composting. However, the scheme ran into difficulties because mixed waste was delivered to the park, and nearby residents protested the deterioration of their environment. Now, another NGO, Sukuki Exnora, has taken over the scheme, in which the waste delivered is market waste (largely organic), and the location has been shifted to a less central place in the park. Composting is done on a very small-scale, and cannot be said to have any measurable impact on SWM in the city.

A second initiative was undertaken with the help of the MCH by the SPEQL. It established a composting unit near Kothapet fruit market. The fruit, grain and vegetable waste generated in the fruit market was utilised by the SPEQL for vermicomposting without making any payment either for the garbage or for the site allotted in the market yard. The authorities of the MCH supplied fruit, grain and vegetable waste free of cost. The project originally started as a pilot, but is now working on a larger scale, after recovering from a fire, which stopped work for several months. Eight people work in the yard currently (both men and women). According to recent information it has 15 beds for composting, able to handle 90 trucks of waste. Organic waste is combined with neem oil cake and cow dung to make it more useful. Compost is bought both by government nurseries and the public. Training on vermicomposting is also provided through this channel.

A number of other actors in Hyderabad also carry out composting activities. These include the military at their farms in Bowenpally (dairy and agricultural activities), who carry out vermi- composting utilising cow dung slurry and farm waste. In the city, cow dung has been used for composting by nurseries. The organic manure generated through composting is mixed with red soil and used at the time of planting.

---

7. The latest information indicates that the plant is processing 200 tons daily, although it is designed to put through 700 tons per day.

The horse dung from the Hyderabad race club, which has 600 horses at Malakpet, is transported daily by a truck to the Tekron Industry at Medchal and composted and used as manure for mushroom crops. The park service utilises leaf and flower droppings for composting, reworking it in their parks (five parks surveyed). The authorities of some of the star hotels in the city, *viz*. Viceroy and Green Park hotels, collect their waste and transport it to their agricultural lands. Food waste is composted by traditional methods and used as manure.

High-income group households in rich localities in Banjara Hills and Jubilee Hills of the city are using their kitchen and garden waste for composting by traditional methods, utilising it in their own gardens as organic manure.

10.4. URBAN-RURAL LINKAGES: REUSE OF DUMP SOLID WASTE BY FARMERS

Farmers from several villages near one of the Hyderabad dumps (Mansurabad) were interviewed about their use of organic solid waste from the dumpsites[8]. Large and medium farmers, growing commercial and horticultural crops, used mixed solid waste from dumpsites, together with other organic and chemical fertilizers. Farmers used the mixed urban waste mainly on irrigated land where commercial crops are grown. The farmers who used urban organic waste on their land reported that crop yields had increased due to the application of urban organic waste. Benefits found were a loosening of the soil, increase in soil fertility, increased weight of potatoes grown, and the fact that flowers (chrysanthemums) became more shining and remained fresh longer. The advantage of urban organic waste was that it worked for two crops whereas fertilizer and poultry waste were effective for only one crop and could not be used as soil conditioner.

Although many farmers have used used urban organic waste in the past their number is decreasing (see also chapter 11). They faced problems in applying such solid waste picked up directly from the dumpsites due to the following reasons:

- The soil became harder as a result of the application of mixed solid waste over a period of years;
- When wastes were applied on agricultural lands where crops were cultivated, problems of pollution of soil, crop plants, ground and surface wastes emerged gradually;

---

8. The discussions were held with farmers from *Mansanapally(115 farmers) and Chinna Gollapaly (60 farmers),* the former 40 km and the latter 30 km away from the *Mansurabad* dump site (Reddy and Galab, 2000*).* The field survey reveals that the overall proportion of farmers applying the non-segregated urban organic waste is larger in the village nearer the dump site (see the Appendix to this chapter, table App 10.1 and App 10.2). However, there is a clear difference in the use of urban waste according to the size of the holdings, which may be more important than distance.

- The farm workers were unwilling to load and unload the trucks with solid waste due to the presence of blades, nails, other sharp materials and the foul smell;
- The transportation of the solid waste from dumpsites to the villages had become more expensive;
- The farmers felt that the cow dung/poultry waste was cheaper compared to the solid waste; and
- Last but not least, the composition of the waste changed to include much more inorganic waste and inert materials from construction (especially after privatisation of collection).[9]

10.5. COMPOSTING EXPERIENCES IN THE CITY AND THEIR CONTRIBUTIONS TO SUSTAINABLE DEVELOPMENT

In this section, the composting experiences described above are analysed according to their contributions to sustainable development.

Decentralised composting of single-source waste is carried out privately without outside interventions, and seems to be successful, presumably due to the fact that they are composting single source waste. In fact, this private market of single source waste producers and buyers is one which can be usefully explored in the future. Decentralised composting through NGOs and CBOs is heavily promoted, but is still in the experimental stages in the few locations where it occurs These experiments will be analysed according to the criteria developed earlier (Baud *et al.,* 2001).

Sukuki Exnora, an NGO affiliated to the Chennai-based Excellence-Novelty and Radical (Exnora) International, was established in 1998. It started a vermicomposting unit in July 1999. whose general aim was to 'involve the common man in environmental conservation' with special reference to solid waste management[10]. Sukuki Exnora became the contractor for waste collection in several areas of the city: Begumpet, Methodist colony and Minister's colony, providing collection and cleaning services as a part of Hyderabad's drive toward privatisation. On the advice of the Additional Commissioner of the MCH, they took up a project called "Zero Garbage or No Garbage to Landfill Site". As part of the project, Sukuki Exnora took up vermicomposting projects in July 1999 in association with the MCH at Jubilee Hills and Indira Park[11] (see also earlier description). These activities contribute to better resource recovery, as the total waste collected from a neighbourhood is separated, sold or composted. Although a residue remains, which goes back to the municipal dumpsite, access to the waste for recovery purposes is greater than on the street. On the basis

9. The outcomes as indicated by the farmers are based on local experience. There was no attempt in this study to verify local knowledge through experiments. However, current use of solid waste in farming is waning.
10. In 2001, their activities have also expanded to other areas, as indicated on their website.

of the available information, it cannot be said whether the amount of waste going to the dumpsite for final disposal has decreased. The activities are still in a too early stage to comment on this aspect.

Sukuki Exnora started the vermicomposting unit in a corner of Indira Park. They invested an Rs 70,000 in this unit and in the other unit at Jubilee Hills. The government provided them with land, water and power free of cost at Indira Park. The NGO is the contractor for several residential areas, whose collected waste they brought to the park (two lorries of garbage daily). Because it consisted of mixed garbage, the environmental conditions deteriorated. Now market waste is utilized as a basis for composting. This set-up means that investment in this practice is heavily subsidised, and the basic resources for composting and selling are free of cost. Nevertheless, the investment made by the NGO is fairly high by Indian standards.

The process of composting is done in a labour-intensive way. Once the garbage is brought to the unit, the workers separate plastic, paper, glass and metal scrap, and other items like coconut shells, rubber/tyre waste, cloth, and wooden waste. The remaining garbage consists of household waste, leafy and other compostable matter. They allow the biodegradable material to decompose for about 15 to 20 days in pre-treatment beds. The pre-treatment bed is 30 metres long and 2 metres wide and produces 30 to 40 tonnes of garbage. The cow dung slurry treatment is undertaken on the bed, i.e. cow dung is sprinkled on the bed. These beds are covered with black polythene sheets, which make the process of decomposing quicker as they absorb the heat fast. The decomposed biodegradable material is shifted to vermicomposting beds after 20 days. The composting bed is 30 metres long, 12 metres wide and one metre in height. No chemicals are used in this process. Two types of worms are used in the Indira Park, which are put in the bed in different layers. These beds are regularly watered in the morning and evening for about 45 to 60 days. This process leads to harvesting of the manure after 60 days.

There is little seasonal variation in the output over the first period of operation. The harvested quantity of manure was only one ton per month in the six months previous to the interview. They have started selling manure with the brand name "Suvarna Manure" since January 2000, at a rate of Rs 4 per kg, but selling is still a problem. Out of the total quantity of manure, Exnora itself used one ton. There is a buy back agree-

11. Besides vermicomposting units, Sukuki Exnora is associated with socially useful activities like map awareness programmes, GIS, training, advertising, and conservancy services. They have also recently started 'Street Beautifiers Scheme' under which rehabilitation of waste pickers takes place by providing tricycles for collection of household garbage. Apart from this, they have taken up the responsibility of planting, watering and maintenance of 6,300 new plants and also maintenance of 7,700 old plants in circle III covering Kachiguda, Barkatpura, Narayanaguda, Himayatnagar, Vidyanagar, Amberpet and Ramananthapur areas.

ment with the municipal corporation of Hyderabad through the horticultural department of the state government, in case the manure is not sold on the open market. At the time of the interview, approximately 5 tons of manure were ready for sale in the unit. This would mean a potential maximum sales level of Rs 20,000 for six months work. Sukuki Exnora employed four waste pickers in the unit (one woman worker), who were paid Rs 2,000 per month. Workers' salaries were paid by Sukuki Exnora through other sources, as the composting unit has not yet made any sales. These figures indicate that costs were much higher than potential sales levels in the period covered by the study. Thus financial viability was not assured in this activity, even when investment costs were not taken into account.

Employment provided through the NGO was regular, relatively safe and well paid (at Rs 2,000 p/month), with accommodation and protective equipment also being provided. Two workers have undergone training in the composting process for one week when they worked with another NGO with experience. However, it was suggested that the workload was high for the number of workers employed at that time (Dhanalakshmi, 2002).

The organisation activities do not contribute to clean and healthy neighbourhoods. This is one of their most pressing problems, as earlier on, residents made unfavourable comments on the conditions in Indira park, when unmixed waste attracted rodents and the unit smelled. To make the unit more financially viable, more space would be needed to build more vermin-composting beds to absorb the 10 tons of waste coming in daily. This would compound the problem already experienced in the neighbourhood, and suggests clear limitations to such decentralised production of compost.

NGO Decentralised Composting in a Market: the Case of SPEQL

SPEQL set up a vermicomposting unit on a half acre at Kothapet fruit market with a grant of Rs 350,000 from the Government of India through the Department of Agriculture, Government of Andhra Pradesh. The organic waste from the fruit market was used as basic raw material in this unit. Ten sheds were constructed at the cost of Rs 100,000, a lower unit cost than those by Sukuki Exnora. The other infrastructure facilities in this unit were a bore well, a drip system, a storage place, a shredder, and a machine which separates compost from other solid wastes. Therefore, this NGO had a solid financial position at the start.

This unit regenerates the earthworms used in the unit itself. These worms were sold to other people at the rate of Rs 50 per kg.[12] The quantity of compost generated from each bed is about 5 tons every two months (implying a maximum of 7.5 tons per month from three beds). The manure is bought at Rs 2 per kg by households, farmers, fertilizer shops, horticulturists and NGOs, with farmers forming a large proportion of the customers There was no buy back arrangement with the MCH. The amount of organic waste processed through this method is not known, but it has the potential to contribute to more reuse and recycling. Nothing is known about the way the residues are finally disposed of, and whether this happens safely.

This unit employed ten labourers and whenever there was more work, they employed more people. The labour cost per month was Rs 10,000 for the workers Women workers are paid 70 percent of the male wage rates, and there is differentiation among the men as well. These workers were provided with accommodation in the unit itself, but not given any other type of social security. One labourer was given training in the specific skills required for releasing worms and preparing the vermi-beds. This suggests that employment provided was regular for a small group of people, and irregular for another group. Income levels were half of what was paid by Exnora, and there was no information on health and safety features. Provision of accommodation also increased income implicitly.

The expenditure on cow dung was Rs 200 per lorry and electricity charges Rs 1,000 per month. This brought total costs per month to an estimated Rs 12,000 If the maximum production was reached and all the compost sold, the basic income was sufficient at Rs 15,000 per month. These figures suggest that financial viability was possible, but that there was little margin for future investment nor for repayment of initial investments in the unit.

SPEQL was planning to start another unit in the Monda Market at Secunderabad. They have not yet reached an agreement with the government on this aspect. The organisation faced problems from the fruit market committee in 1999. They were asked to vacate the land but due to government intervention, they continued to do composting. However, they had to give some land to the fruit market. The conflicts with the fruit market organisation suggest that local businesses contest the legitimacy of the composting activities.

10.6. CONTRIBUTIONS OF THE PARTNERSHIPS TO ASPECTS OF SUSTAINABLE DEVELOPMENT

In conclusion, an attempt has been made to examine the following issues in case of organic waste management;

---

12. The process of composting was carried out by SPEQL as follows: garbage was unloaded into three pits dug for this purpose. The garbage was dumped into these pits before being laid on the vermi-beds. The large items of waste were cut into small pieces by the shredder in order to make the process of decomposition quicker. The biodegradable material in the pits was allowed to decompose for 15 days. The first layer of a vermi-bed consisted of coconut coir, which absorbed the water to keep the bed moist. Over this layer garbage was spread and cow dung slurry treatment given. After introducing the earthworms, garbage was spread on the bed. This process led to the compost being ready after 60 days.

- identify some of the ways in which actors work together in organic waste management;
- assess the partnerships in terms of resource recovery, socio-economic and environmental health aspects; and
- analysing how existing alliances can be strengthened. The previous sections have highlighted the following types of co-operation between local actors:
- truck drivers providing dumpsite waste to farmers;
- CBOs – NGOs working with local government, and
- single source generators selling organic waste to farmers and horticulturalists. The first and last types of partnerships are market-driven, whereas the second is representative of co-operation between local government and civil society organizations.

Environmental Impacts

The environmental effect of these alliances has been examined in terms of source separation, and promotion of reuse and recycling, and safe final disposal. The alliances which utilised mixed organic waste drawn from the municipal waste flows (truck drivers – farmers, and CBOs – NGOs) have not contributed to better source separation and a higher quality of waste used for recycling. They did contribute to the reduction of the volume of waste in the municipal stream, and have promoted recycling. However, the co-operation between truck drivers and farmers has gone down substantially as it became clear that the mixed urban waste affected soil fertility and worker willingness negatively over a longer period.

Socio-economic Impacts

The socio-economic impacts of these alliances have been assessed in terms of financial viability, employment and income, legal legitimacy and social legitimacy.

All these alliances have problems with respect to financial viability. Farmers obtaining waste from the dumpsites found that costs were increasing and the quality of the urban waste was problematic. Single source generators found their co-operation with buyers of organic waste remain financially viable. However, the vermicomposting undertaken by NGO-local government sponsorship suffers from a lack of demand for the compost made and a lack of social acceptance from immediate neighbours of the units. The social legitimacy of the waste used by farmers was also low, as some farm workers refused to take part in the process.

Generation of employment is not large in the case of the private sector partnerships. It provides – for the group concerned – a relatively regular employment, with fringe benefits.

Environmental Health

The activities of single-source waste generators contribute to clean and health neighbourhoods. The NGOs carrying out composting activities in neighbourhoods are not conducive to cleanliness.

Looking at occupational safety and health aspects, the truck drivers- farmers alliance affected the health of the workers and animals negatively. The NGOs paid some attention to safety aspects. Safe final disposal was not taken into explicit account in these forms of co-operation.

10.7. CONCLUSIONS

One of the main problems in organic waste reuse and recycling is the lack of source separation, which affects the quality of the organic waste routed through the municipal waste stream negatively. This prevents effective use of a major portion of municipal waste generated. Efforts should be put into promoting locally acceptable methods of separation. That there is a potential market for 'pure' organic waste and well-composted waste is indicated by the existing private sector activities of the single-source bulk generators of waste who have little difficulty in having their waste removed through private trucks. The study in chapter 11 also supports this conclusion.

Composting organic waste remains a difficult technical process in the prevailing conditions in Hyderabad. Large-scale public sector composting has proven financially unviable and without a market for the product, while small-scale decentralised composting with the help of NGOs and CBOs remains complex, and not well accepted by local residential communities.

APPENDIX TO CHAPTER 10

Table App. 10.1. Distribution of Farming Households According to Categories and Use of Municipal Solid Waste on their Agricultural Lands at Mansanpally Village

Description of the categories	*Number of farmers who have used municipal solid waste in the past*	*Number of farmers who have not-used municipal solid waste*	*Total number of farmers*
Marginal farmers	3 (20.0)	12 (80.0)	15 (100.0)
Small farmers	11 36.6)	19 (63.4)	30 (100.0)
Medium farmers	20 (40.0)	30 (60.0)	50 (100.0)
Large farmers	14 (70.0)	6 (30.0)	20 (100.0)
All farmers	48 (41.73)	67 (58.27)	115 (100.0)

Note:
i) Marginal farmers – Below 2.50
ii) Small farmers – 2.50- 5.00
iii) Medium farmers – 5.00- 10.00
iv) Large farmers – Above 10.00
Figures in brackets are percentages

Source: Field Survey

Table App. 10.2. Distribution of farming households according to their categories and use of municipal solid waste on their agricultural lands at Chinna Gollapally Village

Description of the Categories	*Number of farmers who have used municipal solid waste*	*Number of farmers who have not-used municipal solid waste*	*Total number of farmers*
Marginal farmers	6 (50.0)	6 (50.0)	12 (100.0)
Small farmers	13 (65.0)	7 (35.0)	20 (100.0)
Medium farmers	4 (40.0)	6 (60.0)	10 (100.0)
Large farmers	10 (55.5)	8 (44.5)	18 (100.0)
All farmers	33 (55.0)	27 (45.0)	60 (100.0)

Note:
i) Marginal farmers – Below 2.50
ii) Small farmers – 2.50- 5.00
iii) Medium farmers – 5.00- 10.00
iv) Large farmers – Above 10.00
Figures in brackets are percentages

Source: Field Survey

Table App. 10.3. Crops grown by the farmers in sample villages according to their status of use of municipal solid waste on their farms

Status of farmer	*Crops grown*	
	Irrigated crops	*Unirrigated crops*
Users of urban solid waste	Paddy, Chrysanthemum flowers, Tuber-taro (itchy potato), brinjal and tomatoes	Sorghum, red gram
Non-users of urban solid waste	Paddy, Chrysanthemum flowers, wheat and brinjal	Sorghum, red gram

Source: Field Survey

Table App. 10.4. Type of land owned by the user and non-user farmers of urban organic waste in the sample villages

Description	*Number*	*Irrigated Land*	*Unirrigated Land*	*Total Land*
Users	12	5.91	2.91	8.83
Non-users	6	2.83	2.08	4.91

Source: Field Survey

Table App. 10.5. Impact of urban organic waste on the yields of the crops grown by the farmers

Reference Period of Impact	*Crop Yields (Quintals per Acre)*				
	Paddy	*Chrysan-themum*	*Tubers-Taro (Itchy Potato)*	*Brinjal*	*Tomatoes*
Before the use of urban organic waste	20.4	10.4	12.5	10.6	8.3
After the use of urban organic waste	29.8	16.0	20.0	19.0	13.5
Total	50.2	26.4	32.5	29.6	21,8

Source: Field Survey

CHRISTINE FUREDY AND RAAKHEE KULKARNI[1]

CHAPTER 11

DEMAND FOR COMPOST FROM URBAN ORGANIC SOLID WASTES IN HYDERABAD

11.1. EXPLORING THE DEMAND FOR COMPOST FROM URBAN ORGANIC SOLID WASTES

As noted in chapter 9, organic waste forms the major component of municipal solid waste in most Indian urban areas. If properly treated and utilised it has the potential of being an important resource for food production. Use of urban organic wastes can also reduce the burden upon municipalities of waste disposal, as well as ameliorating the environmental problems associated with untreated organic wastes.

In the past, urban and peri-urban farmers in India accessed relatively uncontaminated urban organic solid wastes (cf., Nunan, 2001; Furedy, 2002), which was mostly 'mined' from garbage dumps, but the increasing contamination, together with other factors in urban development and management, have led to a decline in organic waste reuse, at a time when pressure for dumping space is increasing.

Following the failure of municipal experiments with mechanical compost plants (built in the mid-1970s), most municipal corporations showed little interest in promoting

1. Ms. Raakhee Kulkarni, then a master's student in the Urban Environmental Management Programme at the Asian Institute of Technology, undertook research on the demand for compost (especially that made from urban organic wastes) in Hyderabad – Secunderabad from February to March 1999. Drs Galab and Sudharkar Reddy, of Centre for Economic and Social Studies in Hyderabad (partners in the EU-INCO project contract no. ERBIC18Ct971052) assisted her in local contacts, surveys of farmers, transportation and translation. Anil Yadav and Umamaheswar Rao, CESS research students, also helped with information from farmers near Hyderabad. The study included an assessment of composting undertakings in Bangalore, where information was supplied by Esha Shah and Almitra Patel. Miss Raakhee thanks the project for the opportunity to undertake the work in Hyderabad and for all assistance provided; she also thanks the informants in Bangalore and Hyderabad who provided information. Ms. Kulkarni was awarded a Master of Science degree for her thesis "Market potential for compost industry in Hyderabad, India: Necessary strategies for public and private cooperation" in August 1999. This report, presented as a contribution to the EU-INCO project, is based on subsequent discussion between Furedy and Kulkarni of the original fieldwork and a selection of the thesis information. This report also includes interpretations and statements not contained in the thesis.

I. Baud et al. (eds.), Solid Waste Management and Recycling, 229-240.

compost making. But some commercial undertakings started up in the 1990s (for example, a plant of Excel Industries in Bombay), along with small projects involving NGOs and community groups (Furedy, 1992). In a recent policy change, composting is now seen as contributing to sustainable waste management and also to job creation and income generation. The Solid Waste Management Rules handed down by the Supreme Court of India and adopted by the Ministry of Environment and Forests in 2000 mandate composting as part of solid waste management in large cities (India, 2000).

Compost products made from city wastes have been of poor quality. This is mainly due to poor management in preventing contamination with glass pieces, sharp objects, plastics and industrial and medical contaminants. For the new municipal solid waste management goals to be achieved, municipalities, the local government, citizens and the private sector must collaborate to overcome difficulties of organisation, technology and product quality.

While a few compost companies have been able to sell compost throughout India and even abroad, the markets for compost products within urban and peri-urban areas are not well developed and little independent research has been done on this subject. The major interest of this study was to explore the market potential for compost from urban organic solid waste in and around the city of Hyderabad – Secunderabad. This is an appropriate city for such a pilot study for several reasons: it has a history of municipal compost-making; near-urban farmers have in the past acquired organic waste for animal feed and land application; recently, some NGOs and community projects have become interested in the potential for compost-making; and, like in a number of Southern Asian cities, the decline of the use of unprocessed urban organic solid waste is contributing to the growing crisis of solid waste disposal in the city.

The specific objects of the research were:

- To explore the market potential for urban-organic-waste-derived compost in and around Hyderabad and suggest how composting can be made more sustainable. The emphasis was upon peri-urban farmers, with some attention to parks, hotels, nurseries, and affluent residences;
- To understand the role of private enterprises in production and marketing of compost in both Hyderabad and Bangalore so as to improve the performance;
- To look at the relationships among the Municipal Corporation of Hyderabad (MCH), NGOs, the private sector, and the public in Hyderabad regarding the prospects for better cooperation in undertaking composting and marketing the product;
- To compare Bangalore and Hyderabad on compost making with a view to aiding its development in Hyderabad.

Composting efforts in Bangalore were included in Kulkarni's research because this city has several compost-making enterprises, and there has been cooperation among the municipality, private enterprise and NGOs in composting from urban organic solid waste. Hence, the Bangalore experiences were seen as relevant for improving the performance in the management of organic wastes in Hyderabad.

11.2. SCOPE OF THE STUDY

This is a management-oriented study, so technical details regarding composting are referred to only in passing. It was not possible to undertake a financial analysis, although we recognise that the pricing of compost and the cost-effectiveness of compost operations strongly influence the demand for the product.

Several barriers to information were encountered that limited this research study, mainly with regard to the availability of data on the nature of solid waste in Hyderabad, the operations of solid waste management, and the costs of management. There were also incompleteness or discrepancies in information supplied by informants. Nevertheless, the information presented in this study provides some insight into the actors in the management and use of urban organic solid waste and into the potential demand for compost in and around a city like Hyderabad.

11.3. SUMMARY OF FINDINGS REGARDING THE DEMAND FOR COMPOST

The respondents

The market potential for the compost was investigated with respect to the following actors: farmers, hotels, nurseries, public parks and households. In all, 21 farmers, 10 hotel managers, 7 nursery managers, 4 public parks managers, and 20 members of high-income households were surveyed in Hyderabad (Banjara Hills, Jubilee Hills and Mahindra Hills).

The farmers selected were chosen on the basis of information from Drs Galab and S. Reddy of the Centre for Economic and Social Studies, Hyderabad. The farmers are located near Nagaram, Shamshabad Mandal, in Ranga Reddy district. This district was chosen because it is one where examples of municipal solid waste use can still be found, in spite of the area being at a considerable distance (about 30-40 km) from the city dumps. The ten hotels (5 star and 3 star) are considered to be typical of their class. The nurseries were contacted through the snowball method, starting with a well-known nursery and following up connections. The households in the elite areas were chosen on the basis of convenience and willingness to respond.

To understand the context of compost making, interviews were conducted with a local NGO undertaking vermicomposting in Hyderabad (Society for Preservation of Environmental Quality of Life-SPEQL) and some public officials. In Bangalore, information was gathered from key informants regarding composting operations of one public sector undertaking, two private companies and three NGOs.

Demand for compost

In this small pilot study, demand was demonstrated among peri-urban farmers, nurseries, hotels, parks and some affluent households. All cultivators responding appreciated the use of organic matter. A majority is prepared to use urban-waste-derived compost. Hence it can be said that there is a demand for such compost. However farmers are willing to procure urban-waste-derived compost only under certain conditions: that it is pure (free of toxic material), and conveniently obtained (delivered to the farm or sold nearby). Farmers are wary of agreeing to a price higher than one rupee per kilogram (40 rupees to the US $). Some other potential users are prepared to pay considerably more.

Farmers

- All 21 farmers use chemical and organic fertilizers ('Organic fertilizers' include animal manures, compost or decomposed solid waste).
- 5 farmers are still using municipal solid waste in their fields. (Most inorganics, constituting 15 percent of the material, are removed from the waste on site. Farmers pay Rs 700 – Rs 1,000 per four ton lorry load and apply about 10 lorry loads to an acre.)
- 3 farmers have discontinued use of municipal solid waste in the past one or two years (reasons given below).
- 10 farmers are using cattle dung, poultry litter, etc., without systematic composting.
- 2 farmers are using regular compost: one obtains it from a private supplier and the other composts his own animal and agricultural wastes.
- 1 farmer is using slaughterhouse waste.
- 85.7 percent of the respondents were ready to buy compost made out of urban organic waste, provided the compost was of good quality. At the same time, they would like to buy it from a place convenient to their fields.
- When asked the price they were willing to pay, most farmers gave no response or said they had no idea what they should pay. The farmers who did respond were not willing to pay more than one rupee per kg.

Farmers made the following comments about using solid waste on their fields:

- It is a good soil conditioner and doubles their yield in some crops (especially the vegetable colocasia, and rice). It is good in both saline and alkaline soils.
- An application of municipal solid waste is good for two years
- Cattle and farms workers have injuries from needles, glass, and other sharp and hard objects from the garbage delivered to the fields.
- Plastic content has increased in the past 3-5 years

- Synthetic cloth, plastic, rubber and other materials remain for years undecomposed.
- Extra expenses are incurred for labour to sort out these manufactured materials.
- Such wastes accumulate in the soils, hardening them and affecting root development of crops.
- Purchase of waste from garbage dumps or from municipal vehicles is no longer popular.

Non-farm respondents

- Households, hotels, nurseries and parks are also ready to buy waste-derived compost.
- Parks officials noted that they are making their own compost with park trimmings, so the Parks Department's need for further compost may be limited.
- The price these respondents are willing to pay ranges from Rs 3 – 5 per kg. A few hotels are willing to pay much more, up to Rs 20 per kg. (Like the farmers, they were uncertain about what would be a reasonable price to pay).

Further comments

- The potential for the compost is considerably high, in the abstract. The demand is for a pure product, convenient to obtain. Potential users, however, have little specific knowledge about compost and its benefits.
- The majority of the respondents did not know what would be a reasonable price to pay for the product they want. Lack of knowledge regarding the market values of compost, the requirements for producing good quality compost, transportation costs, etc., probably all contribute to this uncertainty.
- Of the potential customers considered in this study, farmers producing profitable crops (such as flowers and chillies), 3-5 star hotels, nurseries, and the horticulture department are those most likely to be regular buyers, willing to pay a market price. Hotels were willing to pay the highest price (up to Rs 20 per kg).
- Only a few high-income households with gardens were studied and no firm conclusions can be drawn from their responses.

Conclusions regarding demand for compost

The lack of knowledge about the environmental benefits of composting urban organic solid waste, and uncertainty about how it should be valued and priced, suggests that the market potential could be increased considerably if potential users had access to information about compost and its relevance to their needs. If, in addition, the composting of organic matter was understood as contributing to better urban solid waste management and waste reduction, the willingness to support compost-making should improve.

Other potential users of compost, who should be surveyed in a future study, are:

- Apartment complex managements (for grounds)

- Apartment residents growing plants in their balconies
- Smaller parks and community gardens
- Clubs and rest houses
- Institutions such as schools, colleges, offices etc.

11.4. NGO INTEREST IN VERMICOMPOSTING

Although the study did not explore the demand for vermicompost, information was gathered on the interest in this subject in Hyderabad.

One NGO, the Society for Preservation of Environmental Quality of Life, is very interested. It undertook a successful experimental project with green market wastes at a site leased from the Kothapet Fruit Market in 1996. Supporting funds came from the L. B. Nagar Municipality and fruit, grain and vegetable waste was supplied without charge. Training in vermicomposting was obtained from a workshop of the Centre for Environmental Education in Bangalore. Hence this project had some municipal support, as well as assistance from an NGO in another city.

Vermicompost was produced at the rate of 400 kg every 35 days and sold at Rs 2 per kg. Most of the buyers were farmers growing chillies, tobacco and flowers The Agro-horticultural Society, the Horticultural Department and some residents were also purchasers The success of this experimental project supports the view that there is a demand for compost from urban waste, if it is based on source-separated, uncontaminated organics, and sold at a low cost. (Since SPEQL is a non-profit concern, and they obtained the wastes free, the compost could be sold at a very low price).

This project encountered many difficulties (principally with respect to lack of workers, deficiencies in equipment, the need for more funds, and problems in extending the lease with the market committee). It was interrupted when fire destroyed the site at the Kothapet Fruit Market in 1999. Recently the project work has recommenced (Dhanalakshmi, 2002).

Civic Exnora, an organisation with headquarters in Chennai, launched a compost project a few years ago. It was discontinued, but in 1999, an Exnora branch, Sukuki Exnora, decided to begin again, using the park sites of discontinued Parks Dept experiments (see chapter 10). They contracted with the MCH to use the sites free of rent and to have solid waste delivered to them daily. The project was suspended, however, because of complaints from the public about garbage piled in the parks, with the concomitant odours and rodents. Again, another attempt has begun on a very small-scale in the park (Dhanalakshmi, 2002).

The NGOs interested in vermicomposting in Hyderabad believe there is a demand among hotels, institutions and apartment dwellers with container gardens. It is unclear how large the market might be if the product were to be sold at a price of Rs 5 per kg, as it is in Bangalore, since SPEQL sold their compost at Rs 1-2 per kg. Whether a substantial amount of such compost could be sold in the urban area was not assessed by this study.

11.5. ROLE OF GOVERNMENT

The MCH does not have a strategic solid waste management plan that includes attention to waste minimisation and reduction. However, the basics of solid waste management are being improved with assistance from the centrally funded Housing and Urban Development Corporation (HUDCO). A large grant from HUDCO is geared mainly to the improvement of infrastructure for waste collection and disposal. There is no reference to reduction of organic wastes in that project.

The MCH's efforts at marketing compost produced in a municipal plant failed by the late 1980s. The compost was of low quality and marketing was not vigorously pursued. There was little attempt to understand the potential users' needs. Subsequently the MCH has not been interested in restarting a compost plant or in promoting composting substantially, although the Parks Department briefly experimented with vermicomposting in the 1990s. In general, it appears that composting, whether of source-separated organics or mixed municipal wastes, is not currently a priority for MCH. They are prepared to endorse private undertakings, but have not given attention to the varied supports (in both urban and peri-urban areas) that would be necessary to promote viable composting of urban wastes.

The MCH does not seem to be aware that its stated interest in promoting composting (specifically, vermicomposting) is in potential conflict with its endorsement of a private solid waste pelletisation plant recently started in Hyderabad. This plant requires mixed waste, and so represents an impediment to the principle of separation of organics for composting. Energy production (supported by the central Ministry of Non-Conventional Energy) apparently takes precedence over the needs of peri-urban agriculture and horticulture. At present the plant operates on a small-scale, but, if the process is successful and is expanded, it could undercut support for composting.

There is no public participation with respect to composting in Hyderabad and apparently no plans to educate the public about the benefits of reducing urban waste through separation at source of the organic matter and its diversion to compost production. Nevertheless, the willingness to support NGOs that take up projects in composting shows a positive attitude among some public officials and interest may grow soon, given the impetus to composting of urban wastes in the solid waste management rules of the Ministry of Environment and Forests.

11.6. COMPOSTING IN BANGALORE: LESSONS FOR HYDERABAD

In Bangalore, composting of urban organic solid waste is undertaken in private, semi-government and neighbourhood schemes.

The Karnataka Compost Development Corporation (KCDC) is an autonomous body under the state government. The KCDC dates from 1975, when the Government of India gave grants for compost plants in major cities; it is now a subsidiary of Karnataka Agro Industries Corporation Limited (KAIC). It receives capital cost subsidies from the Government of India, and further support from the Bangalore City Corporation (BCC) and the Karnataka State Cooperatives Marketing Federation. The company makes compost products from mixed municipal waste and markets them to other states as well as Karnataka, at prices ranging from Rs 0.8 to 1.5 per kg. (The proportion of municipal waste and separated market wastes used is not known).

Two private companies engaged in composting of urban waste are Sunrays Composts and Terra Firma Bio-Technologies Inc. The latter, which started production in 1994, specialises in vermicompost and has a research division.

Several NGOs in Bangalore have done project work for 5-8 years and have built up a knowledge of small-scale, decentralised composting and vermicomposting, for example, Waste Wise (Mythri Foundation), Centre For Environmental Education, Clean Environs. They operate on a very small-scale, in public parks, producing less than 50 kg per day. The compost is sold at between Rs 3 and Rs 5 per kg, but the higher priced product does not sell readily. The NGO projects do not recover their costs from the sale of compost (Lardinois and Marchand, 2000).

There appears to be a fairly good awareness among the public of solid waste issues and various local initiatives. Nevertheless, waste generators do not comply with the instructions from NGOs to separate at source. The waste used in these neighbourhood pilot projects, however, is less contaminated than mixed municipal wastes collected by the city vehicles (Lardinois and Furedy, 1999; Lardinois and Marchand, 2000).

The municipal corporation has been supportive of these undertakings. The BCC subsidises the companies, and especially the Karnataka Compost Development Corporation, by supplying municipal waste. The companies appear to be successful and are marketing their products in other states as well as to farmers, parks, forest and horticultural department. The undertakings face problems of contamination (as they are dealing with mixed municipal wastes), rainy season operations, NIMBY (Not in my backyard) reactions from residents, and obtaining consistent and regular supplies of waste.

The trend for corporate undertakings is to have their marketing handled by marketing consultants or firms. The main customers for the private firms are plantations (tea, coffee and areca nut), departments of agriculture, horticulture and forests, and farmers

Hyderabad-Secunderabad should pay attention to these aspects of the Bangalore experience:

- Compost can be made from mixed municipal waste and sold at a price acceptable to farmers and plantations in Southern India. (The quality of the product is unknown.)
- There are a variety of undertakings doing composting and vermicomposting from urban organic solid waste and developing expertise in production and marketing. The private company is producing the best product and making a profit after investing in research and technology for a few years
- Composting undertakings receive a certain amount of subsidy (e.g. free materials, rent-free plant sites and technical advice). The private firm receives the least amount of support.
- Marketing of substantial quantities of compost has been more successful when undertaken by specialised departments or consulting firms.
- Composting companies do not rely solely on local markets, but sell their products in several regions of India.
- Producing source-separated organics for composting requires considerable investment in education and monitoring, as waste generators seldom persist in keeping organic waste free of contamination.
- Small NGO composting experiments can fail if not supported with sufficient staff to allow careful monitoring and efforts to educate waste generators
- Small NGO projects cannot produce and sell enough compost to cover their capital and operating costs.
- Composting of urban organic solid waste still faces problems related to: acquiring sufficient uncontaminated organics, the rainy season, quality control, maintaining an acceptable price for the products and developing markets.
- There is an NGO (Swabimana) that acts as a citizens' forum, with a concern about solid waste management.
- The city government is open to ideas from the public and NGOs and has co-operated with decentralised experiments in waste reduction.

11.7. RECOMMENDATIONS

An important finding in this study is that the farmers who are most inclined to use waste-derived compost and who may be likely to pay a reasonable price for it are those growing higher-income crops such as flowers, tobacco and chillies. This is consistent with observations elsewhere that farmers producing staples such as rice or wheat cannot afford to buy compost produced in ways that reduce contamination (Cointreau, personal communication, 2000). The market development for such compost must be geared to an understanding of these types of farmers

Potential compost users want high quality products. The peri-urban farmers of Hyderabad have demonstrated a declining interest in highly contaminated decomposed wastes acquired from garbage dumps. If widespread separation at source were practised, sufficient high-quality compost could be produced to meet the demand. Vermicompost made from market wastes will be insufficient to supply a significant number of peri-urban farmers and other customers, and will cost too much, if not heavily subsidised. Vermicompost will not be significant in reducing the amounts of solid wastes for disposal. But this product can find a niche market in urban centres.

To produce enough good quality compost will require the co-operation of several stakeholders and much education and promotion to sustain the motivation of waste generators to do the separation adequately. The role of NGOs and community organisations will be important in promoting separation at source. Figure 11.1 presents an overview of the potential stakeholders in the composting business in Hyderabad.

The study recommends a systematic strategy for understanding the demand for waste-derived compost, especially the needs of users There must be an analysis of possible constraints and bottlenecks, and, in this, the experience of undertakings in Bangalore can be helpful.

The strategy for promoting composting of organic waste must be placed in the context of a strategic plan for municipal solid waste management. The municipal officials, especially in the department of the Municipal Officer of Health, need some training in strategic planning, especially with regard to options for waste reduction.

Since the MCH does not wish to engage in centralised composting after the failure of its earlier effort, it needs to enter into agreements with private enterprises and NGOs, if it is to have any role in waste reduction through composting. A degree of government support appears to be essential for the effective, sustainable production and marketing of compost from solid waste (see Dulac, 2001). Some of this support should come from the Ministry of Agriculture and rural extension agencies, since understanding the nature of agriculture and the farmers' needs is essential for an organic waste strategy.

Education and promotion can include demonstration projects, exhibits, workshops, focus meetings, and information brochures. Other cities with experience in waste reduction and composting of organic wastes can be invited to share their knowledge with civic and community leaders in Hyderabad-Secunderabad (see Lardinois and Marchand, 2000; Pitot, 2001).

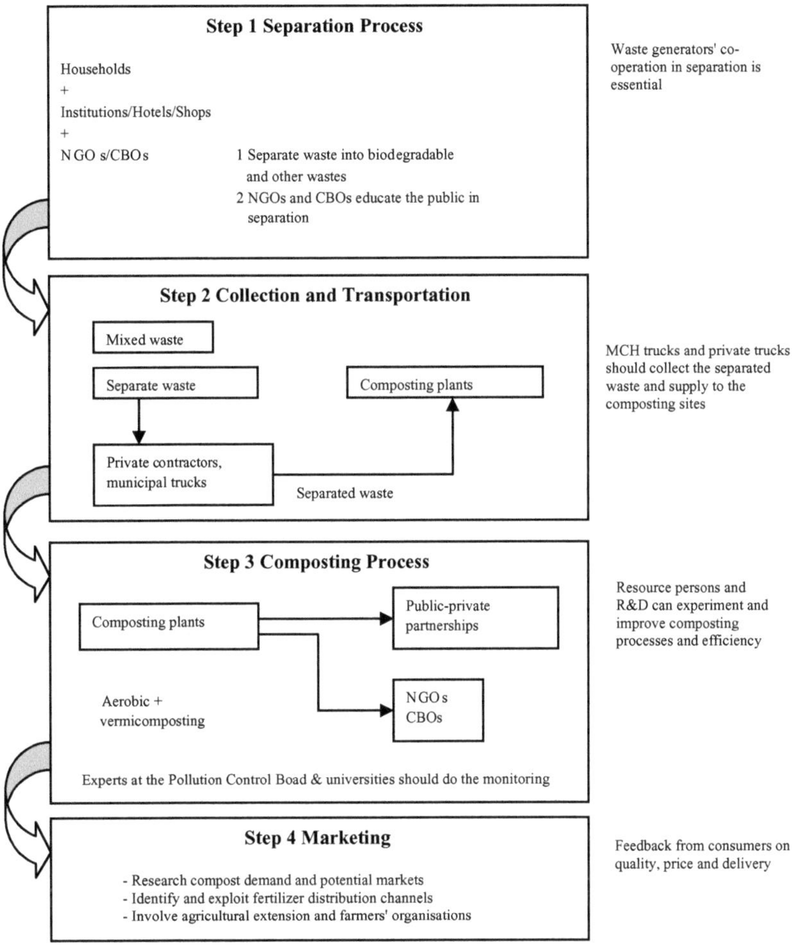

Figure 11.1. Potential stakeholder participation in composting of urban organic solid waste in Hyderabad

It appears feasible to develop markets for waste-derived compost in the urban area of Hyderabad-Secunderabad, but this will take time, since much education and promotion must be undertaken. Technical support and subsidies (e.g. free wastes) appear to be needed to produce a quality product at a price acceptable to a significant number of cultivators.

Farmers cultivating valuable crops, nurseries, plantations, households, institutions, hotels and parks are all potential customers Various stakeholders need to understand how compost making from organic waste can contribute to better solid waste management. Co-operation among all actors can help to develop the necessary behaviours and infrastructure for separation at source, investment in effective compost plants, quality control, and efficient marketing of the product.

THEO C. DAVIES, MOSES M. IKIARA, ANNE M. KARANJA AND CHRISTINE FUREDY[1]

CHAPTER 12
URBAN ORGANIC SOLID WASTE: PRACTICES IN NAIROBI

12.1. INTRODUCTION

Organic waste technically refers to waste containing carbon, including paper, plastics, wood, food wastes and yard wastes. In practice, in municipal solid waste management the term is often used in a more restricted sense to mean material that is more directly derived from plant or animal sources, and which can generally be decomposed by microorganisms. There are three main sources of organic wastes: (a) from rural generation; (b) from separate sources in urban areas, that is, source-separated organics (e.g. green market wastes, canning industry wastes); and (c) from mixed municipal waste (largely organic but significantly contaminated by post-consumer inorganics). Human excreta are not included in urban organic solid waste for purposes of this study.

The organic portion of the municipal solid waste load generated in Nairobi constitutes over 60 percent of the total (Karingi, 1997) and includes green waste, meat, bone and fish remains from markets, hotels, schools, hospitals and other institutions, kitchen wastes, farmyard manure, crop residues and yard trimmings, slaughterhouse remains and a fraction that finds its way into dumpsites. An indeterminate proportion of these wastes is separated and diverted to composting or fed to livestock, but the amount of raw organic waste that is spread on farms is shown to be insignificant. Composting and the use of raw materials like poultry waste and cow dung as manure as well as food waste from hotels, markets and other institutions, as animal feed, continue to gain importance as waste reuse activities that are helping to reduce the organic waste load requiring disposal in Nairobi.

The biodegradable nature of these wastes, if combined with mixed municipal waste that is improperly managed, poses health risks as they provide breeding grounds for for disease-linked micro-organisms, insects and rodents. The stench emanating from rotten organics in heaps of unmanaged solid wastes inconveniences many, while spon-

1. The authors acknowledge the help of the following research assistants for field surveys, under the supervision of Prof. Theo Davies and Dr. Erwin Koster: Kennedy Ogoro; Kevin Khisa; Harrison Kwach; Peter Omondi; Charles Wambugu; Alexander Mugo; Bernard Mwangi and Leah Oyake. Data analyses were carried out by Domisiano Mwabu, with Moses Ikiara, Theo Davis and Anne Karanja.

I. Baud et al. (eds.), Solid Waste Management and Recycling, 241-256.

taneous or deliberate fires contribute further to air pollution. Despite this, it is increasingly being realised that this fraction of the waste stream is a resource, which if managed properly could bring a number of socio-economic and environmental benefits to the community. This chapter addresses farming practices and trends in the use of organic waste in both urban and peri-urban areas of Nairobi. It examines the role of different partnerships and contributions of emerging private stakeholders in fostering this enlightenment in management of organic waste streams in the city of Nairobi.

Initiatives that have been taken by the formal and informal institutions, including local authorities, NGOs and CBOs to convert organic waste into organic manure and compost for marketing and reuse, are reviewed in chapter 4. The concept of an urban farmer is discussed in the light of the development of new production systems that have influenced spatial organisation with the introduction of appropriate crops to meet a changing urban demand.

12.2. TRENDS IN MANAGEMENT OF URBAN ORGANIC SOLID WASTE

According to Peters (1996) organic wastes are not usually scavenged by waste pickers, but are important to street children who often pick through bins to find their next meal. Some of the larger hotels and restaurants in Nairobi also sell their food scraps to farmers for use as pig feed. Organic wastes are also important to the urban agricultural sector as all sorts of livestock, including goats, chickens and occasionally cows, feed on top of waste heaps. This was particularly evident in the past year or two as a result of protracted spells of drought. Pastoralists invaded the City of Nairobi and cattle were feeding on whatever they could find, including wastes from food processing industries, from parks and street plantings, from racecourses, dairies, and so on.

Urban agriculture exists throughout the city on both private and public land. The growing of crops in urban areas is an important survival strategy for the urban poor (especially for those without land holdings) as it reduces the amount of income expended on food (Kettle *et al.*, 1995). Freeman (1991) estimated that one-third of urban households in Nairobi grow crops. A study by Mazingira Institute (1987) estimates that three-quarters of urban farmers consumed all that they produced. Urban agriculture is therefore an important food source for many people, and it should be encouraged. Both crop production, livestock rearing and poultry keeping are practiced at different levels in Nairobi and in the peri-urban areas.

Our recent fieldwork indicates[2] that more than 80 percent of all farms in Nairobi use some form of organic waste produced on site and about a third of the farmers use urban

---

2. The fieldwork on organic wastes conducted under this project consisted of interviews, focus group discussions and surveys of markets, institutions and urban and peri-urban farmers of Nairobi in 1999. For further details, see footnote number 3.

organic waste generated from outside the farm. There is a demand too for the latter kind of organic waste, and a significant number of farms source from this market, though respondents are unable to report accurately on rates of application. Organic waste brought into the farm from outside is, however, mostly animal manure. Most of the organic wastes collected from hotels, markets and other institutions are used for livestock feeding, with a small fraction applied directly to the fields, thus supplementing the role of crop residues as soil conditioners.

Forty three percent of the markets and institutions surveyed reported that some of the organic waste collected from their premises was used as animal feed, largely for pigs. A survey carried out by the Mazingira Institute in 1985 indicated furthermore that 14 percent of animal producers in Nairobi fed their animals on garbage in the wet season compared to 12 percent in the dry season (Mazingira Institute, 1987). Beside the use of organic waste direct from markets, schools, hotels, and other institutions as animal feed and on farms, urban and peri-urban farmers use compost as well. Most farmers rely on family labour, with up to 70 percent of all farms using family labour for applying organic waste. Nevertheless, the use of hired labour for general farming activities is also important.

The 1985 survey by the Mazingira Institute (1987) showed that 35 percent of the farmers in Nairobi used compost, 91 percent of which was from their own sources. Several community-based organisations are involved in composting of organic waste in Nairobi. Farmers do not resort to dumpsites to obtain organic wastes, although in some fringe areas, farmers ask waste companies to deliver municipal waste to their farms. Thus, recycling of urban organic waste including composting and the consumption of large fractions of food wastes resulting from reclamation of some components at the household and institutional levels, are the major determinants of patterns of urban solid waste reuse in Nairobi, though a small seasonal impact may also be present.

12.3. RECYCLING AND REUSE OF URBAN ORGANIC SOLID WASTE

There is no doubt that some recycling and reuse of organic waste is going on in Nairobi. However, the magnitude of these operations is not known with certainty, making it difficult to gauge whether or not they make a significant impact on fluxes in the urban solid waste cycle. Questions such as the quantity of food waste that is diverted and the quantity of dung coming from the urban area need to be addressed.

The actors involved in waste generation include households, markets and institutions, composting CBOs and NGOs offering assistance to CBOs and the farmers who are consumers of compost. The amount of organic waste generated by institutions is enormous and variable. It is difficult to work out precise estimates of quantities because they hardly ever weigh the amount of waste generated. Waste handling by institutions largely involves separation and sorting but many institutions simply dump the wastes together in a bin or some other receptacle to await collection. On average, a market or institution in Nairobi has 51 workers involved in waste handling. The number however ranges from 1 to 400, with 47.4 percent of the markets and institutions having 4 to 15 workers involved in waste handling.

Other aspects regarding the mode of operation of the institutions in the City of Nairobi are covered in chapter 4 of this volume.

12.4. URBAN AGRICULTURE

Provision of most urban services has, since colonial times, been the responsibility of the local government, the NCC in the case of Nairobi. The NCC has largely failed in this responsibility.

Urban-agricultural linkages, revolving around a range of diverse actors from those involved in the generation of urban wastes to reuse, are seen as having the potential to improve SWM. Among the more significant of these actors are: the farmers themselves; private waste handling companies; CBOs; the Ministry of Social Services; NGOs; and institutions such as hotels, markets, hospital and schools.

The urban space

Farming in the city has been categorised into two broad levels: (i) the 'urban' referring to small-scale crop gardens near the central part of the city, often located along roadsides and flood plains, and the high density residential area on the eastern side of the city where there is limited space, and (ii) the 'peri-urban' agriculture, where the land holdings are large enough to allow cultivation and livestock keeping for commercial purposes, as in Karen and Langata. But a phenomenon that is increasingly becoming common in Nairobi is the practice of 'zero grazing' (keeping one or more dairy cattle behind the house) and poultry keeping. Many urban residents are engaging in this practice to supplement food available to households and generate extra income.

Within the central part of the city, flowerbeds and small gardens along the city streets and around buildings are common. Along river valleys, flower and tree nurseries as well as food crops are the major agricultural activities. Food production is considered the most important form of farming in Nairobi. The Mazingira Institute (1994) study established that of the 20 percent of the households that practice urban agriculture, 90 percent use their land for production of subsistence crops.

The urban agricultural production space is derived from the livestock practices that existed before the process of urbanisation. An important characteristic of this notion is that the farming systems are framed within the "urban" concept, with access to infrastructure and public services causing a rise in land prices (Losada *et al.*, 1998). The most important characteristic of the urban concept is its foundation on two fundamental premises: first, the availability of vegetable wastes from the city's large markets and from other food-handling institutions as feed and fodder, as well as for composting; and second, the sustained demand from the local population for the resulting agricultural products. A further consideration is the additional pressure of producing within an urban environment that does not exist within the rural environment.

The peri-urban space

The other agricultural space identified here is the peri-urban, which, according to Losada *et al.* (1998), consists of the last remaining rural spaces in the metropolitan zone and, although there are an urban infrastructure and clear indications of an urban influence, there remains a predominant rural ambiance. It is clear that the influence of the city has affected the manner of agricultural production. In this milieu, a commercial form of farming is practised. The western and northern outskirts of Nairobi, having good agricultural potential as a result of well-drained and fertile soils, support a number of crops such as maize, beans, Irish potatoes, a variety of vegetables, bananas, flowers, fodder and cash crops. Livestock keeping is also practised on both large and small-scales (Mazingira Institute, 1994).

To the north of Nairobi are small subsistence farms and large export oriented horticultural and coffee farms. Cattle ranching is also a common activity and a number of farms rear pigs and poultry. In the western part one finds a mixture of both large and small farms where crops are grown mainly for local markets. Small parts of Kibera Division, such as Karen, have large farms of up to 10 hectares where dairy cattle are raised and fodder crops and other high value horticultural crops are grown.

12.5. FARMERS[3]

As noted above, the great majority of urban (and peri-urban farmers) in Nairobi generally do not use organic waste from the dumpsites, though some composting groups do. However, they use compost and fresh wastes from markets, schools, hotels and other sources. The average farm size in our surveys is 1.2 acres. Farmers applying urban organic solid waste tend to have much smaller holdings (averaging 0.82 acres), compared to conventional farmers, which suggests that farmers with larger holdings are able to generate enough organic waste on their farms, thus obviating the need for

3. Information on farmers was gathered through a survey on the use and application of organic waste, with respondents from 195 family farms selected using a grid sampling design to cover urban and peri-urban Nairobi (including Dandora, Eastleigh, Huruma, Kibera, Kikuyu, Kileleshwa, Langata, Maringo, Mathare, Pumwani and Upper Kabete). For details of the survey design, as well as content and structure of the questionnaires, see Mutoro and Karanja (1998) and Koster (1999).

acquiring waste from external sources. It may also mean that farmers with larger holdings are wealthier and can thus afford to buy manure and fertilizers Urban farming relies largely on family labour with up to 60 percent of all farms using this form of labour for applying organic waste. The average labour cost per farm or household is KSh. 5,687.50 (US$ 74). On average, labour cost is 21.3 percent of the income from farming, which appears rather high.

Problems associated with handling and reuse of urban organic solid waste

Major constraints with respect to the reuse of urban organic solid waste put forward by respondents in our survey were lack of capital, long transportation distances, high production costs and difficulty in storage of waste (Table 12.1).

Table 12.1 Constraints on the use of urban organic waste among urban and non-urban farmes in Nairobi

Constraint	*Percent of respondents (n = 195)*
Difficulty in storage of waste	30.0
Long transportation distances	23.3
Economic	18.4
Too much rains	13.3
Farm too small	10.0
Lack of transport	6.7
Lack of know-how	6.7
Contamination	6.7
Have enough from farm	3.3
Lack of understanding with dealers	3.3
Lack of labour	3.3
Others (unspecified)	6.7

The farmers who do not apply urban organic solid waste say they fear contamination of their soil as they do not know the origin of the organics, which might come from mixed municipal wastes. The only other problems that were cited as significant were difficulty in obtaining the waste, labour unavailability, high costs of sorting, long transportation distances and the associated health risks in handling the waste.

Solutions for tackling the solid waste problem as seen by farmers

The 195 respondents covered in our survey presented wide ranging solutions for combating the solid waste problem. The striking revelation is that privatisation of garbage collection is not viewed as an important solution. Also rather striking is that, in spite of its record of lethargy and apparent failure to deliver services, farmers believe that the NCC is still the best agency to deal with the solid waste problem. Nearly 50 percent of farmers believe that proper collection and transportation of solid waste spearheaded by the NCC, but also involving other stakeholders, is the best solution. They also emphasised the need to encourage the commercial use of organic waste through composting and sale in farming, as well as the need to educate stakeholders on recycling and reuse. Other solutions proposed are detailed in Table 12.2.

Table 12.2 Farmers' solutions for tackling the solid waste problem

Solution	*Percent of respondents*
Proper collection and disposal spearheaded by NCC but involving other stakeholders	46.7
Commercial use of organic waste (composting and sale)	40.0
Recycling and reuse	26.7
Educating people and creating awareness on proper handling	13.3
of waste	10.0
Burning	
Strict anti-dumping policy	6.7
Privatisation of garbage collection	3.3
NCC should pay workers well	3.3

12.6. INSTITUTIONAL FRAMEWORK IN COMPOST HANDLING

The NCC has the sole responsibility for solid waste management in Nairobi (Mwanthi *et al.*, 1997) but its services have been deteriorating, especially in the informal settlements. The inability of the NCC to efficiently handle solid wastes, including the organic waste load, continues to force an advocacy through the local government for the involvement of both large and small-scale private institutions as well as NGOs and CBOs in all segments of the SWM chain, from collection and separation to marketing and reuse. Mwanthi *et al.* (1997) suggested that schools and other institutions of learning, NGOs, religious groups, health workers, womens' groups (e.g. *Maendeleo ya Wanawake*) and landlords should assist the NCC in tackling solid waste management issues. Although the last four to five years have seen the emergence of these groups, an established institutional framework currently exists only in the composting sector, which formally came into being in 1992.

In that year, the Uvumbuzi club, through the encouragement of NGOs, embarked on composting as a means of dealing with the organic fraction of solid waste, primarily from households (especially in low income areas, where these wastes were being randomly dumped). The Uvumbuzi club had two main objectives for the projects it organised: (i) to start an initiative to clean the City, following the continued deterioration in services by the NCC and (ii) to set up the projects as income generating projects, especially in low income areas. The Uvumbuzi Club initially facilitated the formation of 10 groups but a number of other organisations have initiated self-help SWM projects in the City that likewise focus mainly on composting.

Beside composting, the use of raw materials such as poultry waste and cow dung as manure as well as food waste from hotels, markets and other institutions, as animal feed has emerged as waste reuse activities that are helping to reduce organic waste load in Nairobi. In addition, the Mukuru Recycling Centre (MRC), among its varied activities is pioneering the production and use of fuel briquettes as part of their waste conversion strategies. A good proportion of sticks in the municipal waste load is also collected by poor people in the City for use as fuel wood, which also accounts for a reduction of the total organic waste load generated.

Composting groups

Data from our fieldwork reveal that the functional composting groups in Nairobi, which currently number about 20, are generally young, nearly all of them having been established in the 1990s. Nearly 60 percent of these groups were established between 1994 and 1998.

Group membership currently ranges from as low as five members to as high as 62, although previously one of the groups had a membership of 200. Sixty percent of the groups have current membership of between 20 and 50 persons, and there appears to be a gender balance in membership, with women accounting for about 49 percent of the total.

Composting groups were formed in a variety of ways, but mainly through collective initiative by a group of persons in activities that eventually led to composting. The initial spur for the formation of a composting group seems to be economic gain. Seventy percent of the groups surveyed indicated that they were formed as a way of creating jobs through which members could earn sufficient income for improving their living standards. Another important objective relates to the desire to improve the sanitary and general hygienic conditions of their estates. Several composting groups in addition involve themselves in other community activities such as providing public education on improved waste disposal practices, and provision of health services.

Eighty percent of all composting groups had some assistance in getting started, with 50 percent of all the groups citing international organizations such as the United Nations Population Fund (UNFPA) and the United Nations Centre for Human Settlements (HABITAT) as their original sponsors Forty percent of the groups say they received assistance from local NGOs such as the Uvumbuzi Club.

Eighty percent of the composting groups operate from NCC land which has been allocated to them without the payment of rent and some groups, such as the Mukuru Recycling Group have actually reclaimed the land on which they are operating from waste dumps.

All composting groups appear to have acquired their knowledge through training, and indigenous knowledge does not seem to be an important route. The Foundation for Sustainable Development in Africa (FSDA) is an important training institution for composting groups, but the Kenya Institute of Organic Farming (KIOF), Uvumbuzi Group, HABITAT and Kuku Women's Group have also played noteworthy roles in training and capacity building in the field of composting.

Wastes for composting are commonly obtained from households where the composting outfits are located, or from nearby markets, local farms and firms as well as from dumpsites. Ninety percent of compost units do not buy their wastes, but obtain them free of charge.

Composting is a reclamation process, which involves activating and controlling the biological fermentation of organic waste in order to obtain a product which can be used as an agricultural soil conditioner. To obtain such a product, Nairobi composting groups proceed by collecting organic materials, making of a compost pile, and screening. All indicate close monitoring of the composting process to ensure high quality of the end product.

Based on respondents' figures, monthly expenditure on composting varies from as low as Ksh. 50 (US$ 0.65) to as high as Ksh. 35,000 (US$ 455) per group per month, whereas the estimated monthly compost output amounts to 700 kg to 5 tons per month. The Mukuru Recycling Centre, one of the largest and more established of the composting groups, reports a monthly turnover of 3-4 tons of compost produced from 6-8 tons of garbage (Mukuru Integrated Recycling Project, 1998). The MRC maintains at any one time a stock of over 6 tons of compost, which sells at Ksh. 10 (US$ 0.13) per kg. The group uses some of the compost for growing vegetables, which are sold in the market, but the main outlet is through individual farmers, who are their main customers.

Composting groups in Nairobi have a variety of sources of finance, with the most important being funding from local NGOs. Other significant sources of funding include the group's income and members' contributions, funding from international organisations and individual donors Most of the composting groups surveyed maintained that they have received inadequate assistance from public or private institutions. As such, many of these composting groups had not achieved one of their major objectives, that of being able to recompense their workers adequately. Investments in composting were much higher than the earnings and this acted as a disincentive for most of the group members.

Problems associated with composting

Composting has the potential to be income generating but there are a number of constraints facing community groups. The most important of these is probably the procurement of land on which to carry out the activities.

Our survey yielded the following listing of other constraints faced by composting groups during the wet season:
- Lower quality of compost due to erosion and loss of nutrients;
- The compost takes long to dry;
- It is more difficult to collect the waste and turn them in the pile due to wetness;
- Wetness slows down the decomposition process; and
- Difficulty in sieving due to clogging of the compost.

Constraints suffered during the dry season include:
- More water use;
- Longer composting period required;
- Dust produced during sieving; and
- Shortage of markets.

As such, supply cannot meet demand in the rainy season, whereas in the dry season supply may be greater than demand, and there is hardly enough storage space for the compost.

One of the reasons militating against sustainable development in composting operations is the lack of a ready market for the products, which group members attribute to poor marketing strategies. But there is also the wide availability of synthetic fertilizers in the market. In addition, transportation was considered to be a perennial and serious problem in marketing compost, as the long delivery distances involved engender huge expenditure of cash for members, none of whom have their own vehicles. Many of the potential markets for compost are in sites which are largely inaccessible.

Given these transportation difficulties, group members have resorted to the use of handcarts, wheelbarrows and sacks, other means of cheap public transport or request customers to use their own transportation means.

12.7. HEALTH EFFECTS

Health concerns are always mentioned in discussions of waste reuse (e.g. Furedy, 1996; Contreau-Levine *et al.*, 1998; Ratha and Sahu, 1993; etc.) The general effect of urbanisation has been the generation of urban pollution. This has inevitably affected the resources (water, soil and air) and the agricultural environment, reducing the production of traditional cash crops such as maize. This situation has been compounded by the abandonment of agricultural activity in favour of steady work within the city, leading to a vicious circle of deserted farmland, urban expansion and pollution. The extent of the detrimental effects of these pollutants on humans, animals and plants has not yet been documented, nor has enough research been undertaken that is specific for the health implications of organic solid waste reuse in developing cities such as Nairobi. Among the possible reasons for this may be that: the extent of waste reuse is not known; disease outbreaks can rarely be traced to specific practices; there are few experts able to do the research; and developing countries have many other health research targets (Furedy, 1998).

If concern about health risks of waste reuse in urban agriculture increases, there is a range of options to reduce risks to workers and consumers (Furedy, 1996). Minimising the contamination of organic wastes and wastewaters help both health considerations and economic viability. By obtaining pure organics, many public health risks can be reduced and the end product is more marketable.

There are two main procedures for obtaining relatively pure organic wastes:

- Separate collection from special generation points (fruits, vegetable and flower markets; food wastes from restaurants, canteens); and
- Separate collection based on segregation of organics by domestic and institutional waste generators (Furedy, 1998).

Although some studies on the relationship of socio-economic status to morbidity and mortality have been carried out which suggest a potential relationship between environmental agents and lowered health status, the association has not yet been proven, except in some specific, usually occupational related cases (Sexton *et al.*, 1993).

These issues are not easy to investigate. Sexton *et al.* (1993) listed a number of the difficulties involved. These include incomplete knowledge of the aetiology and pathophysiology of many diseases, a latency period of more than 20 years between exposure and clinical effects, lack of surveillance systems and methodologies that would accurately track environmental hazards, and the wide range of non-environmental causes of disease that could be modified by environmental hazards (Mitchell, 1994; Sexton *et al.*, 1993)

Health effects of composting

Despite using various forms of protective gear, more than a third of the composting groups, interviewed reported that group members or employees suffered health problems as a result of their work. These ranged from injury or cuts to headaches, skin ailments, stomach ailments pains and aches in the body, respiratory problems and fungal infections. Only half of the composting groups reported that their group members or employees sought treatment when they got ill as a result of their work. Health checks of composters have not been done, so other effects such as parasitic infections are not known.

The most important problem experienced during storage and application of urban organic waste is bad smell, as cited by almost all the respondents. Other important problems mentioned included flies and cuts from sharp objects.

Trace element- and heavy metal content of urban organic solid waste

The soil has enormous capacity to absorb and transform waste by microbial activity into usable nutrients for plant growth. However, the use of urban organic solid waste has caused concern in recent years because of the possibility of soil contamination by heavy metals and other trace elements. Analysis of samples of municipal compost by Purves and Mackenzie (1973) indicated that they contained up to 100 times as much copper and 300 times as much zinc and lead as uncontaminated, normal soils.

There is hardly any information on trace element and heavy metal composition of urban organic solid wastes in Nairobi, but a number of studies in Hyderabad (e.g. Rao and Shantaram, 1994) have shown that land application of organic waste increases the concentrations of trace elements in soils. These studies have also reported on the concentrations and relative availability of heavy metals in urban solid wastes and the possible implications of the addition of urban solid wastes to soil in relation to heavy metals.

These studies have shown that the long-term effects of such applications to land should be constantly monitored, for although the removal of most heavy metals by plants would be slow their accumulation may present a hazard in the long term. Williams *et al.* (1980), Banin *et al.* (1981) and Emmerich *et al.* (1982) have all advocated a cautions approach to repeated applications of urban solid wastes on land. It is therefore important to assess the impact of long-term application of fresh garbage on soil properties and trace metal concentrations in agricultural fields. Predicting potential hazards associated with land application of garbage is, however, sometimes difficult because of the inherent variability in the composition of wastes and the manures obtained from them. It should be ensured that the land application of wastes does not become an environmental hazard or cause fertility problems for the future (Rao and Shantaram, 1994). Judicious management through proper selection of wastes, soils and crops, and proper reliance on soil and plant testing can avoid most of the potential problems associated with garbage application on land.

12.8. GENERAL PERCEPTIONS OF WASTE WORKERS

Most of the workers in the composting sector see economic gain as the principal benefit that accrues from their work. They also acknowledge the contribution they make towards the attainment of a cleaner environment and thus to the enhancement of environmental quality. Composting groups have a clear vision of what they would like to see in the future. Sixty percent of our respondents spoke of plans to expand their businesses and increase their level of production, constructing more facilities and moving on to new premises. Others spoke of their wish to diversify their activities into provision of better sanitation services, different types of farming and the construction of houses for rental by group members One group indicated that they intend to develop a training institute on agro-forestry and composting, whereas another described plans for starting the operations of the already registered Nairobi Compost Co-ordinating Group.

Motivations for urban farming

The majority of urban farmers have a positive perception regarding the application and use of urban organic waste. The main reasons for applying urban organic waste is its effectiveness as a fertilizer, as a soil conditioner, and for the improvement of soil texture, crop yield and health of plants. Urban organic waste, they contend, has commercial value and is profitable, given its effect on farm productivity. They also attach great importance to some intangible rewards of farming such as knowing that people appreciate what they do, and the benefits for the family. It should be noted that the 'organic waste' they have in mind is farmyard manure, vegetable waste, and compost. No indication of the beneficial effects of each of these waste applications is apparent from the farmers' responses, but it is evident from the pattern of application that the relative proportion of these wastes applied would partly depend upon availability at any given time. A more important reason, however, would be the sometimes prohibitive and unaffordable costs of artificial fertilizers Small-scale community-based composting groups sell ready compost on site to farmers in bags of 20 kg for Ksh. 3.50 (US$ 0.05) to Ksh. 5.0 (US$ 0.065) per kg (Davies *et al.*, 1998), whereas chemical fertilizers could sell for anything up to hundreds of Kenya shillings per kilogram.

Farmers' opinions on use of urban organic waste

Farmers who use urban organic waste believe that it has a positive impact in the long run, especially as a soil conditioner, including improvement in soil texture and fertility. They also think that it gives an improvement in crop and livestock yield as well as enhances water retention capacity.

It was evident in the analysis of our results that the most common observation among organic farmers was that yields of plants increased on the treated areas in comparison with nearby untreated areas. Thus, an increase in total food production is the most common effect of the use of organic waste on agricultural land, as far as food chains are concerned.

12.9. CONTRIBUTIONS TO SUSTAINABLE DEVELOPMENT

Like the federation of pickers in Quezon City, Mexico (Vincentian Missionaries, 1998), urban organic waste handlers in Nairobi are actually serving at least two important functions in society. Through their self-employment initiatives, they absorb part of the otherwise state covered social costs of 'modernisation' such as unemployment and under-employment. Second, they shoulder part of the environmental costs of development by processing waste, which the state would otherwise have to spend money on in terms of organic solid waste transport and disposal.

The various relationships cited give rise to opportunities to find alternative forms of agricultural development for the urban environment, switching from mechanised forms of production that endanger the environment to traditional urban systems of agriculture characterized by 'self-conserving' technologies. Such transition is thought to have the potential of minimising pollution and ecosystem degradation, and of allowing increased income generation, thereby benefiting large sections of the population who have scarce economic resources. An additional objective concerns animal welfare and the search for a method of production which respects the well being of the livestock, whilst maintaining a reasonable standard of living for the producers

Despite some enlightened attitudes relating to sustainable development, the large urban centres are generally regarded as generators of contaminants, destroyers of natural ecosystems and high consumers of external inputs. But this is shown to be too narrow a vision of the urban situation. The growth and development of the urban centre has led to the establishment of new markets, directly influencing traditional systems of production.

In contrast to the blinkered view of the government, the enormous imagination and capacity of the urban farmer has made possible the agricultural production within the urban environment. This phenomenon introduces a very important dimension, demonstrating the capacity of the unprotected sectors (i.e. the urban farmer) to generate alternative solutions in the face of the environmental 'catastrophe' presented by the uncontrolled growth of the city. They have reorganized space, devised new ways of using degraded land, modified conventional agricultural systems in response to the demands of the urban sector and have realised the potential for utilising large volumes of waste as a source of food for animals and/or plants. Awareness is growing in the area of composting, but markets for this product are still poor and not readily available. For now, compost is used mainly in small-scale flower gardens of the Asian and expatriate community in the city. But the potential of composting as part of the waste minimisation strategy and eventually in income generation, is nevertheless huge.

Regarding public health concerns, we must concede that, although there are possibilities that crops of enhanced nutritional quality will result from the transfer of essential nutrient elements from the residues to the edible portion of the crop, there are also possibilities that food and feed crops produced on land treated with organic residues may contain materials from these residues that will be detrimental to the nutritional quality of these crops. The correction of micronutrient deficiencies has become routine agricultural practice, but still little is known about many of the factors controlling these relationships at high levels. An investigation of this subject must surely be advocated.

One of the bottlenecks to sustainable food production in and around developing cities is the lack of communication on urban agriculture among its actual and future practitioners, be they researchers, city farmers, urban planners, consumer organisations, city administrators, national and international support organisations and other stakeholders in urban agriculture.

Advocating proper composting and popularising its use among urban farmers have been used as a case in point. In the absence of a vigorous marketing strategy, and the ready availability of chemical fertilizers, demand for compost continues to be low. Neither is there enough awareness among potential users about its benefits in terms of crop yield and environmental benignity. There is thus a need for reversal of this trend through communication, to facilitate sustainable development in urban agriculture, which will continue to play a major part in food security in developing cities.

The medium of communication must not only be for experts, but should address all types of stakeholders including the city farmers themselves (both male and female). Assuring people's participation and thorough social preparation throughout the programme cycle is a positive step towards self-sustainability.

12.10. CASE STUDY

Some gaps still remain in our knowledge of the potential for urban organic solid waste reuse in Nairobi and whether it can realistically help solid waste management. Questions such as whether or not the informal use of source-separated organics (markets, food processing, etc.) is likely to increase or decrease, underscore the need for further investigations. In attempting to design the blueprint of a questionnaire for our organic farming systems research, we interviewed a subsistence farmer, Grace (a pseudonym), a single woman aged about 60 years, living in Kwawangware, in the outskirts of Nairobi, near her *shamba* (farm).

Grace's *shamba* is a 6-acre clan farm for which she possesses a title deed. She manages it independently, although she has several sons who also have their own families.

Dumping of organic waste on her farm started in 1989. She is paid a non-negotiable amount of Ksh. 50 (US$ 0.65) per lorry load of waste dumped. Grace does the sorting herself and sells appropriate portions, which are used mainly for feeding livestock. She concedes that there is a relatively vibrant market for organic waste in the neighbourhood. She spreads the rest on her farm and claims that she realises an improvement in crop quality from organic waste spread. Most of the crops grown are for family consumption. While we were there, two female neighbours came to recover some organics from the waste on her farm.

Among the problems Grace faces is occasional dispute with the neighbours over the aesthetic quality of the land and the emission of foul odour. But these disputes are often settled in her favour.

12.11. CONCLUSION

It is clear from the foregoing arguments that recovery and utilisation of urban organic waste in Nairobi could provide a very real opportunity to ameliorate some of the problems of waste management in the City. In addition, this may be considered a key strategy to minimise waste since this is a substantial fraction of the total waste load generated in the City.

A strategy for sustainable recovery and utilisation of organic waste must employ a participatory and cyclic process of planning and action to achieve economic, environmental and social objectives in a balanced and integrated manner. The strategy should aim to achieve all three concerns. The process should encompass the definition of policies and action plans, their implementation, monitoring and regular review. Much more effort by city managers is desired in organising the reuse of organics generated by households and institutions so that they do not enter the municipal waste stream. If this is done effectively, it would greatly help solid waste management.

Initiatives by composting groups are supported by various national and international NGOs and donors who finance particular components within the integrated community development programme. By 1997, there were 15 groups in Nairobi involved in composting and other initiatives towards integrated social development of their community. Today, these groups number more than 20 and are mainly concentrated in the low-income areas where they undertake commendable activities in their contribution to the enhancement of environmental integrity. The potential of composting groups to contribute to such awareness in their neighbourhoods is gratifying and compensates somewhat for the present lack of viable and steady markets. Good management, i.e. proper selection of wastes, soils and crops, and proper reliance on soil and plant testing, can avoid most of the identified potential problems from waste-borne microelements.

Lines of communication must always be kept open to all stakeholders so as to drive home the virtues of urban agriculture, but also to raise awareness on the pitfalls of improper application. Knowledge of the source of various toxic elements and their relative strength is necessary for taking effective control measures to prevent pollution in the application of urban organic wastes and compost to agricultural fields. An investigation of this subject must surely be advocated.

PART IV: CONCLUSIONS

JOHAN POST AND ISA BAUD

CHAPTER 13
GOVERNMENT, MARKET AND COMMUNITY IN URBAN SOLID WASTE MANAGEMENT; PROBLEMS AND POTENTIALS IN THE TRANSITION TO SUSTAINABLE DEVELOPMENT

13.1. INTRODUCTION

In this final chapter we return to the core question of the book: to what extent are socio-economic and environmental concerns integrated in the area of SWM, and do they contribute to urban sustainable development? The major results in the domains studied are compared for the two locations in terms of their context, the outcomes and the system concerns they raise. The criteria for sustainable development given in the first chapter provide the guideline for doing so. We also briefly reflect on the implications of the findings for the debate on governance, partnerships, social capital and urban sustainable development. Finally, we present some areas in which we feel there is scope for future research.

13.2. COMPARING RESULTS BY DOMAIN

Solid waste collection: contextual factors

The first research question deals with the impact of the institutional setting on the nature and functioning of activities, actors and partnerships in SWM. The investigations attest to the importance of the local political and administrative setting for effective SWC. In Hyderabad there seems to be a genuine commitment to the public interest and authorities at different levels have all recognized the need to take action. This is apparent, for example, in the issuing of new Municipal Solid Waste Rules by the Central Government in 2000, in the launching of the Andhra Pradesh Clean and Green Campaign in 1998, and in the large-scale privatisation exercise of the MCH. Through privatisation, the MCH has managed to increase spatial coverage as well as the quality of SWC services and has done so at lower costs per ton than previously when the SWC system was entirely based on direct public provision. In Nairobi the local government is largely unaccountable, unresponsive and inefficient, as is apparent in the virtual neglect of the problems of SWC on the part of the authorities. This is closely related to the difficulties in the country's political system, but is aggravated by the continuous political struggle between the (oppositional) NCC and the

I. Baud et al. (eds.), Solid Waste Management and Recycling, 259-281.

central government, which controls funds for local authorities, leading to a virtual paralysis of local government. The privatisation of SWC services in this city has occurred spontaneously in response to local consumer demand but is not accompanied by any safeguards on the part of the (local) administration to keep up health and environmental standards, to ensure co-ordination among actors, and to minimize inequality in service delivery.

The impact of contextual factors is clearly visible in the trial and error process that preceded the adoption of the unit system of privatisation in Hyderabad. The introduction of a rigid system of contract specifications and performance monitoring resulted from the desire to correct earlier shortcomings, notably the sensitivity of the system to corruption and abuse. Furthermore, the current contracting system is largely a product of political pressures by labour interest groups towards the protection of labour rights (especially those of MCH workers) and against attempts at reform. However, contract requirements are such that private initiative is stifled. Technological innovation with cost saving potential cannot be introduced given the specifications of number of workers and types of vehicles and equipment. The typical short duration of contracts also frustrates the willingness to invest and innovate. Furthermore, the authorities actively prevent concentration of power in the hands of a few contractors, allowing them a maximum of no more than three unit areas. Although this increases transaction costs, it also enables the local body to keep firm control on the privatisation exercise and to adequately carry out its regulatory function. At the same time the small size of the total area to be serviced by one contractor prevents cost savings through scale economies (although these are allegedly small in SWC, Cointreau-Levine, 1994: 16). In other words, the specific contracting mode adopted in Hyderabad may fit the particular political-administrative circumstances in this Indian city – a relatively solid state apparatus, a strong concern for public health and powerful labour unions – but it does not allow to reap the full economic benefits of privatisation. Furthermore, the official contract system is so inflexible that it cannot be utilized in slum areas, which remain conspicuously underserved.

Obviously, the private freedom of action is much wider in Nairobi. Here one can observe a degree of differentiation among service providers with relatively big companies dealing with high-and middle-income areas for comparatively high fees, and various small (informal) ones servicing low-income areas at much lower rates. This situation underscores the potential of the private sector to accommodate the needs of various population groups, if at least the potentials of small-scale enterprises are acknowledged.

An important distinction between the Nairobi and Hyderabad experience in SWC relates to the role of residents' CBOs. In the latter city the importance of SWC efforts through CBOs is considerable and collaboration with the local authorities constructive. The support given to the VGDS by the MCH, both politically and financially, is a major reason behind its success. Strict supervision by residential welfare associations has strongly contributed to the success of the scheme. Although involvement of CBOs could be observed in all classes of areas, including slums, their work seems to have the most effect in middle and high-income districts.

In Nairobi the impact of CBOs on SWC is smaller. In the slums CBOs (often with the support of NGOs) endeavour to compensate for the failure of the NCC and the apparent inability or unwillingness of the private sector to extend services to these areas. The (local) government keeps aloof from these CBOs and only provides some moral support by actively participating in their environmental clean-up campaigns. In general CBO-local authority relationships are either non-existent or antagonistic. Such residential welfare associations are increasingly challenging the NCC for its mismanagement and lack of accountability.

Solid waste collection: outcomes

The second research question looks at the contributions of various actors in SWC (on their own or through a partnership) to the goals of urban sustainable development. As the organisation of SWC in both cities differs, one can expect to find variations in terms of socio-economic impacts. A first remarkable observation is that actors in both cities have opted for using labour-intensive and technologically simple modes of SWC (largely working with open vehicles without mechanical lifting devices, manual sweeping), whereas many (top-down and large-scale) privatisation exercises in the developing world stimulated by international agencies (e.g. World Bank) have been based on the application of sophisticated, labour-saving technologies (compaction trucks, multi-lift containers). The methods applied in Nairobi and Hyderabad are both cheap and well adapted to prevailing physical circumstances and have the additional advantage that they allow for integration of small-scale enterprises that do not have large capital assets.

Despite the fact that choices of technology seem to fit local conditions, however, the profitability of its usage in practice cannot be taken for granted. The Nairobi case showed that the economic viability of privatised SWC is seriously impaired, as companies are engaged in a process of open and uncoordinated competition to cater for the needs of scattered clients. There is no legal backing or official support to grant providers exclusive rights to the service in a particular area and to require residents to participate in the service. In the SWM literature this is usually considered a prerequisite for cost-effective servicing. In Nairobi there were quite a number of so-called 'briefcase' companies offering collection services to residents in low-income sections of the city at extremely low fees but without any guarantee of regularity due to the deplorable state of their vehicles. In Hyderabad, in contrast, the economic viability of privatised SWC is more or less ensured as a result of detailed specifications in the service contracts designed by the MCH, including a 10 percent net benefit for the contractor. In actual fact, profit margins can be both slightly higher or lower depending on the balance between savings from small infractions on the one hand and resulting penalties and bribes on the other. Although the contractors were all eager to stay in business, they did object to the rigidity of the unit system. It assumes that the areas served are more or less identical in terms of collection demands simply because they are about equal in size. In actual fact there are differences in the volumes of waste generated as well as in the accessibility and the quality of the roads. The workload clearly differs per area leading to longer working hours for the labourers in 'difficult' areas.

Both the spontaneous privatisation of SWC in Nairobi and the planned privatisation in Hyderabad have resulted in higher levels of employment within the sector, largely by extension of services. So far, and obviously for different reasons, privatisation has not been accompanied by retrenchment of public sector workers In the Hyderabad case this implies that an important reason for local authorities to embark on the road to privatisation, *viz.* reduction of public spending, does not apply. However, in relative terms privatisation is accompanied by lower costs. This is largely due to the fact that labour conditions in the private sector are worse than those in government service (lower wages, higher job insecurity, fewer non-wage benefits and facilities). However, in Nairobi the differences are far less pronounced than in Hyderabad because state employment in Kenya is (no longer) an unquestioned privilege. Low wages are a major reason for frequent absenteeism and low labour productivity of NCC workers in comparison to private sector workers, who are under continuous pressure to perform for fear of being fired. The differences in Hyderabad can also be looked at from the perspective of the new workers in the private SWC companies themselves. They do not compare themselves primarily with the MCH workers, but rather with people from similar backgrounds that face even worse labour conditions.

In both cases there are strong indications of gains in productive efficiency (in terms of costs per ton of collected and disposed waste or in terms of number of workers per ton) when SWC is organised by the private sector. Furthermore, most of the time private operators turn out to be effective service providers and consumers are satisfied with their work. In these respects, therefore, the findings in the two cities corroborate a further transfer of SWC to the private sector.

In terms of allocative efficiency (the degree to which charges cover costs) the situations in Nairobi and Hyderabad are not really comparable. In Nairobi people are charged for waste collection through their water bills, but the majority does not receive any NCC service. Those who have hired the services of private operators are therefore facing a double burden. At the level of the individual private providers allocative efficiency is good, as fees cover the direct expenses incurred. The problem in this city is that the sizeable section of slums in the city is ignored by private companies, and as far as the lower-income areas are concerned, only those situated in the vicinity of the dump can be taken on board at a fee that is both affordable to the consumers and profitable to the entrepreneur. Privatisation of SWC on the basis of the full cost recovery in Nairobi leads to the exclusion of those who cannot afford commercial rates. In addition, one has to remember that the indirect costs of privatised SWC – using up part of the capacity of the municipal dump, environmental externalities – are not taken into consideration.

In Hyderabad, the record of allocative efficiency is very poor. As long as the introduction of service charges is considered politically unfeasible, the entire system will continue to rely on financing from the general municipal budget, and therefore remain subject to political debate (choosing among rivalling priorities). It remains to be seen if the system can be sustained, knowing that SWC expenses already constitute one-fifth of the entire municipal budget and that the city depends on external financing for all major investments in further improving the system. Only the VGDS has a positive score in terms of allocative efficiency. The scheme attests to the willingness of residents to contribute financially to better quality SWC. It also puts into perspective the reluctance on the part of political parties and unions to open the debate on cost recovery in public service delivery.

A special remark should be made on the viability of collection efforts in slum areas. In Nairobi, SWC in these areas is minimal and where it exists it largely depends on active involvement of CBOs. These are usually organized through self-help and youth groups that engage in other community services as well. As most slum residents are not in a position to pay user charges, incomes of CBO-workers largely stem from composting activities and sale of inorganic waste materials. The financial viability of these CBO-activities is somewhat debatable (due to absence of stable markets for recycled and compost products) and most of them continue to depend on financial support by donor agencies. Such dependence on donor support is also noticeable in Hyderabad's slum areas, where NGOs face difficulties in getting residents to pay for SWC services.

In terms of environmental hazards, the situation in Nairobi is considerably worse than in Hyderabad. The collection performance in Nairobi is no more than 25 percent of the total waste generated, while the Hyderabad actors together manage to collect about 70 percent. And although the slum areas in Hyderabad are certainly underserved, garbage dumped in open spaces is at least taken away every now and then. The Nairobi slums are virtually ignored by the authorities when it comes to SWC.

In both cases privatisation has had a favourable effect on the overall environmental health situation due to more effective collection. The spontaneous privatisation of SWC in Nairobi, however, has also induced uncontrolled dumping practices among private operators seeking to reduce transportation costs and to avoid dealing with dump gangs. In Hyderabad the privatisation process is highly regulated and dumping by private contractors occurs in a controlled fashion. The environmental health threats related to prevailing dumping practices are probably higher in Nairobi than in Hyderabad.

Separation of waste streams, especially of hazardous (industrial, chemical, hospital etc.) and ordinary wastes, still leaves a lot to be desired. As far as household waste is concerned, the regulatory and institutional framework for monitoring the SWC process is more developed in Hyderabad and its enforcement capacity more effective (disregarding possible corruptive practices).

A major worry relates to the sheer lack of dumping sites, a phenomenon that is very common among rapidly growing cities in Africa, Asia and Latin America. The new landfill sites that are under consideration in Nairobi and Hyderabad are located at a considerable distance from the city and will increase transportation costs. Under current conditions this is likely to lead to more indiscriminate dumping, especially in Nairobi. The method of open dumping adopted in both cities is also cause for considerable anxiety. Although ecological considerations may have found their way to official documents, they still constitute a marginal item in SWC practices at local level. This is due partly to a lack of awareness at the executive level, and partly to financial constraints. In fact, municipalities have often directed dumping to low-lying areas that are not designated dumps at all. In Hyderabad, the installation of the pelletisation plant is an important step in the right direction. At the same time the MCH indicates that the construction of a proper sanitary landfill is far beyond what the city can afford.

Solid waste collection: system concerns

In the study we also paid some attention to the coordination within a particular sub-sector as well as the interaction with the other sub-sectors of the SWM system.

A major strength of the system of SWC applied in Hyderabad is that all residents living in planned areas of the city receive a basic level of SWC services – collection through vantage points at walking distance from the houses – regardless of wealth. On the other hand, private providers are unable to tailor their services to the specific demands of their clientele because they are tied to the contract system. Those areas desiring a higher level of services have to organize this themselves. In the recent past this has been done through the VGDS, which offers a house-to-house collection service in return for payment of a small user fee. The VGDS operates entirely independently of the basic collection system. Therefore, opportunities for better coordination and cost saving, which could be realized if all activities in a particular area came under one firm's responsibility, are lost.

In Nairobi, the lack of coordination is a much bigger problem. There is a situation of open competition among a variety of operators, ranging from respectable firms to brief-case enterprises. Furthermore, a regulatory framework to oversee private service operators as well as an official monitoring institution are absent. Although the lack of organisation may offer attractive opportunities for informal ingenuity and entrepreneurship, it also constitutes an immense threat to the public interest.

A system weakness that could be observed in both cities comes from the lack of coordination within the SWM sector. In policies the emphasis is always on collection and disposal, while the potentials of reducing waste flows through recycling and reuse are hardly recognised, if at all. In the realm of SWC this is most apparent in the absence of official guidelines and facilities to enhance waste separation at source. In Nairobi, no such thing as a coherent SWM policy exists. The limited input of the NCC seems to be dictated by particularistic and political motives, albeit under the disguise of environmental health promotion. There is hardly any attempt to move beyond SWC and to integrate collection with the recycling and reuse business. Both the positive and negative externalities of collaboration across the boundaries of the three domains are purely coincidental. In Hyderabad, a coherent policy on SWM is also absent. The authorities seem to be preoccupied with conventional concerns for environmental health, and service efficiency and effectiveness. In their privatisation policy no attention is given to the recycling and reuse of (in)organic waste materials. Nevertheless, conditions to merge perspectives seem to be more favourable. In the VGDS, supported by the MCH, an attempt is being made to combine classic SWC concerns with broader socio-economic and environmental goals.

In general a certain degree of antagonism could be observed between waste collection workers (either in private or public service) and waste pickers searching for valuable materials. The latter are accused of making the work of collectors and sweepers more demanding because of littering around collection points related to waste sorting and because they remove valuable waste components that could otherwise be recovered by collection crews. The more organized and regulated the SWC system, the more the livelihood of waste pickers is jeopardized. The fewer the opportunities for waste pickers to sort and trade waste, the larger the volume of mixed waste that reaches the final disposal site.

Recycling and reuse of inorganic waste: contextual factors

First of all, it has to be kept in mind that the statements below are confined to the trade and recycling of common materials from households, shops and institutions. Industrial recyclables, for example, were not taken into consideration.

Different regulatory frameworks affect collection, trade and waste recycling as they are carried out in commodity chains dominated by enterprises using waste as raw material in both cities. These concern local by-laws on SWM, labour and factory regulations, and the regulations for imports of raw materials.

The studies confirm that local authorities have not introduced any 'greening' of local government by-laws on SWM. They remain firmly rooted in the public health perspective, in which there is no room for waste material recovery and recycling. No elements of the 'waste management' hierarchy have been introduced through local government in either city. Neither is source separation by households part of the regulatory framework. This means that current activities in that sector remain outside the government purview, and take place solely out of economic considerations. The danger is that when these reasons disappear, the recycling sector will also be reduced (as occurred in Europe in the 1960s and early 70s). If so, it will take more to rebuild it afterwards, when an environmental perspective becomes more accepted.

The reasons for trading and recycling waste materials currently lie in the economic context of both cities, of too little employment with a living wage, and enterprises producing products which do not require high quality raw materials. This situation has existed over a long period of time in India, and has gradually emerged more recently in Nairobi, as the economy deteriorated there. The activities take place in the context of an economy characterized by enterprises and employment ranging on a continuum from formal to very informal. This context makes the survival activities of waste pickers and itinerant buyers possible, as they do not adhere to labour and enterprise regulations. Such activities are likely to become uneconomical if the regulatory context for production and employment changes, although such changes are unlikely in either country in the near future.

National regulatory frameworks for raw material imports have generally made it difficult for entrepreneurs to obtain virgin materials, such as plastic in Kenya and good quality paper in India. However, after the two countries started opening their economies in the 1990s, import barriers were reduced, with the result that alternative sources have become available in both countries. This has negatively affected the recycling of domestic waste materials (plastics, paper): in Kenya, the plastics market has collapsed. The effects are felt in Kenya more than in India, as the market for waste materials is less developed. It does not preclude the international trade in waste materials. If they were not hazardous, this would be an alternative market, as van Beukering has suggested (2001).

Privatisation was the last major change in the regulatory context. Although the changing way of collecting waste should have affected access to waste for recycling in both cities, the studies found no major closing off of access because of this process. In both cities, the private sector waste collection companies also earn from their waste trading activities.

Reuse, recovery and recycling of inorganic waste: outcomes

Similar materials form the basis of the recovery and recycling chains in both Hyderabad and Nairobi, glass, paper, plastic and metal being important categories in both cities. However, the complexity of the commodity chains differs substantially between the two cities. In Nairobi, the level of specialisation is more limited than in Hyderabad. This reflects the lower levels of demand and profitability for recycled materials there, which do not allow wholesalers to specialise in one material but require them to spread risks by dealing in several types of materials, whereas in Hyderabad, wholesalers commonly specialise.

Differences in profitability are also reflected in the structure of demand for waste materials among recycling units. In Nairobi, only one large-scale company (several hundred workers) buys any given material, thus gaining a monopoly over pricing and amounts traded. This has led to large fluctuations in price, and a current dip in demand for waste materials from Kenya itself. The only exception is scrap metal, in which small enterprises also have a role in the recycling. However, Asian traders are currently driving up prices of scrap metal, which may lead to a lack of profitability among small metal recyclers In Hyderabad, in contrast, a large number of mainly small companies (7 out of 10 have less than 50 workers) buy waste materials for producing new goods. They show an average profit level of 10 percent.

Although the recycling enterprises operate on the formal side of the law, the wholesalers supplying them usually operate on a borderline. In both Nairobi and Hyderabad, they require licences to operate, but can also manage through informal payments to inspectors and police monitoring activities. The itinerant buyers and waste pickers supplying the wholesalers and dealers operate completely informally, and are vulnerable to harassment and 'unlawful payments' by police and others because of this. This reduces the profitability and productivity of their activities.

In both cities, recycling enterprises using recovered materials do so on the basis of technical and economic considerations. In both cities, waste materials are combined with virgin materials to produce higher quality end products. Reductions in the price of virgin imported plastic have led entrepreneurs in Nairobi to change the composition of their inputs. Similarly, in Hyderabad, imports of higher quality paper from abroad can lead to a smaller demand for domestic waste paper. In the Nairobi case, the vulnerability of the trading and recycling chain is greater than in Hyderabad, because of the monopoly of one recycling enterprise there per material. The size and differentiation in the market for waste materials in India provides more of a buffer for such fluctuations. This implies also that these factors remain important for the level of productivity within these enterprises. As there is no external subsidy (which could be the case if the contributions to environmental aspects were recognised), it is the trade-off between cost and quality which makes entrepreneurs decide whether or not to keep using secondary materials.

Employment generated in both cities throughout the commodity chain differs sharply in quality. The street and dump pickers in both cities collect and sell waste at a survival level, obtaining cash and goods in kind from their picking. This occupation has emerged in large numbers in Nairobi since 1985, whereas it has existed for much longer in India. In both cities, men are dominant among street pickers, women among dump pickers Women earn less than men in their activities. The pickers are illegal in both cities, and have no protection from harassment of police and populace. Their main form of social security consists of loans from the dealers to whom they sell their materials. Itinerant buyers are also usually informal, but tend to earn higher incomes in both cities than waste pickers do. Their ties with shopkeepers providing goods or capital also tend to give them a little more security. For both categories working conditions are unsafe and unhealthy, as they do not take protective measures while working. Among pickers and itinerant buyers, there is little upward mobility to employment at the next step in the commodity chain. In fact, there is increasing competition for waste materials being experienced in both cities.

Employment in enterprises is confined to people employed by dealers and recycling units. Dealers in Nairobi employ mainly family labour, as do small dealers in India. The differentiation among dealers is much greater in India than in Nairobi, although in both cities they form the link between recycling units and the pickers and itinerant buyers The small dealers work mainly with unpaid family labour, whereas the large wholesalers employ casual and contract labour in India for sorting activities. This is often work done by women and children. In the recycling units, employment also consists of contract work in both cities. The companies in Nairobi employ more workers than those in India, but they are the only ones in their sector. In India, on average, the number of workers employed in the recycling units is 50, of which almost forty percent are women. Again, even in the recycling units, safety and health measures are absent for workers Only more permanent workers are provided with medical insurance and other non-monetary benefits.

The legitimacy of the trade and recycling sector is low as regards the collection and trading activities, and pickers and small dealers experience the disadvantages of such lack of legitimacy. They are harassed by police and inspectors, and are forced to make higher overhead costs because of bribes they have to pay. The recycling enterprises are part of the registered formal economy, which provides greater legitimacy to their production activities. However, neither the government nor the entrepreneurs themselves officially recognise their contributions to environmental sustainability.

Turning to the environmental aspects of waste recovery and recycling, the studies show that substantial amounts of waste are recovered. The two cities are difficult to compare by amounts of waste materials recovered, as in Nairobi waste is obtained from all over the country and in Hyderabad it is obtained from local and regional sources. However, the amounts show that on a weekly basis several tons of the main types of material (glass, metal, plastic) are recovered in both localities.

The studies show that waste recovery is linked not only to domestic or municipal waste streams. Wastes from institutions and enterprises are also preferred sources, because they provide higher quality unmixed waste. Therefore, efforts to increase waste segregation at source in institutions and enterprises remains as important a channel for resource recovery as increasing source separation by households. It also suggests that attempts to limit private access to waste streams are misplaced in promoting greater resource recovery.

Recycling and reuse of inorganic waste: system concerns

The major system concerns relate to the lack of coordination between waste recovery and recycling efforts and the municipal SWM collection system. This should be of growing concern, as in both cities, lack of space for disposal of waste is already a big problem which local authorities are unable to solve. Efforts to increase the effectiveness of collection only exacerbate the issue of disposal, whereas increasing the extent of waste recovery and recycling can make a substantial contribution to reducing waste flows.

There is a serious problem in promoting co-operation and partnerships between the private sector and local government, as long as the regulatory framework for SWM does not include goals for waste recovery and recycling. At the moment, the contestation remains that such trading and enterprise activities remain on the border of or on the wrong side of the law concerning tax and employment regulations, making entrepreneurs, traders, itinerant buyers and pickers very reluctant to become more formalised than they are. Nevertheless, if waste recovery is to be increased, and the quality of waste maintained and improved, the regulatory framework has to be changed to provide additional incentives to both waste generators to segregate waste, and to waste recycling enterprises to invest in expanding production and improving the quality of products made. Finally, the issue of balancing open imports of raw materials with promoting recovery of local waste materials as input needs to be put on the agenda in such a way that the quality of production is not affected, and the extent of materials recovered is maximised.

Another problem of coordination stems from the fact that waste recycling is not limited to city boundaries: in Kenya it extends to the national level in plastics, and in India it has sub-national regional links, as well as international linkages. This implies that linking such systems to municipal SWM may not always be an easy combination, as the trading and recycling companies may work at different scale levels than the local municipality does.

Reuse and diversion of organic waste: contextual factors

Organic waste makes up a major portion of municipal waste in both cities, as elsewhere in developing countries. Therefore, the patterns of recovery are a major concern both for SWM and in promoting sustainable development.

The diversion and reuse of urban organic waste in Hyderabad and Nairobi consist of a peculiar mix of private ingenuity, community activism, and donor supported social welfarism. The (local) authorities have largely stayed aloof in this aspect of the SWM system. But although official policies for enhancing organic waste recovery are sadly missing, the MCH has at least attempted, in the 1970s and 1980s, to make use of urban organic waste as a resource. Unfortunately, experiences with large-scale municipal composting have been disappointing and hence enthusiasm to continue investing in such operations has waned. Currently, the MCH does provide some help to small pilot-scale composting activities (providing them with land, sometimes with market wastes) albeit more as a political gesture than an attempt to reduce municipal waste flows. So far, the MCH has not taken any initiative to enhance waste separation through sensitising campaigns, technical facilitation and/or regulatory provisions. However, this situation may change in the near future as a result of the recent promulgation of the national directive for solid waste management in large cities, making separation at source and composting mandatory. But although the national context seems to have become more sensitive to the idea of addressing urban organic waste problems, the MCH, for the moment, has put its trust in a private attempt for energy recovery from (organic) waste[1]. This move acts as a disincentive to a more vigorous policy towards the utilisation of urban organic waste in (peri)-urban agriculture.

In Nairobi, the overall political situation has obstructed any attempt at coherent policy formulation, not to say implementation. Except for occasional support to composting groups (providing them with land and waste) the authorities have not done anything to make better use of the organic waste potential. What takes place in terms of waste diversion is largely through collaboration of non-public actors In fact, an important observation in both cities is that composting projects primarily arise from CBO-NGO collaboration rather than stemming from public sector policy interventions. Most NGOs firmly believe in the potentials of community action and facilitate such processes financially and technically. However, in Nairobi and Hyderabad they have not really managed to take the idea of neighbourhood composting beyond the level of experimentation and to turn it into a cornerstone of official SWM policy.

1. Although this idea has gained some popularity in policy circles, most energy from solid waste projects so far have failed (Furedy, personal communication).

Reuse and diversion of organic waste: outcomes

The reuse and diversion of urban organic waste is currently receiving a great deal of attention in the literature as the ultimate way of reducing volumes of solid waste in cities in Africa, Asia and Latin America. In cities with a sizeable urban and peri-urban agricultural sector (such as Nairobi) the potential seems to be best. The findings indicate that it is essential to make a distinction between various types of urban organic waste: pure urban organic waste, urban organic waste mixed in municipal waste, and compost. As far as the first is concerned, bulk generators can easily find appropriate outlets and the municipalities do not have to interfere. Within this category of pure urban organic waste we can further differentiate between animal manure and food and fruit leftovers (mainly used as animal feed or for fertilizer production). In the Kenyan capital agricultural applications of pure urban organic waste predominantly concern animal dung, but goats are fed market wastes. In Hyderabad both types of usage are extensive.

Mixed urban organic waste (coming from the municipal waste stream) is not much used in Nairobi in (peri-)urban farming. Hyderabad did have a tradition of farmers using decomposed matter from mixed municipal solid waste but it seems to have become less popular in the past five years Many farmers around the city have discontinued using this type of urban organic waste on their fields, at least in the area taken up in our study, because of mounting negative effects (health hazards for workers and animals, obstruction by workers, lower yields due to soil contamination). Besides, transportation costs for this particular type of low quality agricultural resource soon become prohibitive, implying that demand is highly localized, and consequently restricted.

The core problem inhibiting wider usage of urban organic waste in and around the two cities is the lack of waste separation. In the course of time this has become an increasingly serious handicap because of higher levels of contamination of waste (more non-biodegradable components, more hazardous substances entering the waste stream). This problem also affects composting undertakings which, in both cities, are currently dominated by small-scale units. The production of good quality compost is difficult. It not only requires appropriate technical skills and facilities, but also depends on the availability of an adequate uncontaminated feedstock. However, not all composting units succeed in getting sufficient access to pure organics. This is especially troublesome for neighbourhood composting units that depend on local waste collection efforts rather than the purer supply by bulk generators (such as markets and food processing companies).

In both cities compost was predominantly used in specific niche markets (non-food crops, notably floriculture and mushroom cultivation where it can successfully compete with chemical fertilizers). The major reason for this is that most local farmers either cannot afford the products (the small and marginal farmers) or have adequate alternatives (especially manure). In addition, farmers in both cities seemed to be insufficiently informed about the potentials of compost as soil conditioner/fertilizer and, hence, about what would be a fair price for them to pay. One should not forget that most composting activities take place in small-scale, neighbourhood units that lack adequate marketing and promotion channels to enable them to fully exploit the market potential. The lack of official backing and interest is also frustrating the expansion of the composting sector.

For the time being most neighbourhood composting groups, even the more sophisticated vermicomposting unit in Hyderabad, are unprofitable undertakings. In both cities the activity depended heavily on donor funding, not only to cover initial investments, but in many cases also for keeping businesses running. Although this may partly be justified by the positive social or environmental spin-off (creating employment and training opportunities for a category of deprived people, raising community awareness on solid waste problems, etc.), the long-term survival of these schemes is not guaranteed as external support is likely to be phased out. In Hyderabad the continuity problem was further compounded by the negative attitudes among neighbouring residents complaining about vermin and stench in community composting experiments.

Reuse and diversion of organic waste: system concerns

The overall impact of the reuse and diversion of urban organic waste from households on reducing volumes of mixed municipal waste going to the dumpsites is limited in Nairobi and Hyderabad. Without an active policy towards separation of different types of waste at source the likelihood of progress in this respect is low. The two cities do not have an incentive structure – for example through differentiated rates for the collection of mixed and separated waste – that encourages households to participate in separation at source. Neither do they have the capacity to sensitise people on the environmental gains of such schemes. This is yet another piece of evidence of the lack of coordination and integration within the SWM system. Potential savings in collection and disposal costs could be utilised (partly) to subsidise waste separation, reuse and composting schemes, but responsible city authorities have not yet adopted a more comprehensive outlook that would allow for a reconciliation of brown and green agenda issues at local level. The prevailing bias towards service efficiency and effectiveness at the cost of long-term environmental concerns – thereby making systems of solid waste disposal artificially cheap (especially due to the practice of open dumping) – results in an underestimation of the potential savings of reducing waste volumes. Consequently, it seems less likely that the cities will want to subsidise initiatives that contribute to waste reduction. Nevertheless, large cities in India are now required to divert organics from dumps through composting and other techniques, so they will have to grapple with the dilemmas of organic waste management. The successes and failures in India in the coming years will provide guidance to smaller cities and other countries.

13.3. GOVERNANCE, PARTNERSHIPS AND SOCIAL CAPITAL

The research project touched upon several issues that are subject of debate in academic and policy circles. The results enable us to underscore or qualify several of these ideas, notably three strongly interwoven topics: partnerships, social capital and (environmental) governance. Both in Nairobi and in Hyderabad several activities within the SWM system are moulded around partnerships. The nature of these arrangements partially depends on the activity concerned. Sometimes an activity is organised on a purely commercial footing – for example the trade in recyclables – despite the fact that it may have important externalities in terms of environmental health, reduction of waste flows or social legitimacy. At the other extreme there are activities that belong to the category of purely collective goods for which no market exists, like for example street cleansing or final disposal of residual waste. In that situation the public sector usually takes prime responsibility not only in organising for its delivery but also in financing it. Obviously, the basic orientation of a partnership influences the balance between various interests. A distinguishing characteristic of partnerships, however, is that they directly or indirectly serve the public interest, which distinguishes them from exclusively commercial relationships (Gonzalez *et al.*, 2000; Baud *et al*, 2001). A major question, therefore, is what provisions have been made to safeguard the public interest within various types of partnerships. Sometimes this leads authors to state that in partnership arrangements at least one of the partners should be public (Peters, 1998).

In the literature there is an undeniable bias in the direction of partnerships that have been planned by the authorities in a top-down manner. Their emergence can be traced back to new conceptions of (urban) public management and local governance that are based on the idea that each actor has its own set of comparative advantages. The assumption is that the common cause is best served by utilising these potentials, and, therefore, the core challenge is to find the most appropriate mix of capabilities. In these partnerships the public sector is always the key player. However, partnerships can also emerge from the bottom-up through collective action, for example through CBOs engaged in waste collection, recycling and reuse. Collective action refers to the organisation of shelter, basic services, employment and security by local communities themselves without any government assistance and usually against considerable odds (Baud, 2000; Mitlin, 2001). Communities can do a lot on their own to improve their living and working environment, but many actions require them to establish partnerships with external actors that can bring in essential complementary resources (Lee, 1998). Our study on SWM in Nairobi and Hyderabad has clearly shown that such partnerships are relevant and need to be taken into consideration in addition to public-private arrangements. In the subsequent analysis we will reflect on various factors that help to explain differences in the types and functioning of partnerships in SWM within the two cities under scrutiny.

Although both the Kenyan and the Indian government have set out on the road to liberalisation and deregulation of the economy and aim at transferring conventional state responsibilities to the private sector, the situation on the ground is very different. In Kenya, the state suffers from a crisis of legitimacy. It is continuously accused of mismanagement and corruption, and virtually unable to satisfy elementary public sector obligations. Private and civil society actors are reluctant to engage in partnership relationships with such an undeserving partner. The Nairobi case shows that an important requirement for successful collaboration across the public-private divide is missing (Evans, 1996; Ostrom, 1996; Mwangi, 2001). Norms of mutual trust that provide the basis for exploiting the potential for synergy are not established. Private operators try to avoid dealing with the authorities and their representatives and if they are required to do so (to obtain a business license, to get access to the dump site etc.) they try to arrange affairs informally. Similarly, residents have lost faith in the authorities and seek alternative ways of satisfying collective needs. In Nairobi, the social capital that constitutes the basis for mutually beneficial state-society collaboration has been severely eroded.

In Hyderabad one can also observe some hesitation on the part of public and private actors to work together on the basis of trust. In the privatisation of SWC the authorities have learnt to be constantly alert to infractions and abuse by contractors, while the latter feel the MCH is unduly squeezing them through the inflexibility of contracts and rigid application of the system of deductions. Nevertheless, the local state's legitimacy and reliability goes almost unquestioned. A feeling of mutual trust and respect is developing between private contractors that are well performing and a council that lives up to its financial obligations. Furthermore, the MCH not only accepted the VGDS as an add-on to its own basic SWC service, it also supported the initiative financially and logistically. In other words, both public and private actors in Hyderabad are working towards the strengthening of social capital, i.e. the social underpinnings of governance (Stoker, 1998).

This does not mean to say that the transition to governance – from public control to power sharing in collective affairs – comes as a matter of course in the Indian context. In fact, political commitment to the idea of partnering is still problematic in both cities studied. Furthermore, the administrative structure and culture (attitudes, working procedures) are still not well adapted to the idea of working together across the public-private divide (requiring, especially, more open, participatory modes of planning and policy-making). In Nairobi considerable pressure is exercised on the authorities to enter into partnerships – for example through the internationally sponsored Nairobi Urban Slum Development Project – but prevailing attitudes among government officials and employees are still largely negative. The political climate is still permeated by authoritarianism and reluctance on the part of the old establishment to alter the power structure. Actual commitments to the idea of partnerships, therefore, are few and seem to be motivated primarily by the desire to gain access to foreign funds rather than a genuine belief in the potentials of partnering. In the absence of a regulatory framework for partnering, the private-private arrangements that sprang up in the city have collectively developed 'regulatory rules' and market-efficiency themselves. This is a difficult process, in which much can go wrong, as conflicts and lack of trust among the firms show.

In Hyderabad, both political and administrative opposition to the transfer of traditional public sector responsibilities to the market or civil society continues to be quite strong. The country's history of state-led development and the fact that many of its major institutions (political parties, trade unions) are imbued with socialist ideas translates into slow implementation of reforms that challenge the state's leading role. This is shown by the strong path-dependence of institutional reform, i.e. the difficulty in adapting not only official rules and regulations, but also the mind set and capacities of those responsible for implementing them (cf. Leach, Means and Scoones, 1999). The partnership idea may be gradually gaining ground in Hyderabad, but at the same the authorities clearly indicate that they want to remain firmly in the driver's seat.

One also has to take the capacity of the public sector to properly fulfil its role in partnership arrangements into account. It is often claimed that privatisation requires the 'guiding hand of the state' to become effective (Batley, 1996; Post, 1999). When responsibilities are passed on to the private sector, safeguards must be built in to ensure appropriate standards, achieve coordinated provision, ensure a competitive environment and avoid monopoly control of essential services by private providers which are not publicly accountable, and to minimise corruption and inequity (Rondinelli and Iacono, 1996; Burgess et al, 1997). The move from direct provision to co-ordinating and supervising private delivery implies that local bodies have to completely reorganize and reorient their administrative machinery with much more emphasis on contract management and performance monitoring. In addition, the entire regulatory framework has to be modified to suit the new division of tasks and responsibilities. In Nairobi these pre-conditions are certainly not fulfilled, and therefore no adequate basis exists on the public side to develop partnerships with the private sector. In Hyderabad the local state is much more competent and better equipped to adequately discharge of its supervisory responsibilities. Although the first steps on the privatisation track produced mixed results, the local body quickly learned from its mistakes and adapted its role accordingly. The authorities have created a situation in which private firms can operate efficiently and effectively. However, their desire to keep everything under control also implies that the freedom of movement of private partners is seriously curtailed.

The comparative strength or weakness of the public partner is also important from another perspective. A major characteristic of partnerships is that each partner can bargain on its own behalf and take decisions without having to consult with other authorities (cf. Hordijk, 2001). An important hindrance in the Kenyan capital is that the NCC is incapable of acting as a principal. Its chronic lack of financial means, related to the weakness of local government in Kenya, together with its antagonistic relationship with the central government disqualify the NCC in this respect. The MCH, on the other hand, is sufficiently solvent and independent to be considered a robust partner. Therefore, the level of uncertainty is comparatively low for those wishing to engage with the MCH.

Similar to the public sector, private commercial and community actors also have to qualify as a partner. One of the requirements of partnerships is that each of the participants brings something to the arrangement that the others are either missing or could only provide at prohibitive costs (complementarity) (Evans, 1996; Peters, 1998). The first problem is that the potential may be there, but the authorities fail to recognize it. Very often there is reluctance on the part of the authorities, for example, to deal with large numbers of informal operators (such as waste pickers). This aversion partially stems from expected transaction costs. More important, however, is that the elusiveness of these activities is at odds with the enforcement of rules and regulations (including sanitary codes and health standards) and could make effective sanctions in case of malpractice difficult to enforce. Official attitudes towards such undertakings are still overwhelmingly hostile, especially when they relate to the most non-formalised activities in the waste sector, which are socially stigmatised as dirty, unhealthy, chaotic and illegal (Baud *et al.*, 2001). Currently, there is a total lack of any form of partnership between small private recycling enterprises, waste pickers, dealers and the local authorities in the two cities studied. This relationship is characterized by extreme avoidance and antagonism. This is a pity, because cooperation could lead to improvements by increasing levels of resource recovery, as well as higher (and more protected) levels of employment for waste pickers/itinerant buyers/dealers currently working in informal forms of employment. The way to cut across this divide does not emerge from the case studies, and there is little evidence from elsewhere (Baud *et al.*, 2001) to suggest that there are methods of developing partnerships. Nevertheless, in view of environmental considerations, it remains important to look into this issue further.

Within inorganic waste recycling commodity chains, (commercial) partnerships are well-established in both cities, although those in India tend to be of a longer duration. They build up social capital by contributing to the continuity and regularity of the economic activities carried out. They also provide important informal forms of social security, through the loans given by dealers to pickers and itinerant buyers Although the employment created within these commodity chains is not protected by labour legislation in either city, efforts can be made to increase safety and health aspects, rather than regulating the private sector in such a way that entrepreneurs are inclined to discontinue their business.

At the same time, and despite the enthusiasm among many academics and policy-makers about the potential of community involvement in urban governance and environmental management (Mitlin, 2001; Hardoy *et al.*, 2001), CBOs are not automatically perceived as potential partners by local governments. Although there are many examples of community action in SWM in our two cases, and many positive aspects of community involvement (participation in clean-ups, organizing for political pressure, community-based projects providing work to unemployed youngsters) most of the time it concerns one-time initiatives or actions that are strictly confined to neighbourhood level or both. Linking up with the local authorities is difficult. In Nairobi 'collaboration' with CBOs is restricted to tolerance of some international donor action and an occasional clean-up campaign. Decisive hindrances are that many CBOs operate in unrecognised slums and that their interventions do not always satisfy existing standards. In Hyderabad official attitudes seem to be somewhat more positive. The MCH was willing to collaborate with CBOs at neighbourhood level through its Voluntary Garbage Disposal Scheme (which also indirectly engaged them with a sizeable group of waste pickers). However, the support was not only on a one-off basis (providing a tricycle free of charge), it also suffered from a middle-class bias. In the slum areas the VGDS could only be implemented thanks to donor activism. Our findings corroborate conclusions drawn elsewhere about the ponderous relationships between CBOs and local authorities in urban environmental management (Lee, 1998; Hordijk, 2000). Part of the explanation probably lies in the lack of coordination within the SWM system by local authorities, notably the bias towards effective and efficient collection of waste and the neglect of recycling and reuse. If more emphasis was put on the latter aspects, collaboration with the community (and the informal private sector) would probably receive higher priority.

However, the potential of community action should not be exaggerated. Our two cases show that such action often depends on NGO support and/or the devotion of un(der)paid workers, both of which constitute uncertain (non-structural) elements. This was particularly true for the community composting initiatives in both cities.

The comparative strength of entrepreneurs and residents in dealing with the authorities partly depends on their level of organisation (Schenk, Baud and Bhuvaneshwari, 1998). Most private sector collection enterprises in our study are self-employed small or micro-business units, each lacking the power to put pressure on governments (Haan *et al.*, 1998). In Hyderabad, the level of organisation among collection enterprises was better than in Nairobi. This resulted in greater effectiveness in challenging the authorities (cf. the strategy to manipulate bids prior to the adoption of the unit system) or in negotiating with them (presenting demands). Similarly, comparatively well-organized middle-class communities managed to win official support for the upgrading of the service in their areas through the VGDS. In Nairobi the absence of any sort of entrepreneur interest group partly explains why no by-laws have been designed to accommodate privatised collection services. Networking among resident organisations seems to be a promising avenue for exercising pressure on governments to make them more responsive and accountable (cf. the 'We Can Do It movement' in Nairobi). Fostering these forms of social capital can help to turn authorities into more reliable actors and/or partners

13.4. TOWARDS MORE SUSTAINABLE SOLID WASTE MANAGEMENT: THE TRADE-OFFS

This study was designed to provide a more integrated analysis of the dynamics of the SWM system than is usually given in the field of urban planning and management. It goes beyond the focus on specific issues, such as privatisation of SWC, livelihoods of waste pickers, or potentials of composting groups. It has also analysed in a qualitative manner the performance of SWM systems beyond the customary focus on either technical, financial, socio-economic or environmental aspects. This approach follows from the concept of integrated sustainable waste management as developed by researchers from WASTE (Lardinois and van de Klundert, 1995; van de Klundert and Anschütz, 1999). Obviously, we could only partly meet the requirements of such an integrated approach, which is ambitious because it not only takes on board most stakeholders, solid waste sub-sectors, and aspects of sustainable development, but also seeks to link it to other systems (e.g. health, infrastructure) and to encompass various scale levels. Nevertheless, the study enables us to comment on a number of important areas of contention when it comes to making SWM more sustainable.

A first basic issue preventing the SWM sector from increasing contributions to sustainable development is related to the existing segmentation of state responsibilities for the different aspects included (e.g. the sectoral basis of government organisation). This means that mandates for policy initiatives and actions integrating environmental health, ecological, and socio-economic concerns are not readily given priority. Coordination across government departments would be an alternative, but is notoriously difficult to accomplish in actual practice. This situation is particularly vexing in the case of integrating ecological concerns in current patterns of solid waste collection, transportation and disposal. Collection is mandated to local Public Health Departments, with a focus on public health in the conventional sense (rather than on a more encompassing agenda of environmental health). In contrast, environmental issues are officially mandated to provincial or national departments, focusing on large-scale (national or global) environmental issues, usually in the context of natural resource management. Although some attention is given to urban environmental issues within these departments, this is mostly oriented towards industrial pollution. This means that there is no clear mandate for any level of government to reduce household waste flows, to promote waste separation and to maximize recycling and reuse of resources used by households and institutions. It must be said, however, that in India the new SWM rules attempt to solve this issue by offering a more encompassing mandate to local authorities.

Existing practices in recycling waste materials are predominantly based on economic motives. Furthermore, much of these activities take place in an informal setting, making it difficult to draw them into any system in which the government plays a role. Where there is some regulation of these activities, they come under the jurisdiction of Departments of Economic Affairs, whose main mandate is financial regulation, and Departments of Employment and Social Affairs, who enforce labour standards. The former are just beginning to discover the importance of environmental regulation (partly to satisfy requirements emerging from international agreements, partly to accommodate increasing pressures by national environmental movements), whereas the latter mainly focus on safety and health issues at firm level as part of their mandate. Together, these factors confound any attempt towards a more integrated SWM policy.

In short, the entire debate on sustainable development notwithstanding, the actual impact it has had on urban development policies is disappointingly meagre (Mattingly, 2000). This is particularly true for the ecological sustainability aspects of sustainable development (the green agenda). Requirements for promoting such an integrated policy include an obligatory national framework (as set by the new SWM rules in India), and the process requirements set out by Satterthwaite and McGranahan (2000) to reconcile the green and brown agenda: a wide usage of open and participatory methods of designing environmental policies, national policies and guidelines that support urban development, and a good knowledge of the state of the environment.

A second, related issue, is the conflict between scale levels for which a mandate is given, which prevents integration of different aspects of sustainable development. In order to promote reuse and recycling of resources, a wider area than the city often needs to be included, preferably a regional or even national context. This is also vital if the environmental impacts of waste disposal on the absorption capacity of local and regional sinks are to receive adequate attention (Satterthwaite, 1997). It implies the need to develop a new regulatory framework that goes beyond the idea of the single city and requires local authorities to at least coordinate activities with neighbouring or higher level authorities.

Thirdly, the segmentation of domains leads to fluctuations in the extent to which resource/materials recycling takes place. The macro-economic context affects the relative prices for domestic and imported resources/materials: if imported virgin materials are available at equal or lower prices than domestic used materials, there will be no effective demand for resources recovered from waste. Although little is known yet about the macro-economic context of pricing materials, this is an important area for further research.[2]

A final issue concerns conflicts in financing different aspects of sustainable development in SWM. It is virtually impossible to attach an accurate price tag to each of the goals of integrated SWM as well as to allocate the costs fairly. Usually, there are only two actors that contribute financially to SWM: the local authorities (albeit sometimes with the help of central grants) and the waste generators Cost recovery can be used as a method to increase financial viability through service charges, but in an urban context with large numbers of low-income households, it is unlikely that the total costs can be retrieved this way. Many poor residents cannot be charged according to their use of services, but ignoring them will have important detrimental effects on overall public health and ecological sustainability. Therefore, the local government will have to accept the necessity of subsidizing the system from general revenues in order to meet its statutory obligation to provide adequate collection services and to keep up public health standards. However, severe budgetary constraints usually urge them to compromise on considerations of equality and total coverage, at the expense of residents living in unplanned areas, spontaneous settlements or slums (with negative spill-over effects for serviced areas). Furthermore, they will not be keen to invest in systems for increased waste separation and recycling out of environmental considerations. Therefore, various priorities in integrated SWM are excluded. The experiences with the privatisation of solid waste collection in Hyderabad and Nairobi may serve to illustrate this point once more. The negative trade-off of privatisation is that environmental concerns are not included in the current regulations for private collection enterprises, as private contractors usually do not carry out waste separation. Efficiency and effectiveness gains in waste collection through privatisation, therefore, result in an increase of waste being offered for final disposal, thereby reducing environmental sustainability (i.e. waste that could also have been diverted in case promotion of recycling and reuse had been part of the agreement). This loss, however, is not translated into monetary terms nor put on anyone's doorstep.

13.5. AREAS FOR FUTURE RESEARCH AND ACTION

The studies have brought out a number of issues on which further research needs to be done: both to explore the analytical questions, as well as to examine the possibilities for improving practice.

A major area for further research relates to how the 'waste management hierarchy' and ideas concerning integrated sustainable waste management can gain more recognition and acceptance by governments (both local and national) in developing countries. The question concerns not only the role of research in analysing the trade-offs inherent in the multiple goals of an approach integrating socio-economic and environmental aspects in SWM, but also further study into how research results can be fed more

---

2. An exception is the work by van Beukering (2001) regarding international trade flows in waste materials, although it does not cover the competition from virgin materials.

effectively into policymaking and administrative processes. The latter is a discussion which has been going on elsewhere in more general terms (cf. Stone, 2001; RAWOO, 2001; Baud 2002; Hall 2002), but can be usefully applied to urban basic services as well.

A second important area is the ways in which partnerships and 'up-scaling of partnerships' can be promoted, and the conditions needed to do so effectively. Although the study has brought out some elements of such potentials, contrasting situations in which the informal waste recovery sector has been able to realize partnerships still need further comparative study. Particularly the issue of three-way partnerships, in which NGOs can and do play an intermediary role between local government and groups of people working in informal employment or non-recognized activities, is an area to be explored further.

Finally, future studies need to analyse how the linkages between urban SWM concerns and the region in which the city is located can be made more effective. Such linkages concern several areas. They include possible changes in the regulatory framework, which force local authorities to coordinate actions that have environmental implications beyond city boundaries. It also concerns linkages with the surrounding region needed to promote organic waste reuse and composting in partnership with rural farmers, and the problems of using urban organic waste in peri-urban agriculture and how to reduce them. Further study on trade flows of various waste materials within and between countries can contribute to our understanding of integrated sustainable waste management in the future.

METHODOLOGICAL APPENDIX

A variety of methodologies were used to carry out the study at the two locations. The following paragraphs describe the choices and limitations of the study, the methodologies used to collect data at the local level, and an assessment of planned versus actual methodologies used.

The fieldwork methods used are grouped according to the three themes taken up in the study: the institutional arrangements for collection and transportation of solid waste, the system of trade and recycling of inorganic waste, and the system of trade and recycling of organic waste.

RESEARCH METHODOLOGIES FOR INSTITUTIONAL ARRANGEMENTS

Theme 1 concerned the institutional arrangements for collection and transportation of solid waste. The main aspects covered in the study were:

- *overall organisational structure:* including operational characteristics of public and private sector organisations involved in generating and dealing with solid waste collection (including NGOs and CBOs), labour conditions of employees, financial aspects and division of costs among stakeholders, operational efficiency and effectiveness, and geographical spread of activities;
- *individual employee level:* personal information on contract and working conditions, and aspects of waste handling (storage, sorting, selling of waste fractions by individual employees, collection, transportation, intermediary disposal);
- *household-level* generation of waste, perceptions of waste, actual dealing with waste, attitudes toward other waste actors;
- *policy aspects through semi-structured interviews with key people in waste activities and partnerships* on notably policies of privatisation, CBO and NGO involvement and resource recovery at local level.

The main methodologies used to obtain the data on the aspects above were the following. A literature study was carried out to obtain information on the international debates on waste management and resource recovery, as well as the basic data on the situation in Kenya and India with respect to SWM. This was followed by a field study among the various groups generating and dealing with waste. This included bulk generators of waste, a stratified household survey (see below), strategic interviews with officials and private sector entrepreneurs, and a stakeholder workshop to discuss the results of the fieldwork surveys done during the course of the study.

Table 1. Sampling design for field study on theme 1

Actors	*Nairobi*		*Hyderabad*	
	Foreseen	Actual	Foreseen	Actual
Private contractors	10	10	10	20
Private contractor employees	20	34	20	40
Residents:				
High-income:	30	29	30	60
Middle-income:	30	38	30	60
Low-income	30	32	30	60
Slum/illegal	30	29	30	60
Key waste generators:				
Hotels	5	5	12	12
Schools	5	5	15	11
Markets	3	3	10	8
Hospitals	3	3	8	8
Slaughterhouses	-	-	-	2
Key people in org.:				
Politicians:	-	3	-	-
CBOs:	-	15	-	4
NGOs:	-	3	-	-

Note: the total number of slaughterhouses in Hyderabad is 5, and the total number of markets 38.

RESEARCH METHODOLOGIES USED FOR ANALYSING TRADE AND RECYCLING SECTORS: INORGANIC AND ORGANIC WASTE

Theme 2 and 3 concerned the trade and recycling sectors utilizing inorganic and organic waste. For recycling and trade in inorganic materials, the main actors defined were: recycling entrepreneurs, wholesalers, dealers, employees of the different enterprises, itinerant buyers, dump pickers, and street waste pickers (also households and maids, but these were included in the survey for theme 1).

The main aspects covered in the study were:

- viability of the recycling activities at the enterprise/unit/individual level (costs, prices, and turnover, profits);
- volumes of raw materials recovered (and products made);
- forward and backward linkages between pickers, traders, and recycling units in the commodity chain;

- employment and income (including levels of employment and working conditions, income levels, regularity of income, mobility in the sector, both upward and downward; and
- health hazards (physical injuries in the past month among the different groups of workers and pickers), and
- how and why changes in the sourcing of waste materials occurred (including changes in import duties, import policies, changes in suppliers, and changing pricing policies).

There were several limits in the field study for both organic and inorganic waste. They were:
- only important waste fractions that are recycled, were taken into account (paper, plastics, metal and glass);
- no account was taken of interest and depreciation levels in enterprises;
- in terms of health hazards, only physical injuries were considered, as establishing causal relationships between diseases and solid waste was impossible;
- fieldwork (January – April 1999) took place in the peak season of waste production in both locations; questions pertained to recycling levels at that time, with comparisons asked about the most recent past slack season.

The sampling design for the study on recycling of (in)organic waste materials was as follows for both Hyderabad and Nairobi. Because basic overall information was lacking concerning the geographic distribution of the actors and their activities, methods used include
- purposive sampling according to the area of the city: by income levels for residents, and
- by economic activities carried out, and
- snowballing methods within one category of actors

Table 2. Sampling design for field study on inorganic and organic waste recycling

Actor	*Nairobi*		*Hyderabad*	
	Foreseen	Actual	Foreseen	Actual
Street picker	60	68	60	120
Dump pickers	60	74	60	72
Itinerant buyers	30	14	30	60
Dealers	30	31	30	55
Wholesalers	–	1	10	25
recycling units-entrepreneurs	60	33	60	10
Workers	60	34	60	60

– The number of recycling enterprises in Nairobi in the different waste materials was smaller than expected; therefore, the number of workers was also reduced.

Organic waste recycling methodology

For organic waste recycling specifically, categories of actors identified were large bulk generators of organic waste, collectors and those composting organic waste (CBOs and NGOs, or private enterprises, farmers), and users of organic waste, in this case, confined mainly to urban and semi-urban farmers

Although the framework used for the inorganic waste trade and recycling commodity chain was the same for organic trade and recycling, specific aspects were covered that pertain only to organic waste. These aspects concerned the way

- organic waste materials are used by farmers;
- the ways organic waste (from urban sources) fit into the pattern of organic waste use; and
- composting initiatives were undertaken by NGOs/CBOs.

The main limitations of the fieldwork on organic waste trade and recycling were that only specific waste generators were included: households in general, whose sum is a large waste generator, and bulk waste generators, such as hotels, educational institutions, hospitals, and markets for produce. The main category of users in the urban-rural context was assumed to be that of farmers; only categories of farmers applying organic waste were compared to a control group of farmers with similar crops and animal stocks who have never used organic waste. In-between categories were avoided, as a stratified sample taking in-between categories into account would have been too large to manage in the time and budget concerned. However, three small studies were commissioned on organic waste use and recovery. The first was a paper by M. Put, on the use of organic manures and inorganic fertilizers in four villages south of Hyderabad (1999). The second was a field study among 195 urban and peri-urban farmers in Nairobi by A. Karanja and E. Koster. The third study was done in Hyderabad by Rakhee Kulkarni among different categories of users of organic waste, with the help of the CESS researchers involved in the project (see chapter 11).

Sampling design for field study

The sampling design for the study on recycling of organic waste materials in Hyderabad and Nairobi makes use of the field surveys carried out for the other themes as far as waste generators is concerned, for composting different categories of users were identified, the main ones being CBOs/NGOs and farmers, and specified buyers of compost. For energy recovery, the companies involved in this method were interviewed. The main methods used to conduct field surveys among farmers include:

- purposive sampling according to the area of the city: by income levels of residents, and economic activities carried out there, and
- the snowballing method within one category of actors In Nairobi, this led to a survey among 195 farmers in/around Nairobi, selected using a grid pattern laid over the city and its surrounding areas for 100 farmers (50-50 urban and peri-urban),

and through purposive sampling in the SWM sector to identify users of organic waste.

In Hyderabad, bulk generators of organic waste were identified and interviewed (see sample above under theme 1). Centralized composting activities were limited to one private company, and decentralized initiatives were also included in the interviews done (2). A small study was done among farmers in two villages 30-40 kms away from the city dump still in use to collect information on their use of urban mixed organic waste (in total 175 farmers with different size holdings).

Assessment of methodology

There are several areas in SWM, which remain difficult to trace with a reasonable degree of accuracy. The composition of waste remains a vexed point in case. Although figures are often presented in the literature as if very accurate, when checked they usually stem from estimates made on very shaky grounds.

When studying inorganic waste recycling, one has to keep in mind that any particular city is not a closed system, and therefore waste recovered and used, is not necessarily limited in its source to the city itself. Therefore, very little can actually be said about total amounts of waste recovered and the impact on city level waste flows.

However, in studying the system of waste recovery and recycling, the recycling units are very dominant in the value chain for inorganic waste. Therefore, in future studies it is to be recommended to have a relatively large sample of such units to be able to trace the differentiation among them, and to use the method of tracing back from them up the chain.

Organic waste reuse and recovery also remains a difficult area to study. In addition to studying mixed urban organic waste flows, single source producers and the ways such sources of organic waste are used, are likely to be equally interesting. It is recommended that such sources are included in future studies as a matter of course.

BIBLIOGRAPHY

Adarkwa, K.K. and Post, J. (eds) (2000), *The Fate of the Tree, Planning and Managing the Development of Kumasi*, Thela Thesis and Woeli Publishing Services, Amsterdam and Accra.

Ali, M. (1993), *Final Report on Private Sector Involvement in SWM in Karachi*, DFID Report, London.

Ali, M., Olley, J. and Cotton, A. (1999), 'Public Sector Delivery of Waste Management Services: Case from the Indian Sub-Continent', *Habitat International*, Vol. 23(4), pp. 495-510.

Allison, M. and Harris, P. (1996), *A Review of the Use of Urban Waste in Peri-urban Interface Production Systems. Overseas Development Administration, the Peri-urban Interface Production System*, Natural Resources Systems Programme, Renewable Natural Resources Research Strategy, African Studies Centre, Coventry University, Coventry.

Anand, P.B. (1999), 'Waste Management in Madras Revisited', *Environment and Urbanization*, Vol. 11(2), pp. 161-77.

Andhra Pradesh Pollution Control Board (1992), *Waste Disposal Practices in Selected Major Hospitals of Twin Cities of Hyderabad and Secunderabad* (Submitted by Induss. Environ Consultants), Hyderabad, India.

Anschütz, J. (1996), *Community-based Solid Waste Management and Water Supply Projects: Problems and Solutions Compared, A Survey of the Literature*, UWEP working document 2, WASTE, Gouda.

Ariyo A. and Jerome, A. (1999), 'Privatization in Africa; an Appraisal', *World Development*, Vol. 27(1), pp. 201-13.

Arrossi, S. (1994), *Funding Community Initiatives*, Earthscan Publications, London.

Arroyo Moreno, J., Rivas Rios, F. and Lardinois, I. (1997/99), *Solid Waste Management in Latin America, the Role of Micro- and Small-scale Enterprises and Cooperatives*, Urban Waste Series 5, (english version), WASTE, Gouda.

Aseto, O. and Okello, J. (1997), *Privatisation in Kenya*, Basic Books, Nairobi.

Asomani-Boateng, Raymond (1999), *Planning and Managing Urban Organic Solid Waste in an African City: Linking Organic Solid Waste Composting to Urban Cultivation in Accra, Ghana, West Africa*, Doctoral thesis, Univ. of Waterloo, Waterloo, Canada.

Asomani-Boateng, Raymond and Haight, Murray (1999), 'Assessing the Performance of Mechanized Centralized Composting Plants in West Africa: the Case of Teshi Nungua Composting Plant in Accra, Ghana', *WARNER Bulletin*, Vol. 69, Nov. 1999, pp. 4-6.

Banin, A., Navort, J., Noiy and Yoles, D. (1981), 'Accumulation of Heavy Metals in Arid Zone Soils Irrigated with Treated Sewage Effluents and their Update by Rhoades Grass', *Journal of Environmental Quality*, Vol. 10, pp. 536-42.

Baron, E. and Castricum, D. (1996), *Second-hand Life: Small-scale Recycling Enterprises in the Cono Sur, Lima, Peru*, MA thesis, University of Amsterdam, Amsterdam.

Bartone, C.L., Leite, L., Triche, T., and Schertenleib, R. (1991), 'Private Sector Participation in Municipal Solid Waste Services: Experiences in Latin America', *Waste Management and Research*, Vol. 9, pp. 495-509.

Bartone, C., Bernstein, J., Leitmann, J., and Eigen, J. (1994), *Toward Environmental Strategies for Cities, Policy Considerations for Urban Environmental Management in Developing Countries,* UMP Paper no. 18, World Bank, Washington DC.

Batley, R. (1996), 'Public-Private Relationships and Performance in Service Delivery', *Urban Studies*, Vol. 33(4-5), pp. 723-51.

Baud, I.S.A. (2000), *Collective Action, Enablement, and Partnerships: Issues in Urban Development*, Inaugural Lecture, 27 October, Free University, Amsterdam.

Baud, I.S.A. (2002). ' North-South Research Partnerships in Development Research: an Institutions Approach', in *North-South Research Cooperation*, KNAW, Amsterdam.

Baud, I.S.A., Grafakos, S., Hordijk, M. and Post, J. (2001), 'QOL and Alliances in Solid Waste Management: Contributions to Urban Sustainable Development', *Cities*, Vol. 18(1), pp. 1-10.

Baud, I.S.A, Huysman, M. and Schenk, H. (1996), 'Solid Waste Management in Three Indian Cities: Linkages Between Formal and Informal Systems', in P.A. Erkelens and G. van der Meulen (eds), *Urban Habitat: the Environment of Tomorrow*, Eindhoven University Press, Eindhovern, pp. 353-62.

Baud, I.S.A. and Schenk, H. (1994), *Solid Waste Management: Modes, Assessments, Appraisals and Linkages in Bangalore,* Manohar Publishers, New Delhi.

Beall, J. (1997), *Households, Livelihoods, and the Urban Environment: Social Development Perspective on Solid Waste Management in Faisalabad, Pakistan*, PhD thesis, London School of Economics and Political Science, London.

Belevi, H., Leitzinger, C., Binder, C., Montagero, Strauss, M., Zurbrügg, C. (2000), *Material Flow Analysis: A Planning Tool for Organic Waste Management in Kumasi, Ghana*, Unpublished internal document, Dept. Of Water and Sanitation (SANDEC), Swiss Federal Institute for Environmental Science and Technology (EAWAG), Duebendorf, Switzerland.

Berkes, F. and Folke, C. (1998), *Linking Social and Ecological Systems, Management Practices and Social Mechanisms for Building Resilience*, Cambridge University Press, Cambridge.

Beukering, P.J.H. van (1994), 'An Economic Analysis of Different Types of Formal and Informal Entrepreneurs, Recovering Urban Solid Waste in Bangalore, India', *Resources, Conservation and Recycling*, Vol. 12, pp. 229-52.

Beukering, P.J.H. van (2001), *Recycling, International Trade and the Environment: An Empirical Analysis*, PhD thesis, Vrije Universiteit, Amsterdam, the Netherlands.

Beukering, P. van and Duraisppah, A. (1996), *The Economic and Environmental Impacts of the Waste Paper Trade and Recycling in India: A Material Balance Approach*, CREED Working Paper Series no. 10.

Beukering, P.J.H. van and Duraiappah, A. (1998), 'The Economic and Environmental Impact of Waste Paper Trade and Recycling in India: A Material Balance Approach', *Journal of Industrial Ecology*, Vol. 2(2), pp.23-42.

Bhuvaneshwari, R. (1994). *Approaches to Urban Solid Waste Management in Indian Cities, a study of NGO/CBO initiatives in ten Indian cities*. Nayadamma Memorial Science Foundation, Madras.

Birkbeck, (1978), 'Self-Employed Proletarians in an Informal Factory: the Case of Cali's Garbage Dump', *World Development*, Vol. 6(9-10), pp. 1173-85.

Birley, M.H. and Lock, K. (1999), *The Health Impacts of Peri-urban Natural Resource Development*, Trowbridge, Cromwell Press, U.K.

Bliek, Julie van der (1992), *Urban Agriculture: Possibilities for Ecological Agriculture in Urban Environments as a Strategy for Sustainable Cities,* ETC Foundation, Leusden.

Blore, I. (1999), *Reclaiming the Wasteland, Systems of Markets and Governance of Household Waste in South Asia*, Occasional Paper 30, School of Public Policy, University of Birmingham, Birmingham.

Blore, I. (2000), *Corruption, Contracts and Competition: the Transition Economics of Waste and their Political Consequences*, Paper to Bradford Conference on Environmental Resources: Conflict, Co-operation and Governance, May 17-18, Bradford.

Blumenthal, U.J., Strauss, M.D., Mara, D. and Cairncross, S. (1989), 'Generalized Model for the Effect of Different Control Measures in Reducing Health Risks from Waste Reuse', *Water Science and Technology*, Vol. 21, pp. 567-77.

Blumenthal, U.J., Mara, D.D., Peasey, A., Ruiz Palacios, G. and Storr, R. (2001), 'Reducing the Health Risks of Using Wastewater in Agriculture', *Urban Agriculture Magazine*, Vol. 1(3), pp. 26-9.

Bose, A. and Blore, I. (1993), 'Public Waste and Private Property: An Enquiry into the Economics of Solid Waste in Calcutta', *Public Administration and Development*, Vol. 13, pp. 1-15.

Broekema, J. (1998), *Vast Afvalmanagement in Bangalore; Een Onderzoek naar het Formele Vast Afvalmanagement in Bangalore met Speciale Aandacht voor de Rol van de Private Sector (Solid Waste Management in Bangalore; a Study on Formal Solid Waste Management in Bangalore with Special Concern for the Role of the Pirvate Sector)*, unpublished master thesis, Department of Geography and Planning University of Amsterdam, Amsterdam.

Bromley, R. (1978), 'The Urban Informal Sector' Critical Perspectives on Employment and Housing Policies, special issue *World Development*, Vol. 6(9/10), pp 1031-1200.

Brook, Robert and Julio Davila (eds) (2000), *The Peri-Urban Interface: A Tale of Two Cities*. University of Wales and University College London. Bethesda (Wales) UK.

Brunt, L.P., Dean, R.B. and Patrick, P.K. (1985), 'Composting', in Michael Suess (ed), *Solid Waste Management: Selected Topics*, World Health Organization, Copenhagen, pp. 37-77.

Burgess, R., Carmona, M. and Kolstee, T. (1997), *The Challenge of Sustainable Cities, Neoliberalism and Urban Strategies in Developing Countries*, Zed Books, London.

Chakraborty, Satyesh (ed) (2000), *A Research Report on Informal Economy of Solid Waste Disposal in the City of Calcutta*, Ekta Ecological Foundation, Kolkata (unpublished report).

Chaturvedi, B. (1998), *Public Waste Private Enterprise, an Enquiry into the Issue of Integrating Rag Pickers into Formal Solid Waste Management Systems*, Heinrich Boll Stiftung, Berlin.

Chikarmane, P., Deshpande, M. and Narayan, L. (n.d.), *Study of Scrap Collectors, Scrap Traders, and Recycling Enterprises in Pune*, ILO and SNDT Women's University, Bombay (unpublished report).

Coad, Adrian. (1997) (ed.), *Lessons from India in Solid Waste Management*. WEDC, Loughborough, UK.

Cointreau, S. (1984), *Integrated Resource Recovery. Recycling from Municipal Refuse*, WB Technical Paper no. 30, WB, Washington DC.

Cointreau-Levine, S. (1994), *Private Sector Participation in Municipal Solid Waste Management in Developing Countries, Vol. 1: The Formal Sector*, Urban Management Programme Policy Paper no. 13, World Bank, Washington.

Cointreau-Levine, S., Listorti, J. and Furedy, C. (1998), 'Solid wastes', in J. Herzstein, W.B. Bunn III, L.E. Fleming, J.M. Harrington, J Jeyaratnam and I.R. Gardner (eds) *International Occupational and Environmental Medicine*, First Edition, Mosby Year Book Inc., St. Louis, Miss., USA, pp. 620-32.

Davies, T.C., Mutoro, B.A. and Karanja, A.M. (1998), *Enabling Strategies for Environmental Management of Solid Waste in Growing Multi-Million cities: An Integrated Economic and Environmental Assessment of Solid Waste Systems in Nairobi City, Kenya, and Hyderabad, India*, A Preliminary Report, Moi University, Kenya (unpublished report).

Devas, N. (1999), *Who Runs Cities? – the Relationship between Urban Governance, Service Delivery and Urban Poverty,* Theme Paper 4 of the Urban Governance, Partnership and Poverty Programme, University of Birmingham, Birmingham.

Dhanalakshmi, R. and Iyer, S. (1999), *Solid Waste management in Madras City,* Pudhuvazhvup Pathippagam Ltd., Chennai, India.

Dhanalakshmi, R. (2002), *Field Notes in Hyderabad*, Report prepared for EU-INCO project no. ERBIC 18Ct 970152.

Dillinger, W. (1994). *Decentralisation and its Implications for Urban Service Delivery.* UMP no. 16, UMP, World Bank, Washington DC.

Djabatey, Raphael Lawer (1996), *Space, Land-use Planning and the Household Economy: The Role of Urban Agriculture in the Accra Metropolitan Area, Ghana*, Doctoral thesis, Dept. of Geography, University of Toronto.

Drakakis-Smith, D. (1995), 'Third World Cities: Sustainable Urban Development 1', *Urban Studies*, Vol. 32(4-5), pp. 659-77.

Drechsel, Pay and Agmar Kunze (eds) (2001), *Waste Composting for Urban and Peri-urban Agriculture: Closing the Rural-Urban Nutrient Cycle in Sub-Saharan Africa*, CABI Publishing for International Water Management Institute, Wallingford, Oxon, UK.

Drescher, Axel W. (1994), 'Gardening on Garbage: Opportunity or Threat?', *ILEIA Newsletter*, Vol. 10(4), pp. 20-1.

Drèze, J. and Sen, A. (1997), *India, Economic Development and Social Opportunity,* Oxford University Press, New Delhi.

Dulac, Nadine (2001), 'The Organic Waste Flow in Integrated Sustainable Waste Management', Document in Arnold van de Klundert, Maria Muller, Anne Scheinberg, Nadine Dulac, Justine Anschutz and Lane Hoffman (2001), *Integrated Sustainable Waste Management: A Set of Five Tools for Decision-makers*, WASTE, Gouda.

Dunnet, Simon (1998), *Urban Waste and Energy in Developing Countries. A Workshop Report*, International Technology Development Group, Rugby (unpublished report).

Edwards, Peter (1992), *Reuse of Human Wastes in Aquaculture: a Technical Review,* Water and Sanitation Rep. 2, World Bank, Washington, D.C.

Eerd, M. van (1995), *Gender Related Labour Market Fragmentation in the Informal Recycling Sector in Bangalore*, Master Thesis, University of Amsterdam, Amsterdam.

Elliot, L. (1998), *The Global Politics of Environment,* MacMillan Press, London.

Emmerich W.E., Lund, L.J., Page, A.L. and Chang, A.C. (1982), 'Movement of Heavy Metals in Sewage Sludge-Treated Soils', *Journal of Environmental Quality,* Vol. 11, pp. 174-179.

EPTRI (1997), *State of the Environment for Hyderabad Urban Agglomeration*, Government of Andhra Pradesh, Hyderabad, India.

Esho, (1997) quoted in UNCHS, 1998.

Escobar, A. (1995), *Encountering Development: The Making and Unmaking of the Third World*, Princeton University Press, Princeton.

Environment and Urbanization, (1998), *Special Issue – Sustainable Cities Revisited.* Vol. 10(2) pp 3-208.

Environment and Urbanization, (1996), *Special Issue, Future Cities*, Vol 8, No 1, pp 3-154.

Etuah-Jackson, I., Klaassen, W.P. and Awuye, J.A. (2001), 'Turning Municipal Waste into Compost: the Case of Accra', in Pay Drechsel and Agmar Kunze (eds), *Waste Composting for Urban and Peri-urban*

Agriculture: Closing the Rural-Urban Nutrient Cycle in Sub-Saharan Africa, CABI Publishing, Wallingford, Oxon, UK, pp. 84-95.

Evans, D. (1996). 'Government Action, Social Capital and Development: Evidence on Synergy', *World Development*, Vol. 24(6):, pp. 1119-32.

FAO (Food and Agriculture Organization), Environmental Technology Centre and Resource Centre for Urban Agriculture and Forestry, (2000), FAO-ETC/RUAF Electronic Conference 'Urban and Periurban Agriculture on the Policy Agenda'. Available at: http://www.ruaf.org.

Fernandez, A. (1993). 'Public-private Partnerships in Solid Waste Management' in: *Regional Development Dialogue*, Vol. 14(3), pp. 3-23.

Freeman, D.B. (1991) A City of Farmers, Informal Urban Agriculture in the Open Spaces of Nairobi, Kenya, McGiil Queeen's University Press, Montreal & Kingston.

Frijns, J., Kirai, P., Malombe, J., and Vliet, B. van (1997), *Pollution Control of Small-scale Metal Industries in Nairobi,* Dept. of Sociology, WAU, Wageningen, the Netherlands.

Furedy, Christine (1990), *Social Aspects of Solid waste Recovery in Asian Cities,* Environmental Sanitation Review, no. 30, Environmental Systems Information Centre, Asian Institute of Technology, Bangkok.

Furedy, Christine (1992), 'Garbage: Exploring Non-Conventional Options in Asian Cities', *Environment and Urbanisation,* Vol. 4(2), pp. 42-53.

Furedy, Christine (1993), *Pilot Survey of Householders and Source Separation in Bangalore* (unpublished).

Furedy, Christine (1995), 'One World of Waste: Should Countries Like India Solve Solid Waste Problems Through Source Separation?', in E. Tepper and J.R. Wood (eds), *Enriched by South Asia: Celebrating 25 Years of Scholarship,* Vol. 2, Social Sciences, Canadian Asian Studies Association, Montreal, pp.87-107.

Furedy, Christine (1996), 'Solid Waste Reuse and Urban Agriculture: Dilemmas in Developing Countries – the Bad News and the Good News', Paper presented at Joint Congress of the Association of Collegiate Schools of Planning and Association of European Schools of Planning, Toronto, July (unpublished). Available at: www.cityfarmer.org/Furedy.html

Furedy, Christine (1997a), 'Reflections on Some Dilemmas Concerning Waste Pickers and Waste Recovery', *Source Book for UWEP Policy Meeting,* WASTE, Gouda (unpublished report).

Furedy, Christine (1997b), 'Household-level and Community Actions for Solid Waste Management and Recycling in Asian cities', in A. Fernandez, K. Oya and D. Dungate (eds), *Recycling in Asia: Partnerships for Responsive Solid Waste Management*, United Nations Centre for Regional Development, Nagoya, pp. 13-25.

Furedy, Christine (1997c), 'Socio-Environmental Initiatives in Solid Waste Management in Southern Cities: Developing International Comparisons', *Journal of Public Health,* Vol. 27(2), pp. 142-56.

Furedy, Christine (1998), 'Sustainable Consumption and Municipal Solid Waste Reduction in Developing Countries of Asia', in UNCHS *Promoting Sustainable Consumption in Asian Cities. Report of a Regional Conference,* United Nations Centre for Human Settlements, Nairobi, pp. 86-107.

Furedy, Christine (2001), 'Reducing Health Risks of Urban Organic Solid Waste Use', *Urban Agriculture Magazine*, Vol. 1(3), March, pp. 23-5.

Furedy, Christine (2002), 'Organic Waste at Low Cost: Dilemmas of a Transition Period', *Urban Agriculture Magazine*, Issue 6, March. Available at: www.ruaf.org.

Furedy, Christine and Ghosh, Dhrubajyoti (1984), 'Resource-conserving Traditions and the Creative Use of Urban Wastes: the Sewage-fed Fisheries and Garbage Farms of Calcutta', *Conservation and Recycling*, Vol. 7(2-4), pp. 159-65.

Furedy, Christine, Virginia Maclaren and Joseph Whitney (1999), 'Waste Reuse for Food Production in Asian Cities: Health and Economic Perspectives', in Mustafa Koc, Rod MacRae, Luc J.A. Mougeot, and Jennifer Welsh (eds). *For Hunger-proof Cities*, International Development Research Centre and Center for Studies in Food Security Ryerson Polytechnic University, Totonto, pp. 136-145.

Furedy, Christine and Alison Doig (2002), 'Socially-responsive Energy from Urban Solid Wastes in Developing Countries", in Grover, V.I., V.K. Grover and W. Hugland (eds), *Recovering Energy from Waste: Various Aspects*, Enfield (NH) USA: Science Publishers Inc. pp. 307-314.

Gatheru, W. and R. Shaw (eds) (1998), *Our Problems, Our Solutions: An Economic and Public Policy Agenda for Kenya*, Institute of Economic Affairs, Nairobi.

Gonzalez, J., K. Lauder and B. Melles (2000). *Opting for Partnerships, Governance Innovations in South Eastern Asia*, Institute on Governance, Ottawa.

Government of India, Ministry Of Environment and Forests (2000), *Notification: Rules to Regulate the Management and Handling of the Municipal Solid Wastes*, Government of India Gazette, 25 September 2000, New Delhi.

Grover, V.I., V.K. Grover and W. Hogland (eds) (2002), *Recovering Energy from Waste: Various Aspects.* Science Publishers Inc., Enfield (NH) USA.

Gugler, Josef, (1997), 'Life in a Dual System Revisited: Urban-Rural Ties in Enugu, Nigeria, 1961 – 1987', in Josef Gugler (ed), *Cities in the Developing World: Issues, Theory, and Policy*, Oxford University Press, U.K.

Gutkind, Peter C.W. (1967), 'The Energy of Despair: Social Organisation of the Unemployed in Two African Cities: Lagos and Nairobi', IDS, University of Nairobi, Nairobi (unpublished report).

Haan, H.C., Coad, A. and Lardinois, I. (1998), *Municipal Solid Waste Management. Involving Micro- and Small Enterprises: Guidelines for Municipal Managers*, International Training Centre of the ILO, SKAT, WASTE, Turin, Italy.

Haan, H.C., Coad, A. and Lardinois, I. (eds) (1999), *Solid Waste Management in Latin America- the Role of Micro- and Small-scale Enterprises and Cooperatives,* WASTE/GTZ/SKAT/ILO, Geneva, Zwitserland.

Hardoy, J., Mitlin, D. and Satterthwaite, D. (2001), *Environmental Problems in an Urbanizing World*, Earthscan Publications, London.

Hake, Andrew (1977), *African Metropolis: Nairobi's Self-Help City*, Chatto & Windus, London.

Hall, A. (2002). Development and Use of the Innovation Systems Framework in India: a northern Perspective', in *North-South Research Cooperation*, KNAW, Amsterdam.

Hardoy, J.E., Cairncross, S. and Satterthwaite, D. (1990), *The Poor Die Young: Housing and Health in Third World Cities*, Earthscan Publications, London.

Hardoy, J.E., Mitlin, D. and Satterthwaite, D. (1992), *Environmental Problems in Third World Cities*, Earthscan Publications, London.

Hardoy, J.E., Mitlin, D. and Satterthwaite, D. (2001), *Environmental Problems in an Urbanizing World, Local Solutions for City Problems in Africa, Asia and Latin America*, Earthscan Publications, London.

Harriss, J., Hunter, J. and Lewis, C. (1997), *The New Institutional Economics and Third World Development*, Routledge, London.

Harris, P.J.C., Allison, M., Smith, G., Kindness, H.M. and Kelly, J. (2001), 'The Potential Use of Waste Streams Products for Soil Amelioration in Peri- urban Interface Agricultural Production Systems', in Pay Drechsel and Agmar Kunze (eds), *Waste Composting for Urban and Peri-irban Agriculture:*

Closing the Rural-Urban Nutrient Cycle in Sub-Saharan Africa, CABI Publishing, Wallingford, Oxon, UK, pp. 1-28.

Harrison, Kwach Ouma and Paumard, A. (2000), *Mukuru Recycling Centre*, UNCHS, Nairobi (unpublished report).

't Hart, D. and Pluijmers, J. (1996), *Wasted Agriculture: The Use of Compost in Urban Agriculture*, WASTE, Gouda.

Helmsing, A.H.J. (1993), 'Small Enterprise and Industrialization Policies in Africa: Some Notes', in A.H.J. Helmsing and Th. Kolstee (eds), *Small Enterprises and Changing Policies: Structural Adjustment, Financial Policy and Assistance Programmes in Africa*, IT publications, Exeter, UK.

Helmsing, A.H.J. (2000), *Decentralisation and Enablement, Issues in the Local Governance Debate*, Inaugural Address, Faculteit Ruimtelijke Wetenschappen, Universiteit Utrecht, Utrecht.

Hoornweg, D. and Thomas, L (1999), *What a Waste: Solid Waste Management in Asia*, Working Paper 1, Urban and Local Government, World Bank, Washington DC.

Hoornweg, Daniel, Thomas, Laura and Otten, Lambert (1999), *Composting and its Applicability in Developing Countries*, Urban Waste Management, Working Paper Series 8, World Bank, Washington D.C.

Hoornweg, Daniel (2000), Personal Communication with C. Furedy.

Hordijk, M.A. (1999), 'A Dream of Green and Water: Community Based Formulation of a Local Agenda 21 in Peri-urban Lima', *Environment and Urbanization*, Vol. 11(2), pp. 11-31.

Hordijk M.A.(2000). Of Dreams and Deeds, the Role of Local Initiatives in Community-based Urban Environmental Management, A Case Study from Lima, Peru, Thela Thesis, Amsterdam.

Hordijk, M.A. (2001). Combining capitals, the Assets of Community-based Organisations and Local Government in Neighbourhood Environmental Management inLima, In: I.S.A. Baud, A.J. Dietz, L.J de Haan and J. Post (eds) *Re-aligning Government, Civil Society and the Market: new challenges in urban and regional development, essays in honour of G.A de Bruijne,* AGIDS, Amsterdam, pp 109-130.

Hunt, C. (1996), 'Child Waste Pickers in India: the Occupation and its Health Risks', *Environment and Urbanization,* Vol. 8(2), pp. 111-9.

Huntington, H.G. (1977), *An Empirical Study of Ethnic Linkages in Kenyan Rural-Urban Migration*, Discussion paper no. 206, IDS, University of Nairobi, Nairobi.

Huysman, M. (1994), 'Waste Picking as a Survival Strategy for Women in Indian Cities', *Environment and Urbanization,* Vol. 6(2), pp. 157-74.

Huysman, M. and Baud, I.S.A. (1994), 'Solid Waste Recovery, Reuse and Recycling: Formal and Informal Aspects of Production and Employment in Indian Cities', in I. Baud and H. Schenk (eds), *Solid Waste Management: Modes, Assessments, Appraisals and Linkages in Bangalore,* Manohar, India, pp. 6-23.

IBSRAM (International Board for Soil Research and Management) (2000), *Municipal Organic Waste Recycling for Urban and Peri-urban Agriculture in Africa and Asia*, Available at City Farmer website: www.cityfarmer.org.

India Supreme Court (1999), *Solid Waste Management in Class 1 Cities in India. Report of the Committee Constituted by the Hon. Supreme Court, March 1999,* New Delhi, Government of India, pp. 6-23.

Japan International Cooperation Agency (JICA) (1998), *The Study on Solid Waste Management in Nairobi City in the Republic of Kenya*, March 1988. Done together with CTI Engineering Co. Ltd. and Environmental Technology Consultants Co. Ltd, Nairobi.

Jones, G.A. and Ward, P.M. (1994), 'The World Bank's "New" Urban Management Programme: Paradigm Shift or Policy Continuity', *Habitat International*, Vol. 18(4), pp. 33-51.

Jong, P. de (1999), *Organizing Waste Reduction in the Dutch Waste Sector*, PhD thesis, University of Amsterdam, Amsterdam.

Jordens, S. (1996), *Plastic Recycling Enterprises in Bangalore*, MA Thesis, University of Amsterdam, Amsterdam.

Kajese, K. (1991), 'African NGO Decolonisation: a Critical Choice for the 1990s', *Critical Choices for the NGO Community: African development in the 1990s*, Centre of African Studies, University of Edinburgh, Edinburgh.

Kanyinga, K. (1995), 'The Politics of Development Space in Kenya', in J. Semboja and O. Therkildsen (eds), *Service Provision under Stress in East Africa*, Centre for Development Research, Copenhagen and James Curry, London.

Karanja, A. (1999), 'Women in the Urban Environment: Solid Waste Management in the Unplanned Settlement Areas of Nairobi', *Perspectives*, Vol 2(1), January, Daystar University, Nairobi, Kenya.

Karanja, A. Mumbi, (2003), Assessing solid waste management practices in Nairobi: actors, partnerships and contributions to sustainable development. PhD dissertation (in preparation), Institute of Social Studies, The Hague, The Netherlands.

Karingi, K., (1997), 'Managing Organic Solid Wastes in Nairobi: Alternative Approaches for Economic and Environmentally Sustainable Urban Management', *Water and Sanitation News*, Vol. 4(2), pp. 5-7.

Keivani, R and Werna, E. (2000) Refocusing the Housing Debate in Developing Countries From a Pluralist perspective, in: *Habitat International* Vol. 25: 191-208.

Kenya, Government (1986), *Economic Management for Renewed Growth,* Sessional Paper no. 1, Government Printer, Nairobi, Kenya.

Kenya, Government of (1988), *1989-1993 Development Plan,* Government printer, Nairobi, Kenya.

Kenya, Government of (1992), *Small Enterprise and Jua Kali Development in Kenya*, Sessional Paper no. 2, Nairobi, Government Printer, Nairobi, Kenya.

Kenya, Government of (1996), *1997-2001, Development Plan,* Government Printer, Nairobi, Kenya.

Kenya, Government of (2000), *Interim Poverty Reduction Strategy Paper for the period 2000-2003*, Government Printer, Nairobi, June 2000.

Kenya, Government of (2001), *1999 Population and Housing Census: Counting Our People for* Development, Vol. 1 (*Population Distribution by Administrative Areas and Urban Centres*) and Vol. 2 (*Socio-Economic Profiles of the Population*), January 2001.

Kerkum, N. (1991), 'Hunting for Waste: the Fringe of the Growth Pole', in J.J.F. Heins, E.N. Meijer and K.W. Kuipers, *Factories and Families, a Study of a Growth Pole in South India*, Manohar, Delhi.

Kettle, B., Muirhead, B., Abwunza, J., Daly, G., Malombe, J., Morley, D. and Ngua, P. (1995), *Urban Poverty and the Survival Strategies of the Urban Poor in Nairobi*, Final Report, Toronto, Faculty of Environmental Studies, York University (unpublished report).

Khatkhate, D.R. (1997), 'India's Economic Growth: A Conundrum', *World Development*, Vol. 25(9), pp. 1551-9.

Khouri, N., Kalbermatten, J.M. and Bartone, C.R. (1994), *The Reuse of Wastewater in Agriculture: a Guide for Planners*, UNDP-World Bank Water and Sanitation Program, Washington D.C.

Kiango, S. and Amend, J. (2001), 'Linking (peri) Urban Agriculture and Organic Waste Management in Dar es Salaam', in Pay Drechsel and Agmar Kunze (eds), *Waste Composting for Urban and Peri-urban agriculture: Closing the Rural-Urban Nutrient Cycle in Sub-Saharan Africa*, CABI Publishing, Wallingford, Oxon, UK, pp 115-28.

Kibwage, J.K. (1996), *Towards the Privatization of Household Waste Management Services in the City of Nairobi*, M.Phil. Thesis, Moi University, Kenya (unpublished report).

King, K. (1996), *Jua Kali Kenya: Change and Development in an Informal Economy, 1970-1995*, East African Educational Publishers, Nairobi.

Klundert, A. van der and Lardinois, I. (1995), *Community and Private (Formal and Informal) Sector Involvement in Muncipal Solid Waste Management in Developing Countries,* Background paper for the UMP Workshop in Ittingen, 10-12 April.

Klundert, Arnold van der, Muller, Maria, Scheinberg, Anne, Dulac, Nadine, Anschu?tz, Justince and Hoffman, Lane (2001), *Integrated Sustainable Waste Management: A Set of Five Tools for Decision-makers*, WASTE, Gouda.

Koc, M., R. MacRae, L.J.A. Mougeot and J. Welsh (eds.) (1999), *For Hunger-proof Cities*, International Development Research Centre, Ottawa, and Centre for Studies in Food Security, Ryerson Polytechnic University, Toronto.

Koster, E. (1999), Report on Farming in Nairobi, Internal report for EU-INCO project ERBIC18ct971052.

Krugman, P. (1997), *Development, Geography, and Economic Theory*, MIT Press, Cambridge, MA.

Kulkarni, R.S. (1999), *Market Potential for Compost Industry in Hyderabad, India: Necessary Strategies for Public and Private Cooperation*, Master of Science thesis, Asian Institute of Technology, School of Environment, Resources and Development, Bangkok, Thailand.

Kwach, O. H. and Paumard, Antoine (2000), *Mukuru Recycling Centre*, UNCHS (Habitat) Report, July 2000, Nairobi.

Lamba, D. (1994), *Nairobi's Environment: A Review of Conditions and Issues*, Mazingira Institute, Nairobi.

Lapid, D., Munez, L.U., and Bongon, L.L.I. (1996), *Community Participation in Urban Solid Waste Management in Metro Manila and Metro Cebu, the Philpippines*, UWEP Case Study Report, WASTE, Gouda (unpublished report).

Lardinois, Inge and Furedy, Christine (eds) (1999), *Source Separation of Household Waste Materials: Analysis of Case Studies from Pakistan, The Philippines, India, Brazil, Argentina and the Netherlands*, Urban Waste Series 7, Gouda: WASTE Consultants.

Lardinois, Inge and Arnold van de Klundert (eds) (1994), *Organic Wastes--Options for Small-scale Resource Recovery*, Urban Solid Waste Series no. 1, Technology Transfer for Development and WASTE Consultants, Amsterdam.

Lardinois, Inge and Marchand, Rogier (2000), *Technical and Financial Performance at Integrated Composting-Waste Management Project Sites in the Philippines, India and Nepal*, Paper for Internet Conference on Material Flow Analysis of Integrated Bio-Systems, March-October. http://www.ias.unu.edu/proceedings/icibs/ic-mfa/lardinois/paper.html.

Le, Thi Huong (1995), *Urban Waste Derived Compost in Hanoi, Vietnam: Factors affecting Supply and Demand,* Asian Institute of Technology, Bangkok. (Unpublished masters thesis).

Leach, M., Means, R., Scoones, I., (1999). 'Environmental Entitlements: Dynamics and Institutions in Community-Based Natural Resource Management'. *World Development,* 27 (2), 225-247.

Lee, Y.F. (1997),'The Privatisation of Solid Waste Infrastructure and Services in Asia', *Third World Planning Review*, Vol. 19(2), pp. 139-62.

Lee, Y-S. (1998), 'Intermediary Institutions, Community Organizations, and Urban Environmental Management; The Case of Three Bangkok Slums', *World Development*, Vol. 26(6), pp. 993-1011.

Leftwich, A. (1994), 'Governance, the State and the Politics of Development', *Development and Change*, Vol. 25(2), pp. 363-86.

Lewcock, C.P. (1994), *Case Study of the Use of Urban Waste by Near Urban Farmers of Kano, Nigeria*, Natural Resources Institute, Project no. A0354.

Lewcock, C.P. (1995), 'Farmer Use of Urban Waste in Kano', *Habitat International*, Vol. 19, pp. 225-34.

Lewis, B.D. (1998), 'The Impact of Public Infrastructure on Municipal Economic Development: Empirical Results from Kenya', *Review of Urban and Rural Development Studies*, Vol. 10(2), pp. 142-55.

Lock, Karen and Veenhuizen, Rene van (2001), 'Editorial', *Urban Agriculture Magazine*, Vol. 1(3), March, pp. 1-5. Available at: http://www.ruaf.org.

Losada, H., Martinez, H., Vieyra J., Pealing R., Zavala, R. and Cortes, J. (1998), 'Urban Agriculture in the Metropolitan Zone of Mexico City: Changes Over Time in Urban, Suburban and Peri-Urban Areas', *Environment and Urbanization*, Vol. 10(2), pp. 34 -54.

MacGranahan, G. and Satterthwaite, D. (2000), 'Environmental Health or Ecological Sustainability? Reconciling the Brown and Green Agendas in Urban Development', in C. Pugh, (ed.), *Sustainable Cities in Developing Countries*, Earthscan Publications, London, pp. 73-90.

Maqsood Sinha, A.H.M. and Nurul Amin, A.T.M. (1995), 'Dhaka's Waste Recycling Economy: Focus on Informal Sector Labour Groups and Industrial Districts', *Regional Development Dialogue*, Vol. 6(2), pp. 173-95.

Masinde, K.M. Catherine (1996), 'Small Enterprise Development: Production and Distribution in Kenya's Motor Industry', in D. McCormick and Paul Ove Pederson (eds), *Small Enterprises: Flexibility and Networking in an African Context*, Longhorn, Kenya.

Mazingira Institute (1987), *Urban Growth and Reform*, Report of Mazingira Institute, Nairobi, Kenya (unpublished report).

Mazingira Institute (1994), *Nairobi's Environment. A Review of Conditions and Issues*, Report Prepared by Davinder Lamba, Nairobi, Kenya (unpublished report).

Mathur, O.P. (1996), 'The Implications of Decentralisation for Municipal Finance', in K. Singh and F. Steinberg (eds), *Urban India in Crisis*, HSMI/IHS, New Age International Publishers, New Delhi, pp. 261-76.

Mathur, O.P. (1998), 'Regional Meet on Devolution of Functional and Financial Powers to Urban Local Bodies', *Urban India*, Vol. 18(2), pp. 127-39.

Mattingly, M. (1999), 'Management of the Urban Environment', in A. Atkinson, J.D. Dávila, E. Fernandes and M. Mattingly, *The Challenge of Environmental Management in Urban Areas*, Ashgate, Aldershot, pp. 105-113.

McCormick, Dorothy (1991), 'Success in Urban Small-Scale Manufacturing: Implications for Economic Development', in P. Coughlin and G. Ikiara (eds), *Kenya's Industrialisation Dilemma*, Heinemann Kenya ltd, Nairobi, pp. 335-61.

MCH (1997), *Municipal Corporation of Hyderabad*, Municipal Corporation of Hyderabad, Hyderabad, India.

MCH (1998a), *Municipal Corporation of Hyderabad: Report on Activities*, Municipal Corporation of Hyderabad, Hyderabad, India.

MCH (1998b), *Budget Estimates for the year 1998-99 and Revised Estimates 1997-98*, Municipal Corporation of Hyderabad, India.

MCH (1999a), *Privatization of Conservancy Services in Hyderabad Municipal Corporation*, Municipal Corporation of Hyderabad, India.

MCH (1999b), File no. 2997/H/Health/99 Health Section, Municipal Corporation of Hyderabad, India.

Mitchell, F.L. (1994), 'Environmental Justice – A Significant Factor in Environmental Contamination Situations', *Occupational and Environmental Medicine*, Vol. 8, p. 57.

Mitlin, D. (1999), *Civil Society and Urban Poverty,* Theme Paper 5, Urban Governance, Partnership and Poverty Programme, University of Birmingham, Birmingham.

Mitlin, D. (2001) Civil Society and Urban Poverty, Examining the Complexity, *Environment and Urbanization*, 13 (2): 151-173.

Mitlin, D. and Satterthwaite, D. (1994), *Cities and Sustainable Development*, Earthscan Publications, London.

Mockler, Margaret (1998), *Community-based Solid Waste Management in Indonesia*, World Bank Background Paper, World Bank, Jakarta.

Mougeot, Luc (1999), 'For Self-reliant Cities: Urban Food Production in a Globalizing South', in Koc, M. R. MacRae, L.J.A. Mougeot and J. Welsh (eds), *For Hunger-proof Cities: Sustainable Urban Food Systems*, Ottawa, International Development Research Centre, pp.11-25.

Moyo, S.S. (1998), *Privatization of Municipal Services in East Africa*, UNCHS, Nairobi.

Mukuru Integrated Recycling Project (1998), 'Jumuiya ya Mukuru', (unpublished leaflet written by the project leaders, Nairobi).

Mutaro, B.A. and Karanja, A.M (1998) 'EU Research Project A Contribution to the Design of Enabling Strategies in Nairobi and Hyderabad: the Case of Urban Solid Waste Management, Nairobi. Research tooh and Methodology'. Eldoret, Moi University, 50 pp. (unpublished report).

Mulei, A. and Crispin Bokea (eds) (1999), *Micro and Small Enterprises in Kenya: Agenda for Improving the Policy Environment*, The International Center for Economic Growth (ICEG), Nairobi.

Mwangi, G.J. (1990), *Solid Waste Management in Nairobi Metropolis*, MSc. Dissertation, Department of Civil Engineering, University of Nairobi, Kenya (unpublished report).

Mwangi, S.W. (2000), 'Partnerships in Urban Environmental Management: an Approach to Solving Environmental Problems in Nakuru, Kenya', *Environment and Urbanization*, Vol. 12(2), pp. 77-92.

Mwangi, S. (2001). '*Local Agenda 21 experiences in Nakuru, Kenya: processes, issues and lessons*', IIED, Urban Environmental Action Plans and Local Agenda 21 Series Working Paper 10.

Mwanthi, M.A., Nyabola, L.O. and Tenambergen, E.D. (1997), 'The Present and Future Status of Municipal Solid Waste Management in Nairobi', *International Journal of Environmental Health Research*, Vol. 7, pp. 345-53.

Nairobi City Council (2000), *Extraordinary Inspection: Summary Recommendations for Implementation*, NCC, Nairobi, August 2000 (unpublished report).

NRC (National Research Council) (1981) *Food Fuel and Fertilizer from Organic Wastes*, Academic Press, Washington DC.

Nunan, Fiona, (2000), 'Urban Organic Waste Markets: Responding to Change in Hubli-Dharwad, India', *Habitat International*, Vol. 24, pp. 347-60.

Nunan, Fiona (2001), 'Rural-urban Interactions: the Purchase of Urban Waste by Farmers in Hubli-Dharwad, India', *Third World Planning Review,* Vol. 23(4), pp. 387-403.

Obirih-Opareh, N. and Post, J. (2001), Quality Assessment of Public and Private Modes of Solid Waste Collection in Accra, Ghana, *Habitat International*, Vol. 26(1), pp 95-112.

Odegi-Awuondo, C. (1994). 'Garbage collection: a survival strategy for Nairobi's Urban Poor', in Odegi-Awuondo, C., Haggai W.Namai, Beneah M. Mutsoto (eds.), *Masters of Survival*, Basic Books, Ltd., Nairobi, Kenya.

Olowu. D. and Smoke, P. (1992), 'Determinants of Success in African Local Governments: an Overview', *Public Administration and Development*, Vol. 12, pp. 1-17.

Omkar, A.C. and Srikant, R. (1996), 'Hyderabad, the City of Garbage', in K.N. Gopi and D. Ravindra Prasad (eds), *Urban Waste Management*, Department of Geography and Regional Centre for Urban and Environmental Studies, Osmania University, Hyderabad, India.

Osborn, Don (2000), Personal communication with C. Furedy.

Ostrom, E. (1996). Crossing the Great Divide: Co-production, Synergy and Development', in *World Development*, Vol. 24(6), pp. 1073-87.

Otieno F.A.O. (1992), 'Solid Waste Management in the City of Nairobi: What are the Prospects for the Future?', *African Urban Quarterly*, Vol.7(1-2), February and May, pp. 142-9.

Oucho, J.O. (1986), 'Rural Orientation, Return Migration and Future Movements of Urban Migrants: A Study of Kisumu, Kenya', *African Urban Quarterly*, Vol. 1(3-4), pp. 207-19.

Pacheco, M. (1992), 'Recycling in Bogota: Developing a Culture for Urban Sustainability', *Environment and Urbanisation*, Vol. 4(2), pp 74-79.

Parker, J.C. and Torres, T.R. (1994), *Micro and Small-Scale enterprise in Kenya: Results of the national baseline survey*, GEMINI technical Report 75, GEMINI, Nairobi.

Patel, Almitra (2000), Personal communication with C. Furedy.

Pedersen, J.D. (2000), 'Explaining Economic Liberalization in India: State and Society Perspectives', *World Development*, Vol. 28(2), pp. 265-82.

Peltenburg, M., de Wit, J. and Davidson, F. (2000). 'Capacity Building for Urban Management: Learning from Recent Experiences', *Habitat International*, Vol. 24(4), pp. 363-73.

Peters, K.A. (1996), *Community-based Waste Management for Environmental Management and Income Generation in Low-income Areas: a Case Study of Nairobi, Kenya,* M.A. Thesis, York University, Ontario, Canada (unpublished report).

Peters, J.G. (1998), 'With a Little Help from our Friends: Public Private Partnerships as Institutions and Instruments', in J. Pierre, (ed), *Partnerships in Urban Governance: European and American Experience*, MacMillan Press, London.

Pierre, J. (1998), *Partnerships in Urban Governance: European and American Experience*, MacMillan Press, London.

Pitot, Hanns-André. (2001), 'Source Separation of Organic Waste and Composting Proves Feasible in a Delhi 'basti'', *ASEP Newsletter*, September, 4-5, 12,15.

Poerbo, H. (1991), 'Urban Solid Waste Management in Bandung: Towards an Integrated Resource Recovery System', *Environment and Urbanisation*, Vol. 3(1), pp 60-9.

Post, J. (1997), 'Urban Management in an Unruly Setting: the African Case', *Third World Planning Review*, Vol. 19(4), pp. 347-66.

Post, J. (1999), 'The Problems and Potentials of Privatising Solid Waste Management in Kumasi, Ghana', *Habitat International*, Vol. 23(2), pp. 201-16.

Post, J., Broekema, J. and Obirih-Opareh, N. (2001), *Trial and Error in Privatisation, Experiences with Solid Waste Collection in Accra (Ghana) and Hyderabad (India)*, paper presented at the Space and Place in Development Geograhpy workshop, 30-31 August 2001; forthcoming in *Urban Studies*).

Purves, D. and Mackenzie, E.J. (1973), 'Effect of Applications of Municipal Compost on Uptake of Copper, Zinc and Boron by Garden Vegetables', *Plant Soil*, Vol. 39, pp. 361-71.

Putnam, R.D. (1993), *Making Democracy Work. Civic Traditions in Modern Italy*, Princeton University Press, Princeton.

Rakodi, C. (1993), 'Planning for Whom?', in N. Devas, and C. Rakodi, (eds), *Managing Fast Growing Cities; New Approaches to Urban Planning and Management in the Developing World*, Longman, Harlow.

Ramamurti, R. (1999), 'Why Haven't Developing Countries Privatised Deeper and Faster?', *World Development*, Vol. 27(1), pp. 137-55.

Rao, K.J. and Shantaram, M.V. (1994), 'Heavy Metal Pollution of Agricultural Soils Due to Application of Garbage', *Indian Journal of Environmental Health*, Vol. 36(1), pp. 31-9.

Ratha, D.S. and Sahu, B.K. (1993), 'Source and Distribution of Metals in Urban Soil of Bombay, India, Using Multivariate Statistical Techniques', *Environmental Geology*, Vol. 22, pp. 276-85.

RAWOO (Netherlands Development Assistance Research Council) (2001). *'North-South Research Partnerships: Issues and Challenges'*, Publication no. 22, RAWOO, the Hague.

Rees, W.E. (1995), 'Achieving Sustainability: Reform or Transition?', *Journal of Planning Literature*, Vol. 9(4), pp. 343-61.

Rees, W.E. (1992). 'Ecological Footprints and Appropriate Carrying Capacity', *Environment and Urbanisation*, Vol. 4(2), pp. 121-30.

Richardson, Glenn and Whitney, Joseph (1995), 'Goats and Garbage in Khartoum, Sudan: A Study of the Urban Ecology of Animal Keeping', *Human Ecology*, Vol. 23(4), pp.433-75.

Rondinelli D.A., McCullough, J.S. and Johnson, R.W. (1989), 'Analysing Decentralisation Policies in Developing Countries; a Political-Economy Framework', *Development and Change*, Vol. 20, pp. 57-87.

Rondinelli, D.A. and Kasarda, J.D. (1993), 'Privatisation of Urban Services and Infrastructure in Developing Countries: an Assessment of Experiences', in J.D. Kasarda and A. McParnell (eds), *Third World Cities: Problems, Policies and Prospects*, Sage Focus Editions, London-New Delhi, pp. 134-60.

Rondinelli, D.A. and Iacono, M. (1996), 'Strategic Management of Privatisation: a Framework for Planning and Implementation', *Public Administration and Development*, Vol. 16, pp. 247-63.

Rosenberg, L. and C. Furedy (eds) (1996), *International Source Book on Environmentally Sound Technologies for Municipal Solid Waste Management*, Compiled by International Environment Technology Centre (IETC) in collaboration with the Harvard Institute of International Development, Osaka, International Environmental Technology Centre, United Nations Environment Program, Osaka/Shiga.

Sachs, W. (ed) (1997), *The Development Dictionary, A Guide to Knowledge and Power*, London, Zed Books.

Safier, M. (1992), 'Urban Development: Policy Planning and Management, Practitioners' Perspectives on Public Learning over Three Decades', *Habitat International*, Vol. 16(2), pp. 5-12.

Samoff, J. (1990), 'Decentralisation: The Politics of Interventionism', *Development and Change*, Vol. 21(3), pp. 513-30.

Satterthwaite, D. (1997), 'Sustainable Cities or Cities that Contribute to Sustainable Development?', *Urban Studies*, Vol. 34(10), pp. 1667-91. (Reprinted in *The Sustainable Cities Reader*, D. Satterthwaite, (ed.), Earthscan Publications, 2000, London)

Schenk, H., Bhuvaneshwari, R. and Baud, I. (1998), 'Perspectives on Waste in Urban India', in A. Kalland and G. Persoon, *Environmental Movements in Asia*, Curzon Press, Richmond, Surrey, pp. 271-285.

Scheu, M. and Bhattacharya, J.K. (1995), 'Reuse of Decomposed Waste', in Adrian Coad (ed), *Lessons from India in Solid Waste Management*, Water, Engineering and Development Centre, Loughborough, UK, pp. 16-7.

Schübeler, P. (1996), *A Conceptual Framework for Municipal Solid Waste Management in Low-income Countries*, UMP Working Paper Series 9, UMP/SDC SKAT, Geneva.

Schuurman, F. (1997), 'The Decentralisation Discourse: Post-Fordist Paradigm or Neo-Liberal Cul-de-Sac', *The European Review of Development Studies*, pp. 150-66.

Sexton, K., Olden, K., Johnson, B.L. (1993), 'Environmental Justice: the Central Role of Research in Establishing a Credible Scientific Foundation for Informed Decision Making', *Toxicology and Industrial Health*, Vol. 9, pp. 685.

Sicular, D. (1992), *Scavengers, Recyclers, and Solutions for Solid Waste Management in Indonesia*, Berkeley, Centre for Southeast Asia Studies, University of California, Berkeley (unpublished thesis).

Singh, K. (1996), 'The Impact of the Seventy Forth Constitutional Amendment on Urban Management', in K. Singh and F. Steinberg (eds), *Urban India in Crisis*, HSMI/IHS, New Age International Publishers, New Delhi, pp. 423-36.

Sivaramakrishan, K.C. (2000), *Power to the People? The Politics and Progress of Dentralisation*, Konark Publishers, New Delhi.

Smit, Jac, Ratta, Annu and Nasr, Joe (1996), *Urban Agriculture: Food, Jobs, and Sustainable Cities*, Habitat II Series, UNDP, New York.

Snel, M. (1997), *The Formal and Informal Sector of Solid Waste Management in Hyderabad*, M.Phil thesis, University of Sussex, UK.

Southall, R. (1999), 'Re-Forming the State? Kleptocracy and Political Transition in Kenya', *Review of African Political Economy*, Vol. 79, pp. 93-108.

Spiaggi, E.P., Biasatti, N.R. and Marc. L.B. (2000), 'Vermiculture for Organic Waste Processing: Mini-livestock in Rosario Argentina', *Urban Agriculture Magazine*, Vol. 1(2), October, p. 36.

Stoker, G (1998). 'Governance as Theory: Five Propositions', *International Social Science Journal* Vol. 50(155), pp. 17-28.

Stoker, G. (2000), *The New Politics of British Local Governance*, McMillan Press, Basingstoke.

Stone, D., Maxwell, S., Keating, M. (2001). '*Bridging Research and Policy'*. Paper prepared for International Workshop funded by DFID, UK, Radcliffe House, Warwick Univ. July 16-17.

Storper, M. (1997), *The Regional World, Territorial Development in a Global Economy*, Guildford Press, New York.

Stren, R. (1993), 'Urban Management in Development Assistance: an Elusive Concept', *Cities*, May, pp. 125-38.

Sudharkar Reddy, S. and Galab, S. (2000), *Urban Organic Waste Recycling: Role of Farming Community, NGOs, CBOs and the Corporate Sector in Hyderabad*, Centre for Economic and Social Studies, Hyderabad, India, (Report prepared for EU-INCO project ERBIC 18CT 970152).

Sudhakar Reddy, S. and Galab, S. (2000a), *Solid Waste Management in Hyderabad City, the Role of Local Bodies and the Civil Society,* CESS, Hyderabad, India (Report prepared for EU-INCO project ERBIC 18CT 970152).

Sudhakar Reddy, S. and Galab, S. (2000b), *Alliances in Solid Waste Management, the Case of Hyderabad,* CESS, Hyderabad, India (Report prepared for EU-INCO project ERBIC 18CT 970152).

Sudhakar Reddy, S. and Galab, S. (2000c), *Solid Waste Management in Hyderabad City, the Role of the Private Sector*, CESS, Hyderabad, India (Report prepared for EU-INCO project ERBIC 18CT 970152).

Sundaram, K.V. (2000), 'From Top-Down Planning to Decentralised Regional and Local Development in India- Vision and Reality: Challenges and Perspectives', *Indian Social Science Review*, Vol. 2(2), pp. 275-300.

Syagga, P.M. (1992), *Solid Waste Management Cycle in Nairobi*, Paper prepared for the Workshop on Urban Management in Kenya, University of Nairobi (unpublished report).

Tukker, A. (2000), 'Life Cycle Assessment as a Tool in Environmental Impact Assessment', *Environmental Impact Assessment Review*, Vol. 20, pp. 435-56.

United Nations Centre for Human Settlements (Undated), chapter *5: Case Studies of Privatization of some Municipal Services*, http://www.unchs.org/unchs/planning/privat.

United Nations Centre for Human Settlements (1996), *An Urbanizing World: Global Report on Human Settlements 1996*, Oxford University Press, Oxford and New York.

United Nations Centre for Human Settlements (1998), *Privatization of Municipal Services in East Africa, a Governance Approach to Human Settlements Management*, Nairobi, http://www.unchs.org/unchs/planning/privat.

United Nations Centre for Human Settlements/United Nations Development Program (1997), *Implementing the Urban Environment Agenda*, Vol. I, Nairobi.

Vincentian Missionaries (1998), 'The Payatas Environmental Development Programme: Micro-Enterprise Promotion and Involvement in Solid Waste Management in Quezon City', *Environment and Urbanization*, Vol. 10(2), pp. 55-68.

WASTE Consultants (1998), 'Pilot projects in Bamako: an overview." *UWEP (Urban Waste Expertise Programme) E-mail Bulletin* no. 12, May 1998 (Available at: www.waste.nl).

Williams, D.E., Vlamis, J., Pukite, A. H. and Corey, J.E. (1980), 'Trace Element Accumulation, Movement and Distribution in the Soil Profile from Massive Applications of Sewage Sludge', *Soil Science*, Vol. 129, pp. 119-32.

World Commission on Environment and Development (1987), *Our Common Future*, Oxford University Press, Oxford.

Werna, E. (1995), 'The Management of Urban Development or the Development of Urban Management? Problems and Premises of an Elusive Concept', *Cities*, Vol. 12(5), pp. 353-9.

Werna, E. (1998), 'Urban Management, the Provision of Public Services and Intra-urban Differentials in Nairobi', *Habitat International*, Vol. 22(1), pp. 15-26.

World Bank (2000), *World Development Report 1999/2000: Entering the 21st Century*, Oxford University Press, Washington D.C.

Yasmeen, Giselle (2001), *Urban Agriculture in India,* Report to International Development Research Centre, IDRC, Ottawa.

Yeung, Yue-man (1985), *Urban Agriculture in Asia*, Food-Energy Nexus Programme Report no. 10., United Nations University, Paris.

Zeeuw Henk de, and Karen Lock. (2001) "Mitigating the Health Risks Associated with Urban and Periurban Agriculture." *Urban Agriculture Magazine*, Vol. 1, no. 3, March, pp. 6-8.

Zurbrugg, Christian and Aristanti, Christina (1999), 'Resource Recovery in a Primary Collection Scheme in Indonesia', *SANDEC News*, Vol. 4, January, pp. 7-9.

The GeoJournal Library

---

1. B. Currey and G. Hugo (eds.): *Famine as Geographical Phenomenon.* 1984 ISBN 90-277-1762-1
2. S.H.U. Bowie, F.R.S. and I. Thornton (eds.): *Environmental Geochemistry and Health.* Report of the Royal Society's British National Committee for Problems of the Environment. 1985 ISBN 90-277-1879-2
3. L.A. Kosiński and K.M. Elahi (eds.): *Population Redistribution and Development in South Asia.* 1985 ISBN 90-277-1938-1
4. Y. Gradus (ed.): *Desert Development.* Man and Technology in Sparselands. 1985 ISBN 90-277-2043-6
5. F.J. Calzonetti and B.D. Solomon (eds.): *Geographical Dimensions of Energy.* 1985 ISBN 90-277-2061-4
6. J. Lundqvist, U. Lohm and M. Falkenmark (eds.): *Strategies for River Basin Management.* Environmental Integration of Land and Water in River Basin. 1985 ISBN 90-277-2111-4
7. A. Rogers and F.J. Willekens (eds.): *Migration and Settlement.* A Multiregional Comparative Study. 1986 ISBN 90-277-2119-X
8. R. Laulajainen: *Spatial Strategies in Retailing.* 1987 ISBN 90-277-2595-0
9. T.H. Lee, H.R. Linden, D.A. Dreyfus and T. Vasko (eds.): *The Methane Age.* 1988 ISBN 90-277-2745-7
10. H.J. Walker (ed.): *Artificial Structures and Shorelines.* 1988 ISBN 90-277-2746-5
11. A. Kellerman: *Time, Space, and Society.* Geographical Societal Perspectives. 1989 ISBN 0-7923-0123-4
12. P. Fabbri (ed.): *Recreational Uses of Coastal Areas.* A Research Project of the Commission on the Coastal Environment, International Geographical Union. 1990 ISBN 0-7923-0279-6
13. L.M. Brush, M.G. Wolman and Huang Bing-Wei (eds.): *Taming the Yellow River: Silt and Floods.* Proceedings of a Bilateral Seminar on Problems in the Lower Reaches of the Yellow River, China. 1989 ISBN 0-7923-0416-0
14. J. Stillwell and H.J. Scholten (eds.): *Contemporary Research in Population Geography.* A Comparison of the United Kingdom and the Netherlands. 1990 ISBN 0-7923-0431-4
15. M.S. Kenzer (ed.): *Applied Geography.* Issues, Questions, and Concerns. 1989 ISBN 0-7923-0438-1
16. D. Nir: *Region as a Socio-environmental System.* An Introduction to a Systemic Regional Geography. 1990 ISBN 0-7923-0516-7
17. H.J. Scholten and J.C.H. Stillwell (eds.): *Geographical Information Systems for Urban and Regional Planning.* 1990 ISBN 0-7923-0793-3
18. F.M. Brouwer, A.J. Thomas and M.J. Chadwick (eds.): *Land Use Changes in Europe.* Processes of Change, Environmental Transformations and Future Patterns. 1991 ISBN 0-7923-1099-3

The GeoJournal Library

19. C.J. Campbell: *The Golden Century of Oil 1950–2050.* The Depletion of a Resource. 1991 ISBN 0-7923-1442-5
20. F.M. Dieleman and S. Musterd (eds.): *The Randstad: A Research and Policy Laboratory.* 1992 ISBN 0-7923-1649-5
21. V.I. Ilyichev and V.V. Anikiev (eds.): *Oceanic and Anthropogenic Controls of Life in the Pacific Ocean.* 1992 ISBN 0-7923-1854-4
22. A.K. Dutt and F.J. Costa (eds.): *Perspectives on Planning and Urban Development in Belgium.* 1992 ISBN 0-7923-1885-4
23. J. Portugali: *Implicate Relations.* Society and Space in the Israeli-Palestinian Conflict. 1993 ISBN 0-7923-1886-2
24. M.J.C. de Lepper, H.J. Scholten and R.M. Stern (eds.): *The Added Value of Geographical Information Systems in Public and Environmental Health.* 1995 ISBN 0-7923-1887-0
25. J.P. Dorian, P.A. Minakir and V.T. Borisovich (eds.): *CIS Energy and Minerals Development.* Prospects, Problems and Opportunities for International Cooperation. 1993 ISBN 0-7923-2323-8
26. P.P. Wong (ed.): *Tourism vs Environment: The Case for Coastal Areas.* 1993 ISBN 0-7923-2404-8
27. G.B. Benko and U. Strohmayer (eds.): *Geography, History and Social Sciences.* 1995 ISBN 0-7923-2543-5
28. A. Faludi and A. der Valk: *Rule and Order. Dutch Planning Doctrine in the Twentieth Century.* 1994 ISBN 0-7923-2619-9
29. B.C. Hewitson and R.G. Crane (eds.): *Neural Nets: Applications in Geography.* 1994 ISBN 0-7923-2746-2
30. A.K. Dutt, F.J. Costa, S. Aggarwal and A.G. Noble (eds.): *The Asian City: Processes of Development, Characteristics and Planning.* 1994 ISBN 0-7923-3135-4
31. R. Laulajainen and H.A. Stafford: *Corporate Geography.* Business Location Principles and Cases. 1995 ISBN 0-7923-3326-8
32. J. Portugali (ed.): *The Construction of Cognitive Maps.* 1996 ISBN 0-7923-3949-5
33. E. Biagini: *Northern Ireland and Beyond.* Social and Geographical Issues. 1996 ISBN 0-7923-4046-9
34. A.K. Dutt (ed.): *Southeast Asia: A Ten Nation Region.* 1996 ISBN 0-7923-4171-6
35. J. Settele, C. Margules, P. Poschlod and K. Henle (eds.): *Species Survival in Fragmented Landscapes.* 1996 ISBN 0-7923-4239-9
36. M. Yoshino, M. Domrös, A. Douguédroit, J. Paszynski and L.D. Nkemdirim (eds.): *Climates and Societies – A Climatological Perspective.* A Contribution on Global Change and Related Problems Prepared by the Commission on Climatology of the International Geographical Union. 1997 ISBN 0-7923-4324-7
37. D. Borri, A. Khakee and C. Lacirignola (eds.): *Evaluating Theory-Practice and Urban-Rural Interplay in Planning.* 1997 ISBN 0-7923-4326-3

The GeoJournal Library

---

38. J.A.A. Jones, C. Liu, M-K. Woo and H-T. Kung (eds.): *Regional Hydrological Response to Climate Change.* 1996 ISBN 0-7923-4329-8
39. R. Lloyd: *Spatial Cognition.* Geographic Environments. 1997 ISBN 0-7923-4375-1
40. I. Lyons Murphy: *The Danube: A River Basin in Transition.* 1997 ISBN 0-7923-4558-4
41. H.J. Bruins and H. Lithwick (eds.): *The Arid Frontier.* Interactive Management of Environment and Development. 1998 ISBN 0-7923-4227-5
42. G. Lipshitz: *Country on the Move: Migration to and within Israel, 1948–1995.* 1998 ISBN 0-7923-4850-8
43. S. Musterd, W. Ostendorf and M. Breebaart: *Multi-Ethnic Metropolis: Patterns and Policies.* 1998 ISBN 0-7923-4854-0
44. B.K. Maloney (ed.): *Human Activities and the Tropical Rainforest.* Past, Present and Possible Future. 1998 ISBN 0-7923-4858-3
45. H. van der Wusten (ed.): *The Urban University and its Identity.* Roots, Location, Roles. 1998 ISBN 0-7923-4870-2
46. J. Kalvoda and C.L. Rosenfeld (eds.): *Geomorphological Hazards in High Mountain Areas.* 1998 ISBN 0-7923-4961-X
47. N. Lichfield, A. Barbanente, D. Borri, A. Khakee and A. Prat (eds.): *Evaluation in Planning.* Facing the Challenge of Complexity. 1998 ISBN 0-7923-4870-2
48. A. Buttimer and L. Wallin (eds.): *Nature and Identity in Cross-Cultural Perspective.* 1999 ISBN 0-7923-5651-9
49. A. Vallega: *Fundamentals of Integrated Coastal Management.* 1999 ISBN 0-7923-5875-9
50. D. Rumley: *The Geopolitics of Australia's Regional Relations.* 1999 ISBN 0-7923-5916-X
51. H. Stevens: *The Institutional Position of Seaports.* An International Comparison. 1999 ISBN 0-7923-5979-8
52. H. Lithwick and Y. Gradus (eds.): *Developing Frontier Cities.* Global Perspectives – Regional Contexts. 2000 ISBN 0-7923-6061-3
53. H. Knippenberg and J. Markusse (eds.): *Nationalising and Denationalising European Border Regions, 1800–2000.* Views from Geography and History. 2000 ISBN 0-7923-6066-4
54. R. Gerber and G.K. Chuan (eds.): *Fieldwork in Geography: Reflections, Perspectives and Actions.* 2000 ISBN 0-7923-6329-9
55. M. Dobry (ed.): *Democratic and Capitalist Transitions in Eastern Europe.* Lessons for the Social Sciences. 2000 ISBN 0-7923-6331-0
56. Y. Murayama: *Japanese Urban System.* 2000 ISBN 0-7923-6600-X
57. D. Zheng, Q. Zhang and S. Wu (eds.): *Mountain Geoecology and Sustainable Development of the Tibetan Plateau.* 2000 ISBN 0-7923-6688-3

The GeoJournal Library

58. A.J. Conacher (ed.): *Land Degradation.* Papers selected from Contributions to the Sixth Meeting of the International Geographical Union's Commission on Land Degradation and Desertication, Perth, Western Australia, 20–28 September 1999. 2001 ISBN 0-7923-6770-7
59. S. Conti and P. Giaccaria: *Local Development and Competitiveness.* 2001 ISBN 0-7923-6829-0
60. P. Miao (ed.): *Public Places in Asia Pacic Cities.* Current Issues and Strategies. 2001 ISBN 0-7923-7083-X
61. N. Maiellaro (ed.): *Towards Sustainable Buiding.* 2001 ISBN 1-4020-0012-X
62. G.S. Dunbar (ed.): *Geography: Discipline, Profession and Subject since 1870.* An International Survey. 2001 ISBN 1-4020-0019-7
63. J. Stillwell and H.J. Scholten (eds.): *Land Use Simulation for Europe.* 2001 ISBN 1-4020-0213-0
64. P. Doyle and M.R. Bennett (eds.): *Fields of Battle.* Terrain in Military History. 2002 ISBN 1-4020-0433-8
65. C.M. Hall and A.M. Williams (eds.): *Tourism and Migration.* New Relationships between Production and Consumption. 2002 ISBN 1-4020-0454-0
66. I.R. Bowler, C.R. Bryant and C. Cocklin (eds.): *The Sustainability of Rural Systems.* Geographical Interpretations. 2002 ISBN 1-4020-0513-X
67. O. Yiftachel, J. Little, D. Hedgcock and I. Alexander (eds.): *The Power of Planning.* Spaces of Control and Transformation. 2001 ISBN Hb; 1-4020-0533-4 ISBN Pb; 1-4020-0534-2
68. K. Hewitt, M.-L. Byrne, M. English and G. Young (eds.): *Landscapes of Transition.* Landform Assemblages and Transformations in Cold Regions. 2002 ISBN 1-4020-0663-2
69. M. Romanos and C. Auffrey (eds.): *Managing Intermediate Size Cities.* Sustainable Development in a Growth Region of Thailand. 2002 ISBN 1-4020-0818-X
70. B. Boots, A. Okabe and R. Thomas (eds.): *Modelling Geographical Systems.* Statistical and Computational Applications. 2003 ISBN 1-4020-0821-X
71. R. Gerber and M. Williams (eds.): *Geography, Culture and Education.* 2002 ISBN 1-4020-0878-3
72. D. Felsenstein, E.W. Schamp and A. Shachar (eds.): *Emerging Nodes in the Global Economy: Frankfurt and Tel Aviv Compared.* 2002 ISBN 1-4020-0924-0
73. R. Gerber (ed.): *International Handbook on Geographical Education.* 2003 ISBN 1-4020-1019-2
74. M. de Jong, K. Lalenis and V. Mamadouh (eds.): *The Theory and Practice of Institutional Transplantation.* Experiences with the Transfer of Policy Institutions. 2002 ISBN 1-4020-1049-4
75. A.K. Dutt, A.G. Noble, G. Venugopal and S. Subbiah (eds.): *Challenges to Asian Urbanization in the 21st Century.* 2003 ISBN 1-4020-1576-3

The GeoJournal Library

76. I. Baud, J. Post and C. Furedy (eds.): *Solid Waste Management and Recycling.* Actors, Partnerships and Policies in Hyderabad, India and Nairobi, Kenya. 2004
ISBN 1-4020-1975-0

KLUWER ACADEMIC PUBLISHERS – DORDRECHT / BOSTON / LONDON

MIX
Papier aus verantwortungsvollen Quellen
Paper from responsible sources
FSC® C105338

If you have any concerns about our products,
you can contact us on
ProductSafety@springernature.com

In case Publisher is established outside the EU,
the EU authorized representative is:
Springer Nature Customer Service Center GmbH
Europaplatz 3, 69115 Heidelberg, Germany

Printed by Libri Plureos GmbH
in Hamburg, Germany